**Studies in Systems, Decision and Control**

Volume 270

**Series Editor**

Janusz Kacprzyk, Systems Research Institute, Polish Academy of Sciences,
Warsaw, Poland

The series "Studies in Systems, Decision and Control" (SSDC) covers both new developments and advances, as well as the state of the art, in the various areas of broadly perceived systems, decision making and control–quickly, up to date and with a high quality. The intent is to cover the theory, applications, and perspectives on the state of the art and future developments relevant to systems, decision making, control, complex processes and related areas, as embedded in the fields of engineering, computer science, physics, economics, social and life sciences, as well as the paradigms and methodologies behind them. The series contains monographs, textbooks, lecture notes and edited volumes in systems, decision making and control spanning the areas of Cyber-Physical Systems, Autonomous Systems, Sensor Networks, Control Systems, Energy Systems, Automotive Systems, Biological Systems, Vehicular Networking and Connected Vehicles, Aerospace Systems, Automation, Manufacturing, Smart Grids, Nonlinear Systems, Power Systems, Robotics, Social Systems, Economic Systems and other. Of particular value to both the contributors and the readership are the short publication timeframe and the world-wide distribution and exposure which enable both a wide and rapid dissemination of research output.

** Indexing: The books of this series are submitted to ISI, SCOPUS, DBLP, Ulrichs, MathSciNet, Current Mathematical Publications, Mathematical Reviews, Zentralblatt Math: MetaPress and Springerlink.

More information about this series at http://www.springer.com/series/13304

Jawhar Ghommam · Nabil Derbel ·
Quanmin Zhu
Editors

# New Trends in Robot Control

Springer

*Editors*
Jawhar Ghommam
Department of Electrical
and Computer Engineering
Sultan Qaboos University
Al Khoudh, Oman

Nabil Derbel
National School of Engineers of Sfax
University of Sfax
Sfax, Tunisia

Quanmin Zhu
Department of Engineering
Design and Mathematics
University of the West of England
Bristol, UK

ISSN 2198-4182 ISSN 2198-4190 (electronic)
Studies in Systems, Decision and Control
ISBN 978-981-15-1818-8 ISBN 978-981-15-1819-5 (eBook)
https://doi.org/10.1007/978-981-15-1819-5



This Springer imprint is published by the registered company Springer Nature Singapore Pte Ltd.
The registered company address is: 152 Beach Road, #21-01/04 Gateway East, Singapore 189721, Singapore

# Preface

The current book *New Trends in Robot Control* publishes the latest developments in the field of robotic and control, presented informally and with high quality. The content of the book is very interesting and attractive as it covers a wide range of technologies and control techniques including advanced robotic systems such as robot manipulators, system network, underactuated robotic system, autonomous drilling system, UAV-type drones, vehicular networking and connected vehicles, and vision process control. The intent is to cover the theory, applications, and perspectives on the state-of-the-art and future developments relevant to the field of robotic and other related areas, as embedded in the fields of control engineering, as well as the paradigms and methodologies behind them. The chapters of this book are designed for graduate students, researchers, educators, engineers, and scientists requiring theories, methods, and applications of mathematical analysis. Each chapter provides an enrichment of the understanding of the research discussed along with a balanced effort in theories, methods and applications, and also contains the most recent advances made in an area of robotic. This book is made of four distinct parts. In total, there are 19 chapters in this book, and they are organized as follows:

The first part of this book focuses on applications of control theory on robot manipulators, and comprises seven chapters:

The chapter "Tuning of Fractional Order Controller and Prefilter in MIMO Robust Motion Control: SCARA Robot" investigates the quantitative feedback theory in combination with fractional order controller to control multivariable systems. A new analytic tuning method of fractional order controller is proposed and aims at ensuring stability and robustness of the MIMO system.

The chapter "Flexible-Link Manipulators: Dynamic Analysis and Advanced Control Strategies" contains advanced studies in dynamic modeling and nonlinear control strategies applied to flexible-link robotic manipulators. The control strategy tackled in this chapter utilizes the classical nonlinear PD with fuzzy and gain scheduling controllers. The robustness of the obtained performance is then further improved involving a novel fractional order fuzzy PD (FOFPD) controller that uses non-integer order differentiator operators in the fuzzy PD controller by fine-tuning the gains of the FOFPD using the particle swarm optimization algorithm. The stability of the closed-loop system is shown to be bounded input-bounded output.

The chapter “Autonomous Trenchless Horizontal Directional Drilling” reports mathematical aspects of drilling and trenches technology and provides an online control scheme for real-time optimization of the drilling parameters and control of steering system to minimize the deviation from the target borehole trajectory of the pilot hole and maximize Rop. Nonlinear model for the drilling process was developed using energy balance equation. The proposed controller design problem is formulated as an optimization problem where an objective function is considered to minimize tracking error and drilling efforts.

The chapter “Stabilization of Second Order Underactuated System Using Fast Terminal Synergetic Control” investigate the control of a class of underactuated mechanical systems using fast terminal synergetic control (FTSC) to guarantee the finite time convergence and the stability of the underactuated system. Hierarchical procedure has been utilized for the design of the finite time type controller.

The chapter “Sliding Mode Fault Diagnosis with Vision in the Loop for Robot Manipulators” investigates the problem of fault diagnosis (FD) for industrial robotic manipulators within the framework of sliding mode control theory. The control scheme for fault detection is complemented by a low-cost vision servoing architecture allowing the design a fault-tolerant control strategy in case of sensor faults. The effectiveness of the proposed FD architecture has been carried out in simulation on a realistic simulator as well as experimentally on a COMAU SMART3-S2 anthropomorphic robot manipulator.

The chapter “$L_1$ Adaptive Control of a Lower Limb Exoskeleton Dedicated to Kids’ Rehabilitation” presents a comparison study of different adaptive controllers developed to control robotic orthoses used for kids rehabilitation. It is shown that among the proposed adaptive controllers, the performance of L1 adaptive control scheme prevails over the other approaches. The developed controllers were validated through different scenarios and simulations.

The chapter “Exoskeletons Control via Computed Torque for Lower Limb Rehabilitation” presents similar robotic system that aids deficient kids ability to walk. An exoskeleton 2-DoF robotic system is therefore considered in this chapter whereof two types of PID controllers have been proposed, in particular the PID based on feedback linearization and the PID adaptive feedback linearization have been tested and compared.

The second part of the book presents most advanced techniques on controlling and navigating autonomous vehicles and includes six chapters:

The chapter “Intelligent Control for an Uncertain Mobile Robot with External Disturbances Estimator” presents a control approach for the trajectory tracking problem of nonholonomic wheeled mobile robot (WMR) that efficiently copes simultaneously with model uncertainties and external torque disturbances. The novelty of this approach is the ability to completely compensate for the uncertainties and external torque disturbances without the requirement of torque measurement. In this approach, a kinematic backstepping controller is proposed to achieve perfect velocity tracking. Then, a robust dynamic adaptive control algorithm with two update laws is developed to estimate and compensate the dynamic uncertainties and the unmeasured external torque disturbances. The design of the update laws uses only position and velocity measurements and is derived from the Lyapunov method.

The chapter "Adaptive Event-Triggered Regulation Control of Nonholonomic Mobile Robots" investigates the design of an adaptive-regulation control of a mobile robot in the presence of model uncertainties with event-based feedback control. The neural network with two layers are utilized to estimate the uncertain dynamics of the mobile robot. The control inputs is then developed with event-sampled measurement update. The Lyapunov's method is utilized to develop an event sampling condition and to demonstrate the regulation error performance of the mobile robot.

The chapter "Optimal Lane Merging for AGV" addresses the generation of an optimal constrained trajectory for lane merging tasks. The developed algorithm ensures a safe and a less fuel consumption for autonomous driving. The security is established with the restriction of both the lateral and the longitudinal AGV's position inside a safe zone while accomplishing a lane change to overtake an obstacle or to follow a lead vehicle.

The chapter "Finite Time Consensus for Higher Order Multi-agent Systems with Mismatched Uncertainties" tackles another aspects of robotic filed, which is networking and connected agents. This chapter presents a finite time distributed consensus problem for a class of robotic system represented by a higher order multi-agent system (MAS). The chapter considers agents when they are potentially affected by mismatched uncertainties and continues the design of a distributed controller combining the integral sliding mode technique and the finite time disturbance observer. Allowing to neglect the effect of mismatched uncertainties during the sliding mode. Stability analysis of the MAS is conducted using strict Lyapunov functions.

The chapter "Path Planning for a Multi-robot System with Decentralized Control Architecture" studies the path planning problem of a group of autonomous Wheeled Mobile Robots (WMR) in a cluttered environment. The problem of coordinating the motion of multiple WMR is tackled from the point of view of finding local minima of artificial potential field imbricated into a decentralized architecture to coordinate the movement of WMR that consists of combining three techniques which are the potential field, the neighborhood system, and the notion of priority between the robots. The approach is validated by simulation resorting to different possible scenarios.

The chapter "Which is Better for Mobile Robot Trajectory Optimization: PSO or GA?", Two optimization approaches for path planning of mobile robot have been proposed and compared. These approaches are the Particle Swarm Optimization and the Genetic Algorithms. The two approaches are then merged together seeking for optimal parametrization of the Dynamic Variable Speed Force Field. Authors' opinion has been provided to decide which of the two methods is the best for path planning of mobile robot.

The third part is concerned about recent developments on autonomous unmanned vehicles and drones, particular interest is brought to control of quadrotor-type aerial vehicles. This part is made of four chapters:

The chapter "Dynamic Modeling of a Quadrotor UAV Prototype" presents two methods to determine the dynamic model of a quadrotor UAV. In a first attempt, a mathematical model is developed to describe the quadrotor dynamics with six degree of freedom, using Euler–Lagrange formalism. The unknown model parameters are then obtained using calculations and experimental tests. In the second attempt, the roll system is estimated using closed-loop identification method and frequency domain analysis. The simulation results are compared with practical responses.

The chapter "Model-Based Fault Detection of Permanent Magnet Synchronous Motors of Drones Using Current Sensors" discusses simple low-cost and effective scheme for the detection of the most common faults in drone actuators which are the permanent magnet synchronous motors (PMSM). The scheme is based on a modeling stage which only requires the current measurements from a fault-less motor. From this, a simplified transfer function of the motor is derived. Then, the output of this model and a healthy motor are used as arguments of simple tests to detect the occurrence of a set of characterized faults in the target motor. The setup of the scheme and the development of the tests are straightforward. The faults considered in this work are inter-turn short-circuit, changes in friction constant and flying off propeller, and other less common faults. Experimental results show that these faults are accurately detected and characterized by the proposed scheme.

The chapter "ENMPC Versus PID Control Strategies Applied to a Quadcopter" investigates the trajectory tracking control problem for an unmanned aerial vehicle (UAV). Two continuous-time strategies of control are presented to solve this control problem. The first technique is based on cascade proportional-integral-derivative (PID) controllers, while the second utilizes the explicit nonlinear model predictive control (ENMPC) technique. Numerical comparison study is conducted to show the effectiveness of these two approaches and to emphasize the advantages of the ENMPC over the cascade PID strategy.

The chapter "Robust and Adaptive State Estimation of UAV Quadrotors with a High Gain Approach" studies the problem of joint states and external aerodynamic disturbances as well as the problem of adaptive estimation of quadrotor systems. Two configurations of the quadrotor are studied: in the first configuration, the dynamics of longitudinal and angular velocities are corrupted by unknown time-varying disturbances and in the second one, the quadrotor system is subject to constant unknown parameters. For the two latter cases of study, only positions and angles are available for measurements. To this end, the problem of observer matching condition is developed to solve both robust and adaptive estimation problems using first-order sliding mode observer in cascade with a high gain observer to reconstruct both the states and the unknown disturbances of the quadrotor system in the case of time-varying disturbance and adaptive observer in cascade with the same high gain observer to reconstruct both the states and the unknown parameters.

The fourth and last part of this book includes two chapters and emphasis on the importance of microsensors and actuators being used in robotics:

The chapter "Compound Fractional Integral Terminal Sliding Mode Control and Fractional PD Control of a MEMS Gyroscope" proposes a compound fractional order integral terminal sliding mode control (FOITSMC) and fractional order proportional-derivative control (FOPD-FOITSMC) for the control of a MEMS gyroscope. In order to improve the robustness of the conventional integral terminal sliding mode control (ITSMC), a fractional integral terminal sliding mode surface is applied. The chattering problem in FOITSMC, which is usually generated by the excitation of fast un-modeled dynamic, is the main drawback. A fractional order proportional-derivative controller (FOPD) is employed in order to eliminate chattering phenomenon. The stability of the FOPD-FOITSMC is proved by Lyapunov theory.

The chapter "Highly Sensitive Polymer/Multiwalled Carbon Nanotubes Based Pressure and Strain Sensors for Robotic Applications" is dedicated to illustrate the recent progress and development in wearable tactile sensors based on polymer nanocomposites for robotic applications, where capacitive pressure sensor and high sensitive and flexible piezoresistive strain sensor were realized. To effectively test these sensors, they were placed in different locations in the hand like fingers and palm for precision and power grasping in addition to gait analysis in a humanoid robot.

The editors are grateful to the contributors for their timely cooperation and patience while the chapters were being reviewed and processed. The editors also thank the editors and supporting staff at Springer for their timely cooperation in bringing out this book. We hope that all the contributors and the interested readers benefit scientifically from this book and find it stimulating in the process.

Al Khoudh, Oman — Jawhar Ghommam
Sfax, Tunisia — Nabil Derbel
Bristol, UK — Quanmin Zhu
April 2019

# Contents

# Part I

# Tuning of Fractional Order Controller and Prefilter in MIMO Robust Motion Control: SCARA Robot

**Mohamed Allagui, Najah Yousfi, Nabil Derbel and Pierre Melchior**

**Abstract** Quite recently, fractional approaches was greatly used in control engineering area. Several methodologies based on fractional controllers tuning has been treated. In this work, a multi-input multi-output (MIMO) quantitative feedback theory (QFT) method is mixed with a fractional order $PD^{\mu}$ controller and a fractional prefilter to govern multivariable systems. Each obtained sub-systems from the MIMO QFT technique is controlled independently. A new analytic tuning method of fractional order controller is developed in the aim to ensure stability and robustness. After the controller tuning a fractional order prefilter is designed and optimized to reach the desired performances. The proposed method effectiveness will be tested and evaluated based on a real robot model.

**Keywords** Control · Fractional calculus · Fractional order controller · Fractional prefilter · MIMO processes · Robotic · Robustness

## 1 Introduction

The works of Laplace, Liouville [12], Abel [1], Riemann [27] and Cauchy gave birth to the mathematical formalism of the non integer calculus (real or complex). This scientific operation concerns at the same time the theory of systems, automatic, control of the electric machines, robotics and signal and image processing. In recent

M. Allagui (✉) · N. Yousfi · N. Derbel
CEM Laboratory in Control & Energy Management, ENIS, University of Sfax, Sfax, Tunisia
e-mail: mohamed07.allagui@gmail.com

N. Yousfi
e-mail: najah.yousfi@enis.tn

N. Derbel
e-mail: n.derbel@enis.rnu.tn

P. Melchior
IMS-UMR 5218 CNRS, University of Bordeaux-IPB-Talence cedex, Bordeaux, France
e-mail: pierre.melchior@ims-bordeaux.fr

 
J. Ghommam et al. (eds.), *New Trends in Robot Control*, Studies in Systems, Decision and Control 270, https://doi.org/10.1007/978-981-15-1819-5_1

years, fractional calculus have been applied in various types of physical systems modeling and systems control. The problem of controller synthesis aimed to provide system stability and to get a desired performance specifications. Fractional calculations were applied in the physical modeling of diverse types of systems, as it is well documented in numerous theories of control or application literature [2, 3]. The problem of synthesis consists in obtaining a controller who insures the stability and answers the task of performance specifications.

From an engineering point of view, fractional order control is a generalization of the classical integer order theory, which brings us back to more adequate modeling and more robust control performance. Indeed, the interest of fractional order control is the optimization of the performances by using the concept of theory of derivatives, integrals and fractional order of systems.

From the historic point of view, they exist in the literature, four main structures of controller with fractional order [4]; The first fractional order control structure is the famous CRONE controller [22], which was proposed by Alain Oustaloup in the early of 1990. The success of this approach is enormous; several variants of this approach are available in our day (1st, 2nd and 3rd generation).

The last fractional order structure is the fractional order $PI^{\lambda}D^{\mu}$ [26] proposed by Igor Podlubny in 1999, this one is a generalization of the PID controller, including a fractional order integration action $\lambda$ and a differentiation action of fractional order $\mu$. It is demonstrated that a best response of this type of controllers is given in comparison with the conventional classical PID ones.

To improve the behavior of the controller PID, the version of fractional order controller is proposed. The most common types of fractional order is the $PI^{\lambda}D^{\mu}$ of fractional order [26], implying an integrator of order $\lambda$ and a differentiation of order $\mu$, where $\lambda$ and $\mu$ are any real numbers.

One of the most important advantages of the controller $PI^{\lambda}D^{\mu}$ of fractional order is its possibility of well controlling the dynamics of the systems of fractional order. Another advantage is in the fact that the controllers $PI^{\lambda}D^{\mu}$ of fractional order are less sensitive to the changes of the parameter uncertainties of a system to be commanded, which gives an improvement of the robustness. It is generally proved that the fractional order $PI^{\lambda}D^{\mu}$ controller, due to the two supplementary tuning variables $\lambda$ and $\mu$, is able to meet more performance criteria and behave more robustly than the traditional PID controller [18, 30]. Several approaches to tune fractional order $PI^{\lambda}D^{\mu}$ exist, with some notable works that use the theory of fractional calculus in controlling both integer order and fractional order dynamical systems [13, 29]. So for these reasons, the fractional $PI^{\lambda}D^{\mu}$ controller is considered in this paper.

Fractional order controller combined with MIMO QFT approach is not widely used in control problem. The aim of this paper consist on designing a new fractional order (proportional derivative PD) controller and fractional prefilter tuning to ensure control of MIMO systems.

To perform this task, a MIMO controller must designed based on MIMO QFT approach which is based on dividing the initial MIMO problem into a set of equivalent subsystems. This MIMO QFT approach [8, 9] is a frequency domain control method that based on Nichols chart.

The progression of this document is presented as follows: Sect. 2, MIMO QFT approach is presented. Section 3, a fractional order derivative proportional controller has been proposed for motion control of MIMO systems. In Sect. 4, the fractional prefilter is developed. The utility of using a fractional prefilter was taken for the MIMO system designed by solving a quadratic inequalities of the upper and lower tracking specifications. So, the validation on a robot model is studied in order to satisfy the various desired objectives including trajectory tracking which is given in Sect. 5. Finally, this work is concluded in Sect. 6.

## 2 MIMO-QFT

The QFT design technique was introduced by Horowitz and Sidi [8, 9]. The QFT design is a frequency technique that uses the Nichols chart to achieve a desired robust performance in specified bands of system uncertainties. The QFT method is an approach which allows controlling uncertain systems based on two degrees of freedom (Controller and prefilter) [9]. The general problem of the QFT procedure is the determination of the controller and prefilter in order to have uncertain systems in the specified low and high bands. The controller allows the reduction of the effects of uncertainties, while the prefilter aims to put the response of the system in the desired domain. In most cases, only the gain of the closed loop transfer function is used to verify the desired specifications.

The specifications aim at putting the temporal answer or frequency inside the wished lower and superior bands. Several types of approaches were combined with this procedure for the command of the MIMO systems. The following figure shows the control structure used in MIMO QFT with two degrees of freedom.

For the MIMO QFT structure, $P$ is an uncertain system to be controlled. It must be square of order $m \times m$ and a minimum phase system $P = \left[p_{ij}\right]_{m \times m}$.

The controller $G(s) = \text{diag}\,[g_{ii}]$, reduces the effect of uncertainty, the transfer matrix of the prefilter $F(s)$, makes it possible to put the temporal or frequency response of the system inside lower and upper desired bands.

From Fig. 1, the transfer function is expressed by:

$$T = [I + PG]^{-1}\, PGF \tag{1}$$

The matrix $P$ must be invertible, then:

$$\left[P^{-1} + G\right] T = GF \tag{2}$$

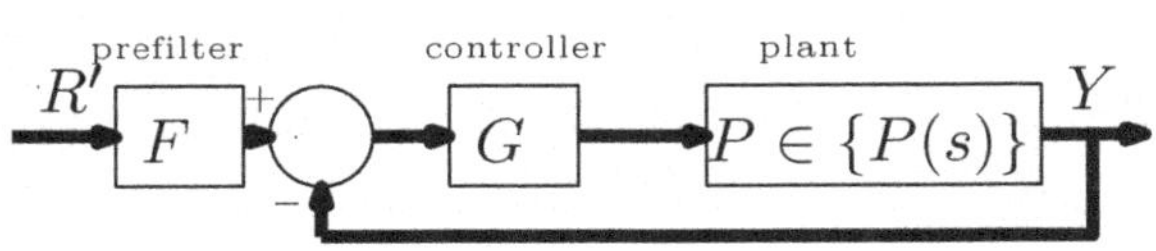


**Fig. 1** Control structure of a MIMO system

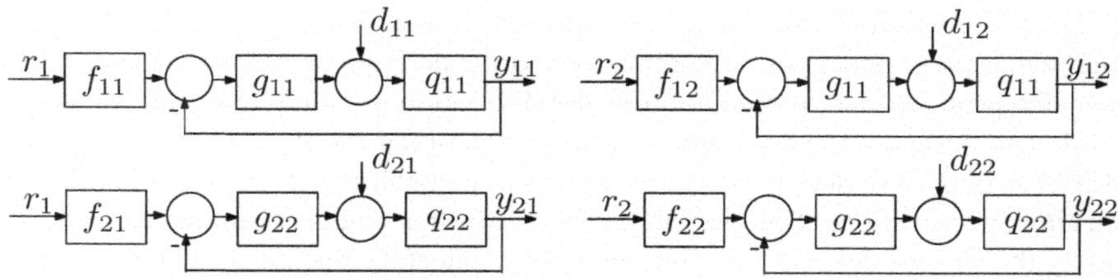


**Fig. 2** Equivalent diagram for 2 × 2 MIMO system

The inverse of the matrix $P$ can be given as follows:

$$P^{-1} = A + B \tag{3}$$

where:

- $A$ is the diagonal part and
- $B$ the anti-diagonal part.

Taking into account Eq. (3), Eq. (2) can be written:

$$T = [A + G]^{-1} [GF - BT] \tag{4}$$

So:

$$T = \begin{bmatrix} t_{11} & t_{12} \\ t_{21} & t_{22} \end{bmatrix} = \begin{bmatrix} \frac{q_{11}}{1+q_{11}g_{11}}\left(g_{11}f_{11} - \frac{t_{21}}{q_{12}}\right) & \frac{q_{11}}{1+q_{11}g_{11}}\left(g_{11}f_{12} - \frac{t_{22}}{q_{12}}\right) \\ \frac{q_{22}}{1+q_{22}g_{22}}\left(g_{22}f_{21} - \frac{t_{11}}{q_{21}}\right) & \frac{q_{22}}{1+q_{22}g_{22}}\left(g_{22}f_{22} - \frac{t_{12}}{q_{21}}\right) \end{bmatrix} \tag{5}$$

In the case of 2 × 2 MIMO system design and based on Eq. (4), 4 equivalent MISO subsystems (see Fig. 2) are extracted from the initial MIMO system [8, 15]. In the given subsystems the transfer function plant is now described by:

$$q_{ij} = \frac{1}{P_{ij}^{-1}} \tag{6}$$

## 3 Fractional Order PD Controller

The issue of differentiation with fractional/on-integer order has been initiated since 17th century and in 19th century, it is well developed [1, 12, 27]. The fractional calculus object consist of evaluating the non integer function differentiation. The fractional order calculus is giving a value of a fractional differentiation of an equation $\frac{d^n y}{dt^n}$ where $n$ is integer, fractional or complex.

The general definition of a fractional integration and differentiation of order $\alpha$ is presented as:

$$ {}_aD_t^\alpha = \begin{Bmatrix} \frac{d^\alpha}{dt^\alpha}, \Re(\alpha) > 0 \\ 1, \Re(\alpha) = 0 \\ \int_a^t (d\tau)^{-\alpha}, \Re(\alpha) < 0 \end{Bmatrix} \tag{7} $$

where $\Re(\alpha)$, a and t are in order the $\alpha$ real part and the operation two bounds. Several presentation of non integer derivative systems are given. The most popular are Grunwald–Letnikov and Riemann–Liouville definitions.

The following expression is related to the Grunwald–Letnikov definition:

$$ {}_aD_t^\alpha f(t) = \lim_{h \to 0} h^{-\alpha} \sum_{j=0}^{\left[\frac{t-\alpha}{h}\right]} (-1)^j \binom{\alpha}{j} f(t - jh) \tag{8} $$

where: $\left[\frac{t-\alpha}{h}\right]$ is the $\left(\frac{t-\alpha}{h}\right)$ integer part and,

$$ \binom{\alpha}{j} = \frac{\Gamma(\alpha + 1)}{\Gamma(j+1)\Gamma(\alpha - j + 1)} \tag{9} $$

$\Gamma$ and $h$ are the gamma function and the step time. The definition of Riemann–Liouville is given by:

$$ \frac{d^\alpha}{dt^\alpha} f = D^\alpha f(t) = \frac{1}{\Gamma(n-\alpha)} \left(\frac{d}{dt}\right)^n \int_0^t \frac{f(\tau)}{(t-\tau)^{1-n+\alpha}} d\tau \tag{10} $$

$n = [\alpha] + 1$, ($[\alpha]$: the integer part of $\alpha$).

There are various approximation methods of the fractional order functions. These approximations [6, 11, 12, 14, 17, 20, 23, 28] are given in both discrete and continuous domains.

The most common form of a fractional order PID controller is the $PI^\lambda D^\mu$ controller [34], involving an integrator of order $\lambda$ and a differentiator of order $\mu$, where $\mu$ and $\lambda$ can be any real numbers. The transfer function of such a controller $g(s)$ has the following form. With $g(s) = (k_p + k_I s^{-\lambda} + k_d s^\mu)$, Where $(\lambda, \mu > 0)$, $k_p$ is the proportional constant, $k_I$ is the integration constant and $k_d$ is the differentiation constant.

Clearly, selecting $\lambda = 1$ and $\mu = 1$, a classical PID controller can be recovered. Using $\lambda = 0, \mu = 1$, and, $\lambda = 1, \mu = 0$, respectively, corresponds to the conventional PD and PI controllers (see Fig. 3). All these classical types of PID controllers

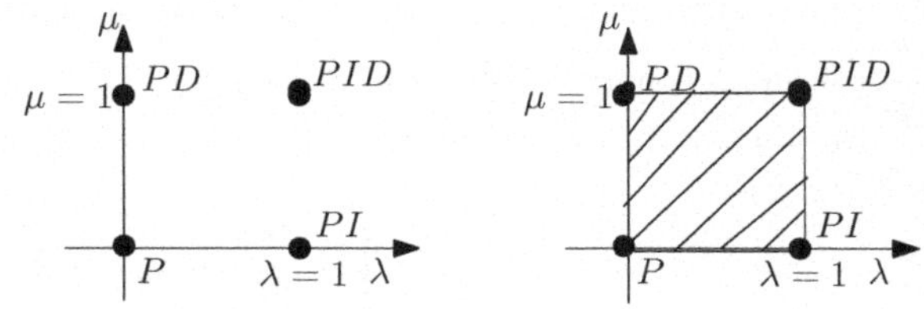


**Fig. 3** Fractional order PID controllers

are special cases of the $PI^{\lambda}D^{\mu}$ controller [24]. It can be expected that the $PI^{\lambda}D^{\mu}$ controller may enhance the systems control performance.

One of the most important advantages of the $PI^{\lambda}D^{\mu}$ controller is the possible better control of fractional order dynamical systems. Another advantage lies in the fact that the $PI^{\lambda}D^{\mu}$ controllers are less sensitive to changes of parameters of a controlled system [24]. In this document, the fractional order $PD^{\mu}$ controller is chosen because of its improved performance compared to the conventional integer controller PD. The $PD^{\mu}$ controller is described by its following transfer function:

$$g(s) = k_p\left(1 + k_d s^{\mu}\right) \tag{11}$$

where $s$ is the Laplace operator; $k_d$ and $k_p$ are the derivative and proportional gains, while $\mu$ is the fractional order differentiation.

The trigonometric function of the fractional order $PD^{\mu}$ controller is written as follows [5, 19]:

$$g(j\omega) = k_p\left[1 + k_d\omega^{\mu}\left(\cos\frac{\pi\mu}{2} + j\sin\frac{\pi\mu}{2}\right)\right] \tag{12}$$

The tuning of the fractional order PD$^{\mu}$ controller consist of determining the values of the three controller parameters, $k_p$, $k_d$ and $\mu$ based on imposed performance specifications.

The parameters of the controller $k_p$, $k_d$ and $\mu$ are calculated according to frequency domain constraints (13), (14). The first two constraints can be written as follows for the open loop system:

$$\varphi\left(H_{open\text{-}loop(j\omega_{gc})}\right) = -\pi + \varphi_m \tag{13}$$

where $H_{open\text{-}loop}(s) = g(s)q(s)$, $\omega_{gc}$ is the crossover frequency and $\varphi_m$ is the desired phase margin. The robustness constraint is allowed by the respect of the following equation [5, 10, 19]:

$$\frac{d\left(\varphi_{open\text{-}loop}\left(j\omega_{gc}\right)\right)}{d\omega_{gc}} = 0 \tag{14}$$

## 4 Fractional Prefilter

To take into account the bandwidth of actuators, the second approach named prefiltering was developed. The regulation of these filters is often empirical, thus the link between the temporal borders and frequency not being able to be established in a general case. The purpose of the generation of movement by basing itself on the use of a fractional filter as prefilters, is to reduce the factor of echo of the subjection upstream to which it is place. While limiting the requests by the instruction of the appropriate modes of actuators, a continuous optimization of both constituent parameters of prefilters is made to minimize the necessary duration to achieve the movement. The fractional filters can be implemented like analog or digital ones, which facilitate its industrial implementation [16, 21]. Fractional prefilter have proven its usefulness in combination with QFT approach to solve the problem of MIMO process in trajectory tracking [15, 25, 32]. In this section a new fractional order prefilter is proposed in combination with PD$^{\mu}$ controller to ensure desired specifications. The non integer prefilter expression is given by:

$$f(s) = \frac{1}{1 + (\tau s)^{\eta}} \tag{15}$$

where $\eta$ is the fractional order and $\tau$ is the time constant.

The two parameters of this fractional order prefilter are deduced from an integral gap optimization. The criterion of integral gap, for a step response, is described as follows:

$$I_c \leq \eta \tau \tag{16}$$

In the case of $n \times n$ MIMO systems, this Eq. (16) is enhanced to be:

$$I_c \leq \eta_1 \tau_1 + \eta_2 \tau_2 + \cdots + \eta_n \tau_n \tag{17}$$

where $\eta_i$ and $\tau_i$ $(for \;\; i = 1 \ldots n)$ are the parameters of designed prefilter to the $i$th subsystems.

The prefilter parameters are obtained while minimizing the criterion (17) and respecting following specifications:

- The performance of rejection of disturbance at the output is allowed by:

$$\left| \frac{1}{1 + H_{open\text{-}loop}(s)} \right| \leq \omega_{d0} \tag{18}$$

  with $\omega_{d0}$ is a positive constant.
- The performance of robust tracking is ensured by both controller and prefilter respecting following equation:

$$|T_L(j\omega)| \le \left| \frac{F(j\omega)H_{open\text{-}loop}(s)}{1 + H_{open\text{-}loop}(s)} \right| \le |T_U(j\omega)| \tag{19}$$

where $T_L(j\omega)$ and $T_U(j\omega)$ named the lower and higher tracking models. The system transfer $T_R(j\omega)$ function is described by:

$$T_R(j\omega) = \frac{g(j\omega)q(j\omega)}{1 + g(j\omega)q(j\omega)} \tag{20}$$

At each frequency $\omega$, the Eq. (19) lead to two inequalities (21):

$$\begin{aligned} |T_L(j\omega)| - |F(j\omega)|\,|T_R(j\omega)| \le 0 \\ |F(j\omega)|\,|T_R(j\omega)| - |T_U(j\omega)| \le 0 \end{aligned} \tag{21}$$

## 5 Results and Discussion

The model of SCARA (Selective Compliance Robot Assembly) robot is one of the most used robots in industries (Fig. 4). Only the first and second axis was considered as shown in Figs. 5 and 6. The SCARA robot linearized model is used in this part to verify the used approach [7]. The MIMO QFT approach is used here to extract the equivalent subsystems. After that, the RGA matrix design is used to obtain the selected pairing (input 1 with output 1 and input 2 with output 2) [30].

The transfer function matrix $P$ of the system has been extracted as:

$$P(s) = \begin{pmatrix} \dfrac{\alpha_2 s + \nu_2}{k\, s\, \Delta(s)} & \dfrac{-(\alpha_2 + \alpha_3 h)}{k\, \Delta(s)} \\ \dfrac{-(\alpha_2 + \alpha_3 h)}{k\, \Delta(s)} & \dfrac{(\alpha_1 + 2\alpha_3 h)\, s + \nu_1}{k\, s\, \Delta(s)} \end{pmatrix} \tag{22}$$

where $\alpha_i$(i = 1, 2, 3) and $\eta_j$(j = 0, 1, 2) are as follows:

- $\eta_0 = \nu_1 \nu_2$
- $\eta_1 = \alpha_2 \nu_1 + \nu_2 (\alpha_1 + 2\alpha_3 h)$
- $\eta_2 = \alpha_2 (\alpha_1 + 2\alpha_3 h) - (\alpha_2 + \alpha_3 h)^2$
- $\alpha_1 = I_1 + I_2 + m_1 x_1^2 + m_2 \left(l_1^2 + x_2^2\right)$
- $\alpha_2 = I_2 + m_2 x_2^2$
- $\alpha_3 = m_2 l_1 x_2$

and

$$\Delta(s) = \eta_2 s^2 + \eta_1 s + \eta_0 \tag{23}$$

With $k, \nu_i, I_i, m_i, x_i, l_1$ and $h$ are orderly arranged as the power amplifier gain, the viscous friction coefficients, the moment of inertia, the mass, the $i$th link, the link 1 length and the value of Cosine of $\delta_2$.

**(a)**

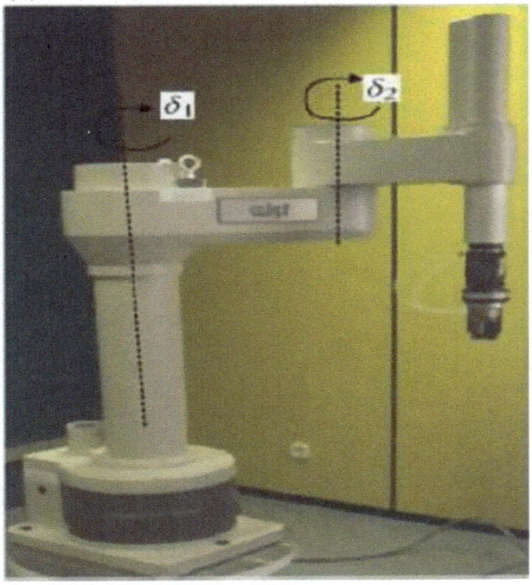


**(b)**

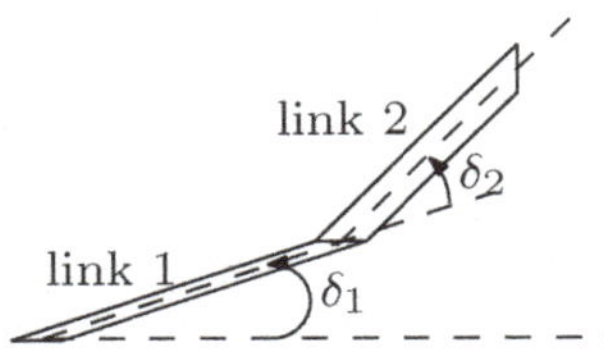


**Fig. 4** **a**, **b**: Configuration of the first and second arms of SCARA robot

**Table 1** Process uncertainties

| | Minimum | Nominal | Maximum |
|---|---|---|---|
| $\alpha_1 k$ | 719 | 766 | 813 |
| $\alpha_2 k$ | 186 | 193 | 200 |
| $\alpha_3 k$ | 134 | 182 | 230 |
| $\nu_1 k$ | 67 | 224 | 381 |
| $\nu_2 k$ | 11.6 | 51.75 | 91.9 |
| $\mu_1 k$ | 344 | 351 | 358 |
| $\mu_2 k$ | 262 | 292.5 | 323 |
| $h$ | −1 | 0 | 1 |

The parametric uncertainties of the plant are given in Table 1.

The closed loop transfer function must be under the specified lower and upper bounds $T_L$ and $T_U$:

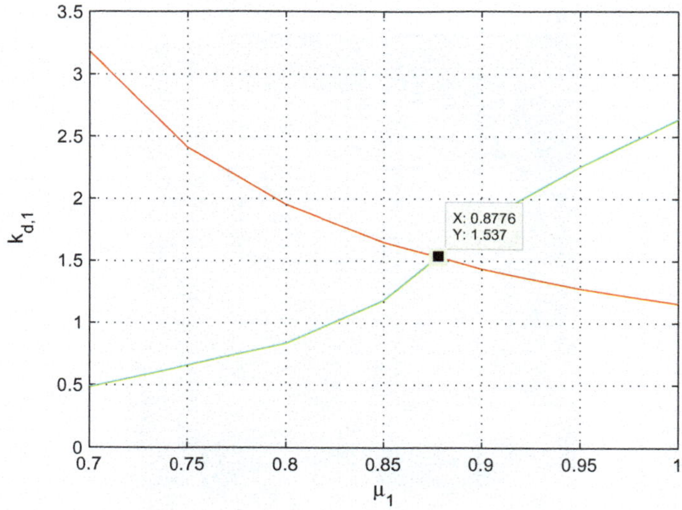


**Fig. 5** $k_{d,1}$ plot function of non integer order $\mu_1$

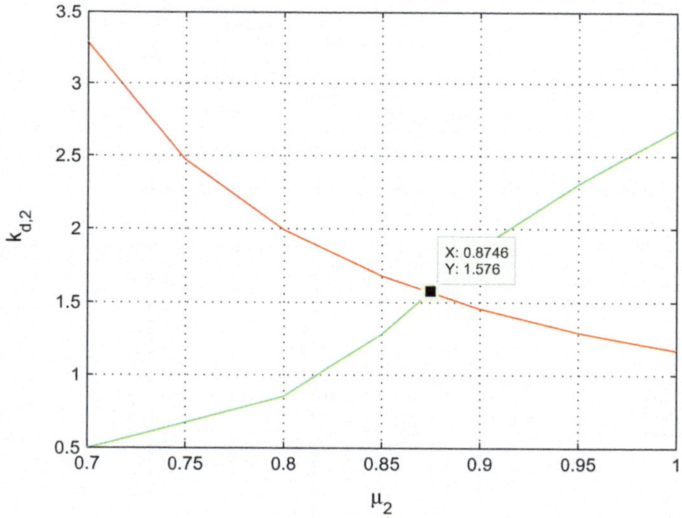


**Fig. 6** $k_{d,2}$ plot function of non integer order $\mu_2$

$$T_L(s) = \frac{2.25}{(s^2 + 4.5s + 2.25)(\frac{1}{10}s + 1)} \tag{24}$$

$$T_U(s) = \frac{12.25(\frac{1}{30}s + 1)}{(s^2 + 5.25s + 12.25)} \tag{25}$$

After using the QFT procedure, the $i$th process transfer function for each subsystems $q_{ii}(s)$ is supposed to be written in this form (26) with $Re$ and $Im$ are respectively real and imaginary parts of the denominator of $q_{ii}$:

$$q_{ii}(\omega) = \frac{1}{Re(\omega) + jIm(\omega)} \tag{26}$$

The controller for each subsystem is given by Eq. (12), so based on (13) and (14), with phase margin $\varphi_m = 70°$ and crossover frequency $w_{cg} = 1$ rad/s, we can obtain following Figs. 5 and 6 by resolving these equations:

$$k_d\omega_{cg}^{\mu} \sin\frac{\pi\mu}{2} - \left[1 + \left[k_d\omega_{cg}^{\mu}\cos\frac{\pi\mu}{2}\right]\tan\left(\varphi_m - \pi + \arctan\frac{Im(j\omega_{cg})}{Re(j\omega_{cg})}\right)\right] = 0 \tag{27}$$

$$\frac{\mu k_d\omega_{cg}^{\mu-1}\sin\frac{\pi\mu}{2}}{1 + 2k_d\omega_{cg}^{\mu}\cos\frac{\pi\mu}{2} + k_d^2\omega_{cg}^2} + \left.\frac{\dot{I}mRe - Im\dot{R}e}{Re^2 + Im^2}\right|_{\omega=\omega_{cg}} = 0 \tag{28}$$

The intersection between these two expressions gives the derivative gain $k_{d,i}$ and the derivative differentiation $\mu_i$ for each MISO loop i.

Once $k_{d,i}$ and $\mu_i$ are determined, it is time to look for $k_{p,i}$. The $k_{p,i}$ is designed in order to make the closed loop response of each subsystems under higher and lower bounds for some frequencies.

Then, the obtained matrix transfer function of the controller is:

$$G(s) = \begin{pmatrix} g_{11}(s) & 0 \\ 0 & g_{22}(s) \end{pmatrix} \tag{29}$$

where:

$$g_{11}(s) = 2800\left(1 + 1.537s^{0.8776}\right), \quad g_{22}(s) = 2000\left(1 + 1.574s^{0.8745}\right)$$

Now, the prefilter is obtained while minimizing the criterion (17) a and respecting the performance specifications given by Eqs. (19) and (20). The prefilter matrix is as follows:

$$F(s) = \begin{pmatrix} f_{11}(s) & 0 \\ 0 & f_{22}(s) \end{pmatrix} \tag{30}$$

where:

$$f_{11}(s) = \frac{1}{1 + (0.804s)^{0.9}}, \quad f_{22}(s) = \frac{1}{1 + (1.128s)^{0.956}}$$

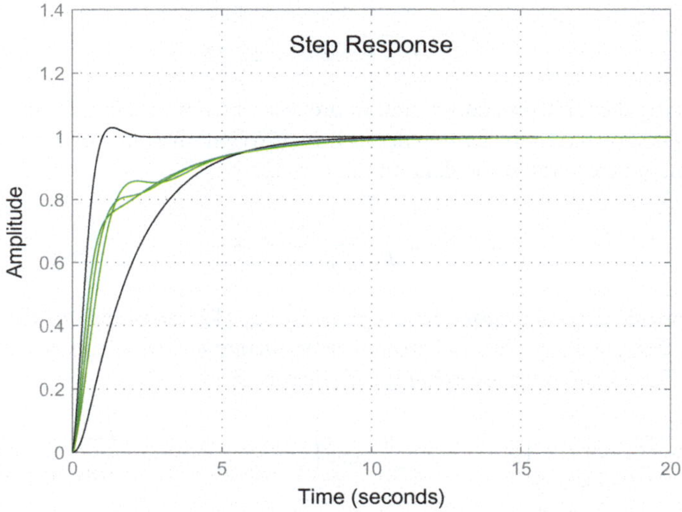


**Fig. 7** 1st closed loop step response (black: desired upper and lower bounds, green: closed loop responses under uncertainties)

Observing Figs. 7 and 8, it is remarked that both closed loop 1 and 2 outputs respected the given tracking specification bound. This figure shows the utility of the proposed design of fractional order and prefilter tuning which is also confirmed by the frequency responses of loop 1 and 2 given by Figs. 9 and 10.

Robust control methods seek to bound uncertain systems. Given a higher and lower bounds, the feedback system with uncertainty meets the desired specifications in all conditions by the robust designed controller. These obtained results in Figs. 7, 8, 9 and 10 confirm the robustness and efficiency of the developed fractional $PD^{\mu}$ controller with the fractional prefilter. I noticed that all uncertain plant step responses have respected our desired tracking specifications. Both in time and frequency domains, the same robustness results are shown. It can be easily noted that our developed fractional order controller and prefilter give a satisfactory performances and are robust under uncertainties of the MIMO plant. Previous developed controllers used in the same context in literature are based on complex designs like CRONE and $H_\infty$ controls [31–33] but our designed method is efficient and simple and can be adapted for different systems in industry.

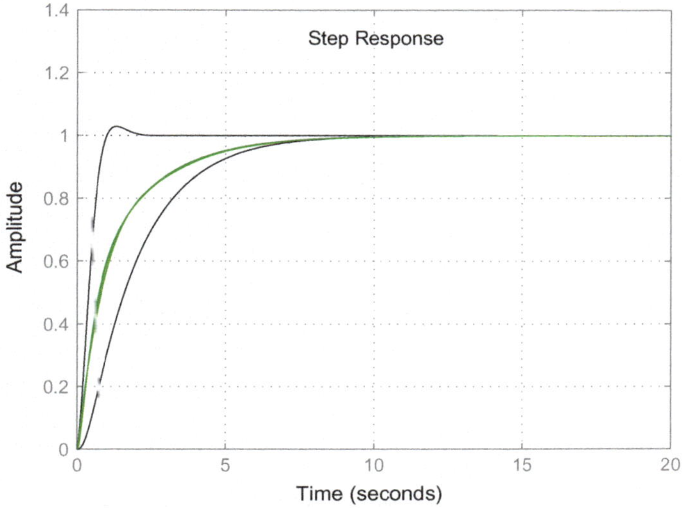


**Fig. 8** 2nd closed loop step response (black: desired upper and lower bounds, green: closed loop responses under uncertainties)

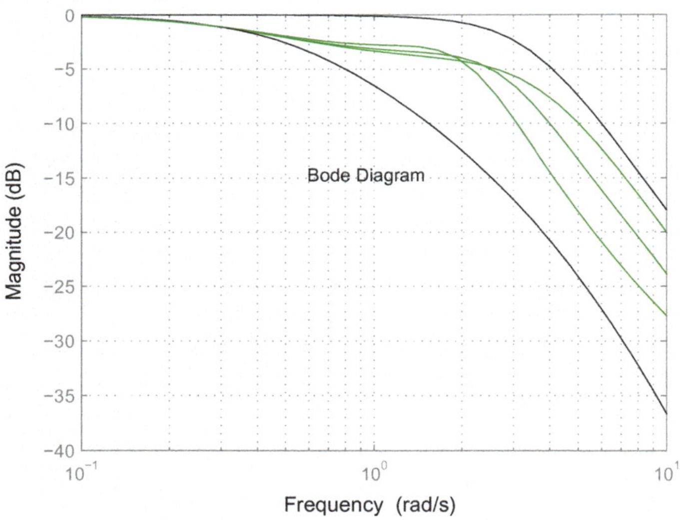


**Fig. 9** 1st closed loop bode response (black: desired upper and lower bounds, green: closed loop responses under uncertainties)

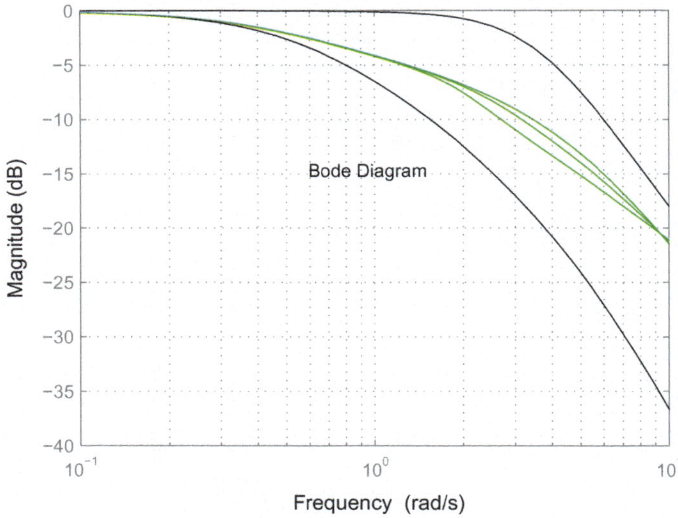


**Fig. 10** 2nd closed loop bode response (black: desired upper and lower bounds, green: closed loop responses under uncertainties)

# 6 Conclusion

In this paper, based on MIMO-QFT control design, a new reliable and simple design of non integer order controller and prefilter is discussed. The MIMO-QFT methodology is based on dividing the original MIMO system on equivalent MISO structure in order to make its control simple. After this step, the controller and prefilter parameters determination is our object. An analytic tuning method for a fractional order PD controller in combination with QFT design and an optimization of fractional order prefilter are presented to solve the problem of MIMO processes. The validation of this method is based on a SCARA robot model of order $2 \times 2$. In future work, we will be interested by the real application of the developed approach and the general form of fractional PID controller will be treated.

# References

1. Abel, N.: Solution de quelques problemes a l'aide d'integrales definies. Gesammelte mathematische Werke **1**, 11–27 (1881)
2. Bagley, R.L., Calico, R.A.: Fractional-order state equations for the control of viscoelastic damped structures. J. Guid. Control Dyn. **14**, 304–311 (1991)
3. Bagley, R.L., Torvik, P.: On the appearance of the fractional derivative in the behavior of real materials. J. Appl. Mech. **51**, 294–298 (1984)
4. Chen, Y., Xue, D.: A comparative introduction of four fractional order controllers. In: 4th IEEE World Congress on Intelligent Control and Automatic (WCICA02) (2002)

5. Dulf, E.-H., Timis, D., Muresan, C.-I.: Robust fractional order controllers for distributed systems. Acta Polytech. Hung. **14**, 163–176 (2017)
6. Euler, L.: Foundations of Differential Calculus. Springer, New York (2000)
7. Garcia-Sanz, M., Egana, I., Villanueva, J.: Interval modelling of SCARA robot for robust control. In: 10th Mediterranean Conference on Control and Automation - MED (2002)
8. Horowitz, I.: Synthesis of Feedback Systems. Academic Press, New York (1963)
9. Horowitz, I.: Survey of quantitative feedback theory (qft). Int. J. Control **53**, 255–291 (2001)
10. Keyser, R.D., Muresan, C.I., Ionescu, C.M.: A novel auto-tuning method for fractional order pi/pd controllers. ISA Trans. **62**, 268–275 (2016)
11. Letnikov, A.: Theory of differentiation of fractional order. Mat. Sbornik **3**, 1–68 (1868)
12. Liouville, J.: Memoire sur quelques questions de geometrie et de mecanique, et sur un nouveau genre pour resoudre ces questions. J. Ec. Polytech. **13**, 1–69 (1832)
13. Luo, Y., Chen, Y.Q., Wang, C.Y., Pi, Y.G.: Tuning fractional order proportional integral controllers for fractional order systems. J. Process Control **20**(7), 823–831 (2010)
14. Mainardi, F.: Fractional relaxation in anelastic solids. J. Alloys Compd. **211**(212), 534–538 (1994)
15. Melchior, P., Inarn, C., Oustaloup, A.: Path tracking design by fractional prefilter extension to square MIMO systems. In: Proceedings of the ASME 2009 International Design Engineering Technical Conference and Computers and Information in Engineering Conference (2009)
16. Melchior, P., Robin, G., L'Hostis, S., Levron, F.: Non integer order movement generation in path planning. In: CESA'98 IMACS-IEEE/SMC Multiconference Computational Engineering in Systems Applications (1998)
17. Metzler, R., Klafter, J.: The random walk's guide to anomalous diffusion: a fractional dynamics approach. Phys. Rep. **339**, 1–77 (2000)
18. Muresan, C.I.: Fractional calculus: from simple control solutions to complex. In: Nonlinear Systems and Complexity, vol. 6, pp. 113–134. Springer, Cham (2014)
19. Muresan, C.-I., Folea, S., Mois, G., Dulf, E.-H.: Development and implementation of an fpga based fractional order controller for a DC motor. Mechatronics **23**, 798–804 (2013)
20. Osler, T.J.: The integral analog of the leibniz rule. Math. Comp. **26**, 903–915 (1972)
21. Oustaloup, A.: Systèmes asservis linéaires d'ordre fractionnaire. Masson, Paris (1983)
22. Oustaloup, A.: La commande CRONE. Hermes, Paris (1991)
23. Oustaloup, A.: La dérivation non entière: théorie, synthèse et applications. Hermes, Paris (1995)
24. Pan, I., Das, S., Gupta, A.: Handling packet dropouts and random delays for unstable delayed processes in ncs by optimal tuning of $pi^{\lambda}d^{\mu}$ controllers with evolutionary algorithms. ISA Trans. **50**, 557–572 (2011)
25. Patil, M.D., Nataraj, P., Vyawahare, V.: Design of robust fractional-order controller and prefilters for multivariable system using interval constraint satisfaction technique. Int. J. Dynam. Control **5**, 145–158 (2016)
26. Podlubny, I.: Fractional order systems and $pi^{\lambda}d^{\mu}$ controllers. IEEE Trans. Autom. Control **44**, 208–214 (1999)
27. Riemann, B.: Versuch einer allgemeinen auffassung der integration und differentiation. Gesammelte Mathematische Werke und Wissenschaftlicher **1**, 331–344 (1876)
28. Ross, B.: Fractional calculus and its applications. In: Proceedings of the International Conference held at the University of New Haven. Springer, New York (1974)
29. Xue, D., Chen, Y.: A comparative introduction of four fractional order controllers. In: 4th IEEE World Congress on Intelligent Control and Automation (2002)
30. Xue, D., Zhao, C., Chen, Y.Q.: Fractional order PID control a DC-motor with elastic shaft: a case study. In: Proceedings of the American Control Conference (2006)
31. Yousfi, N., Melchior, P., Rekik, C., Derbel, N., Oustaloup, A.: Comparison between $h_{\infty}$ and crone control combined with qft approach to control multivariable systems in path tracking design. Int. J. Comput. Appl. **45**, 1–9 (2012a)
32. Yousfi, N., Melchior, P., Rekik, C., Derbel, N., Oustaloup, A.: Path tracking design based on Davidson-Cole prefilter using a centralized crone controller applied to multivariable systems. Nonlinear Dyn. **71**, 701–712 (2013)

33. Yousfi, N., Melchior, P., Rekik, C., Derbel, N., Oustaloup, A.: Path tracking design by fractional prefilter using a combined qft/$h_\infty$ design for tdof uncertain feedback systems. J. Appl. Nonlinear Dyn. **1**, 239–261 (2012b)
34. Zhao, C., Yang, D., Xue, Q.C.: A fractional order PID tuning algorithm for a class of fractional order plants. In: IEEE International Conference on Mechatronics and Automation (2005)

# Flexible-Link Manipulators: Dynamic Analysis and Advanced Control Strategies

**R. Boucetta, S. Hamdi and S. Bel Hadj Ali**

**Abstract** This chapter contains advanced studies in dynamic modelling and nonlinear control strategies applied to flexible-link robotic manipulators increasingly in demand in many fields such as industrial domain, medical intervention, and space exploitation. Taking into consideration the flexibility effect, Hamilton's principle and Euler–Lagrange equations are associated to determine a highly nonlinear and coupled dynamic model. Therefore, the main control goals are to reach a perfect trajectory tracking without vibration impact. That is why, PD, Fuzzy, and gain scheduling Fuzzy PD controllers are applied to a rigid-flexible two links manipulator and then compared in terms of robustness and vibration minimization. A stability study is accomplished using the candidate function of Lyapunov. To improve performances, a robust Fractional Order Fuzzy PD (FOFPD) controller is developed by using non-integer order differentiator operators in the fuzzy PD controller. The gains of the FOFPD are normalized with the Particle Swarm Optimization (PSO) algorithm. The small gain theorem is used to establish the sufficient condition for bounded input-bounded output (BIBO) stability in closed-loop. Simulation results are introduced for each case of control to discuss reached performances.

**Keywords** Rigid-flexible manipulator · Dynamic model · PD control · Fuzzy control · Fractional order fuzzy control · Particle Swarm Optimization

---

R. Boucetta (✉)
Physics Department, Sciences Faculty of Sfax, University of Sfax, Sfax, Tunisia
e-mail: boucetta.rahma@gmail.com

S. Hamdi
MACS Laboratory, National Engineering School of Gabes, University of Gabes, Gabes, Tunisia
e-mail: hamdisalim99@gmail.com

S. Bel Hadj Ali
Preparatory Institute of Tunis El Manar, University of Tunis El Manar, Tunis, Tunisia
e-mail: saloua.belhadjali@enit.rnu.tn



J. Ghommam et al. (eds.), *New Trends in Robot Control*, Studies in Systems, Decision and Control 270, https://doi.org/10.1007/978-981-15-1819-5_2

## 1 Introduction

Flexible robotic manipulators begin to matter in the industrial environment led by needs and demands in production and manufacturing. Their application is also expanded to some critical areas where men cannot intervene properly like space field, medical operations, and radioactive or nuclear plants.

Advantages of manipulators with lightweight arms are multiple: They are light in weight, they require less material and small actuators, they consume lower energy, they are more manoeuvrable and transportable. Moreover, they present a reduced overall cost and can withstand high weight loads in relation to the weight of the robot itself.

However, the major disadvantage of flexible link manipulators is vibration in the free end-point. Their dynamics present non-minimum phase characteristics, high nonlinearity and coupling which make their control a difficult task and to pose a great challenge to establish an adequate command law for vibration cancellation.

In order to develop efficient control laws on a rigid-flexible two-links manipulator, an accurate mathematical model must be properly established. However various methods are used for dynamic modelling of flexible link manipulators, as the Lumped Parameter Method (LPM), Finite Element Method (FEM) and Assumed Mode Method (AMM). Sun, Hong and Xilun [17, 26, 32] used LPM for the dynamic modelling of flexible manipulators. The FEM was used by many researchers such as Baroudi, Bayo, Boucetta and Spong [3–5, 25]. However, the AMM is based on Hamilton's principle and theory of beams of Timoshenko and Euler, where many studies were performed by this method for modelling a flexible manipulator such as those of Hamdi, Fenili and Chanwikrai [7, 12, 15].

To control a rigid-flexible manipulator, different types of controllers are used such as conventional and intelligent controllers. For example, Hussein and Qiu [23, 27] have studied the performance of the PD controller for the manipulator. Fenili [11] developed the State-Dependent Riccati Equation to control a rigid-flexible manipulator. The sliding mode control is studied by Lee and Lochan [20, 21], a boundary control problem for a constrained two-link rigid-flexible manipulator is presented by Cao [6]. An optimal trajectory planning technique for suppressing residual vibrations is performed by Abe and Tarvirdizadeh [1, 28]. However, many researchers are interested in the intelligent control for a flexible manipulator, as He, Irani and Tian [16, 18, 29]. Using neural networks, an adaptive iterative learning algorithm for boundary control of a coupled ordinary differential equation and partial differential equation (ODE-PDE) for two-link rigid-flexible manipulator is developed by Cao [6] and in [2, 13, 34]. Alavander, Gao and Zhang used a fuzzy logic control to study the flexible or rigid-flexible manipulator. The control by fuzzy neural network for manipulator robots is developed by Tian, Tinkir and Gao [14, 30, 31].

On the other hand, a fuzzy logic controller (FLC) has been the most successful and popular of the intelligent controllers in the industry and more particularly for manipulators robots [8, 33]. Furthermore, there are several classes of fuzzy controllers such as fuzzy PI, PD and PID controllers. Kumar [19] exhibited self-tuning abilities, their gains are normalized online with a change in error and rate of change of error. Generally, to obtain a good performance an optimization algorithm is used to determine the gains of the fuzzy controller. There are several types of optimization algorithms such as Particle Swarm Optimization (PSO) [10], Backtracking Search Algorithm (BSA), Genetic Algorithm (GA), etc. Traditionally, the field of control has been dominated with integer order controllers. Recently, the innovation is in the field of fractional calculus, it was evident that fractional order integral and differential can be used in control applications to provide better performance compared to integer order controllers [9, 22, 24].

This chapter is an advanced study in modelling and control of flexible robotic manipulators. Section 2 presents the development of a dynamic model based on Hamilton's principle associated to Euler–Lagrange equations. In Sect. 3, the classical PD controller is illustrated. The fuzzy and the gain scheduling fuzzy PD controllers are explained in Sects. 4 and 5 respectively. Section 6 shows a new type of control such as the Fractional Order Fuzzy PD (FOFPD) controller applied to the rigid-flexible manipulator, compared then to an integer order fuzzy PD (FPD) controller. The gains of these two last controllers are optimized using Particle Swarm Optimization (PSO) algorithm using the Integral of Absolute Error (IAE) as the objective function for this algorithm to follow the desired trajectory of the manipulator.

## 2 Dynamic Modelling

Figure 1 shows the structure of the rigid-flexible manipulator robot system. The control torques applied to the center of the articulation of the rigid link and the flexible link which are driven by a DC motor are respectively $\tau_1(t)$ and $\tau_2(t)$. $m_1$, $L_1$, $I_1$ and $E_1$

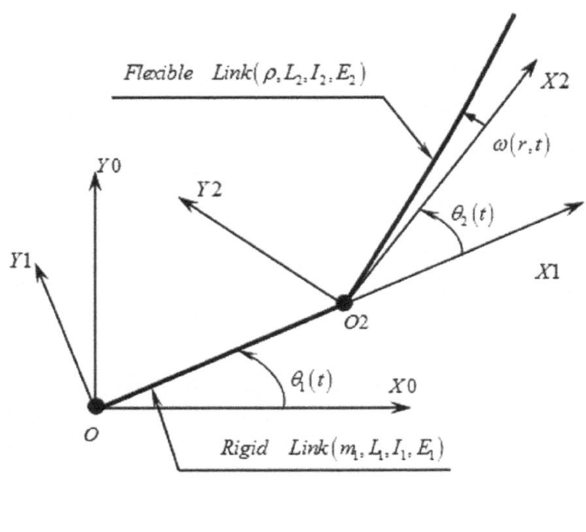


**Fig. 1** Schematic representation of the rigid-flexible manipulator

represent the mass, the length, the moment of inertia of the motor giving the rotation of the arm and the Young's modulus of the rigid link respectively, and $\rho$, $L_2$, $I_2$ and $E_2$ represent the linear density, the length, the moment of inertia of the motor giving the rotation of the arm and the Young's modulus of the flexible link respectively. $\theta_1(t)$ denotes the angular position of the rigid link in the horizontal plane $OX_0Y_0$, $\theta_2(t)$ represents the angular displacement of the flexible link in the horizontal plane $OX_1Y_1$ and $w(r,t)$ is the deflection of the flexible link at position $r$ from the center of articulation measured along the axis $O_2X_2$. $OX_0Y_0$, $OX_1Y_1$ and $O_2X_2Y_2$ present respectively the right-handed inertial and body frames as shown in the Fig. 1.

The motion dynamic equation of the rigid-flexible manipulator is obtained using the Lagrange equation associated to the Hamilton's principle [7], where the Lagrangian is the difference between kinetic and potential energies. The total kinetic energy $T$ of the system is given by the following expression:

$$T = \frac{1}{2}I_1\dot{\theta}_1^2 + \frac{1}{2}m_1\left(\frac{L_1}{2}\dot{\theta}_1\right)^2 + \frac{1}{2}I_2(\dot{\theta}_1+\dot{\theta}_2)^2 + \frac{1}{2}\rho\int_0^{L_2}\Big[\left(w(r,t)(\dot{\theta}_1+\dot{\theta}_2)\right)^2 + \left(r(\dot{\theta}_1+\dot{\theta}_2)\right)^2 + \dot{w}(r,t)^2 - 2L_1\dot{\theta}_1\sin(\theta_2)\,w(r,t)(\dot{\theta}_1+\dot{\theta}_2) + (L_1\dot{\theta}_1)^2 + 2r(\dot{\theta}_1+\dot{\theta}_2)\dot{w}(r,t) + 2L_1\dot{\theta}_1\cos(\theta_2)\left(r(\dot{\theta}_1+\dot{\theta}_2)+\dot{w}(r,t)\right)\Big]dr \quad (1)$$

The potential energy of the system depending on the flexibility of the flexible arm has the following form:

$$V = \frac{1}{2}\int_0^{L_2} EI\left(\frac{\partial^2 w(r,t)}{\partial r^2}\right)^2 dr \quad (2)$$

So, the Lagrangian $L = T - V$ of the rigid-flexible manipulator is expressed as:

$$L = \frac{1}{2}\left[I_1 + m_1\frac{L_1^2}{4}\right]\dot{\theta}_1^2 + \frac{1}{2}I_2[\dot{\theta}_1+\dot{\theta}_2]^2 + \frac{1}{2}\rho\int_0^{L_2}\Big[{L_1}^2\dot{\theta}_1^2 + 2r[\dot{\theta}_1+\dot{\theta}_2]\dot{w}(r,t) + r^2[\dot{\theta}_1+\dot{\theta}_2]^2 + 2L_1\dot{\theta}_1\left(r[\dot{\theta}_1+\dot{\theta}_2]+\dot{w}(r,t)\right)\cos\theta_2 + \left(w(r,t)[\dot{\theta}_1+\dot{\theta}_2]\right)^2 + \dot{w}(r,t)^2 - 2L_1\dot{\theta}_1\sin(\theta_2)\,w(r,t)(\dot{\theta}_1+\dot{\theta}_2)\Big]dr - \frac{1}{2}\int_0^{L_2} EI\left(\frac{\partial^2 w(r,t)}{\partial r^2}\right)^2 dr \quad (3)$$

The motion equations of the manipulator robot system are obtained by deriving the Lagrangian using the Euler–Lagrange equation as follows:

$$Q = -\frac{\partial L}{\partial q} + \frac{d}{dt}\left(\frac{\partial L}{\partial \dot{q}}\right) - \frac{d^2}{dr^2}\left(\frac{\partial L}{\partial q''}\right) \quad (4)$$

where $q = \left[\, \theta_1(t)\ \theta_2(t)\ w(r,t) \,\right]^T$ are the variables of the rigid-flexible manipulator and $Q$ is the external force vector applied to the system.

Substituting Eq. (3) in Eq. (4), three motion dynamic equations of the manipulator are expressed as:

$$\begin{aligned}
&\left\{ m_1 \frac{L_1^2}{4} + \rho L_1^2 L_2 + \rho L_1 L_2^2 \cos(\theta_2) + \int_0^{L_2} \left( \rho w^2(r,t) - 2\rho L_1 \sin(\theta_2) w(r,t) \right) dr \right. \\
&\left. + I_1 + I_2 + \rho \frac{L_2^3}{3} \right\} \ddot{\theta}_1 + \left\{ \int_0^{L_2} \left( \rho w^2(r,t) - \rho L_1 \sin(\theta_2) w(r,t) \right) dr + I_2 + \rho \frac{L_2^3}{3} \right. \\
&\left. + \rho L_1 \frac{L_2^2}{2} \cos(\theta_2) \right\} \ddot{\theta}_2 - \rho \left\{ L_1 \frac{L_2^2}{2} \sin(\theta_2) + \int_0^{L_2} L_1 \cos(\theta_2) w(r,t)\, dr \right\} \dot{\theta}_2^2 \\
&- \rho \left\{ L_1 L_2^2 \sin(\theta_2) + \int_0^{L_2} 2 L_1 \cos(\theta_2) w(r,t)\, dr \right\} \dot{\theta}_1 \dot{\theta}_2 + \left\{ \int_0^{L_2} 2\rho \left( \dot{w}(r,t) w(r,t) \right. \right. \\
&\left. - L_1 \sin(\theta_2) \dot{w}(r,t) \right) dr \} \left( \dot{\theta}_1 + \dot{\theta}_2 \right) + \left. \int_0^{L_2} \rho \left( r + L_1 \cos(\theta_2) \right) \ddot{w}(r,t)\, dr \right\} = \tau_1
\end{aligned} \tag{5}$$

$$\begin{aligned}
&\left\{ I_2 + \rho L_1 \frac{L_2^2}{2} \cos(\theta_2) + \rho \frac{L_2^3}{3} + \int_0^{L_2} \rho \left( w^2(r,t) - L_1 \sin(\theta_2) w(r,t) \right) dr \right\} \ddot{\theta}_1 \\
&+ \left\{ I_2 + \rho \frac{L_2^3}{3} + \int_0^{L_2} \rho w^2(r,t)\, dr \right\} \ddot{\theta}_2 + \int_0^{L_2} 2\rho w(r,t) \dot{w}(r,t)\, dr \left( \dot{\theta}_1 + \dot{\theta}_2 \right) \\
&+ \rho \left\{ L_1 \frac{L_2^2}{2} \sin(\theta_2) + \int_0^{L_2} L_1 \cos(\theta_2) w(r,t)\, dr \right\} \dot{\theta}_1^2 + \int_0^{L_2} \rho r \ddot{w}(r,t)\, dr = \tau_2
\end{aligned} \tag{6}$$

$$\begin{aligned}
&\rho \left( r + L_1 \cos(\theta_2) \right) \ddot{\theta}_1 + \rho r \ddot{\theta}_2 - \rho w(r,t) \left( \dot{\theta}_1 + \dot{\theta}_2 \right)^2 \\
&+ L_1 \dot{\theta}_1^2 \sin(\theta_2) + EI w''''(r,t) + \rho \ddot{w}(r,t) = 0
\end{aligned} \tag{7}$$

According to the assumed mode method, a separation of the variables allows to write the elastic deformation $w(r,t)$ as follows:

$$w(r,t) = \sum_{i=1}^{n} \varphi_i(r) q_i(t) \tag{8}$$

$q_i(t)$ are the modal coordinates and $\varphi_i(r)$ represent the shape modes of the system. To simplify the solution of the problem, the first two vibration modes ($n = 2$) are considered.

Substituting Eq. (8) in the three previous equations, the motion dynamic equation of the rigid-flexible manipulator robot can be written in a matrix form as:

$$H(q)\ddot{q} + C(q, \dot{q})\dot{q} + Kq = u \tag{9}$$

where:

$$H(q) = \begin{bmatrix} h_1 & h_5 & h_6 & h_7 \\ h_5 & h_2 & h_8 & h_9 \\ h_6 & h_8 & h_3 & h_{10} \\ h_7 & h_9 & h_{10} & h_4 \end{bmatrix} \quad C(q, \dot{q}) = \begin{bmatrix} c_1 & c_2 & c_3 & c_4 \\ c_5 & c_6 & c_7 & c_8 \\ c_9 & c_{10} & c_{11} & c_{12} \\ c_{13} & c_{14} & c_{15} & c_{16} \end{bmatrix}$$

$$K = \begin{bmatrix} 0 & 0 & 0 & 0 \\ 0 & 0 & 0 & 0 \\ 0 & 0 & k_3 & 0 \\ 0 & 0 & 0 & k_4 \end{bmatrix} \quad q = \begin{bmatrix} \theta_1(t) \\ \theta_2(t) \\ q_1(t) \\ q_2(t) \end{bmatrix} \quad u = \begin{bmatrix} \tau_1 \\ \tau_2 \\ 0 \\ 0 \end{bmatrix}$$

## 3 PD Controller

The Proportional Derivative (PD) controller is a control law that allows a closed loop control of an industrial process, where the PD controller compares the measured value of the process with set point value of the system. The error signal that is the difference between these two values is used to calculate the new input value which tends to minimize this difference. Therefore, it's necessary to study the stability for the rigid-flexible manipulator system with the PD controller in different reference trajectory tracking cases. The rotations of the rigid and the flexible bodies are assumed to follow a reference trajectory given by a fifth order polynomial between $t = 0\,\text{s}$ and $t = T = 3\,\text{s}$. After this time, the desired output trajectory remains constant at the final value. The initial and final conditions for the desired joints trajectories are: $\theta_{1d}(0) = 0;\ \theta_{1d}(T) = \frac{\pi}{2}$ and $\theta_{2d}(0) = 0;\ \theta_{2d}(T) = \frac{\pi}{3}$.

The torques applied to the manipulator arms are described as follows:

$$\begin{cases} \tau_1 = \tau_{1r} - K_p(\theta_1 - \theta_{1r}) - K_d(\dot{\theta}_1 - \dot{\theta}_{1r}) \\ \tau_2 = \tau_{2r} - K_p(\theta_2 - \theta_{2r}) - K_d(\dot{\theta}_2 - \dot{\theta}_{2r}) \end{cases} \tag{10}$$

where $\tau_{1r}$ and $\tau_{2r}$ denote feedforward control inputs determined from the inverse dynamics solution and $\theta_{1r}$ and $\theta_{2r}$ denote the paths open loop which are the results of the feedforward control, therefore:

$$\begin{aligned}&\left\{m_1\frac{L_1^2}{4}+\rho L_1^2L_2+\rho L_1L_2^2\cos(\theta_{2r})+\int_0^{L_2}\left(\rho w_r^2(r,t)-2\rho L_1\sin(\theta_{2r})w_r(r,t)\right)dr\right.\\&\left.+I_1+I_2+\rho\frac{L_2^3}{3}\right\}\ddot{\theta}_{1r}+\left\{\int_0^{L_2}\rho\left(w_r^2(r,t)-L_1\sin(\theta_{2r})w_r(r,t)\right)dr+\rho L_1\frac{L_2^2}{2}\cos\theta_{2r}\right.\\&\left.+I_2+\rho\frac{L_2^3}{3}\right\}\ddot{\theta}_{2r}+\int_0^{L_2}\rho(r+L_1\cos(\theta_{2r}))\ddot{w}_r(r,t)\,dr-\rho\left\{\int_0^{L_2}L_1\cos(\theta_{2r})w_r(r,t)\,dr\right.\\&\left.L_1\frac{L_2^2}{2}\sin(\theta_{2r})\right\}\dot{\theta}_{2r}^2-\left\{\rho L_1L_2^2\sin(\theta_{2r})+\int_0^{L_2}2L_1\cos(\theta_{2r})w_r(r,t)\,dr\right\}\dot{\theta}_{1r}\dot{\theta}_{2r}\\&+(\dot{\theta}_{1r}+\dot{\theta}_{2r})\left\{\int_0^{L_2}2\rho(\dot{w}_r(r,t)w_r(r,t)-L_1\sin(\theta_{2r})\dot{w}_r(r,t))\,dr\right\}=\tau_{1r}\end{aligned}\tag{11}$$

$$\begin{aligned}&\left\{I_2+\rho L_1\frac{L_2^2}{2}\cos(\theta_{2r})+\rho\frac{L_2^3}{3}+\int_0^{L_2}\rho\left(w_r^2(r,t)-L_1\sin(\theta_{2r})w_r(r,t)\right)dr\right\}\ddot{\theta}_{1r}\\&+\left\{I_2+\rho\frac{L_2^3}{3}+\int_0^{L_2}\rho w_r^2(r,t)\,dr\right\}\ddot{\theta}_{2r}+\int_0^{L_2}\rho r\ddot{w}_r(r,t)\,dr+\rho\left\{L_1\frac{L_2^2}{2}\sin(\theta_{2r})\right.\\&\left.+\int_0^{L_2}L_1\cos(\theta_{2r})w_r(r,t)\,dr\right\}\dot{\theta}_{1r}^2+\int_0^{L_2}2\rho w_r(r,t)\dot{w}_r(r,t)\,dr\left(\dot{\theta}_{1r}+\dot{\theta}_{2r}\right)=\tau_{2r}\end{aligned}\tag{12}$$

$$\begin{aligned}&\rho(r-L_1\cos(\theta_{2r}))\ddot{\theta}_{1r}+\rho r\ddot{\theta}_{2r}-\rho w_r(r,t)\left(\dot{\theta}_{1r}+\dot{\theta}_{2r}\right)^2\\&+L_1\dot{\theta}_{1r}^2\sin(\theta_{2r})+EI\ddot{w}_r(r,t)+\rho\ddot{w}_r(r,t)=0\end{aligned}\tag{13}$$

with $w_r(r,t)=\dot{w}_r(r,t)=\ddot{w}_r(r,t)=\dddot{w}_r(r,t)=0$.

To show the stability of the closed loop system manipulator with reference trajectory, Lyapunov method is used where the candidate Lyapunov function $V$ is selected as:

$$\begin{aligned}V=&\frac{1}{2}\left(I_1+M_1\frac{L_1^2}{4}\right)\left(\dot{\theta}_1-\dot{\theta}_{1r}\right)^2+\frac{1}{2}I_2\left(\dot{\theta}_1+\dot{\theta}_2-\dot{\theta}_{1r}-\dot{\theta}_{2r}\right)^2+\frac{1}{2}\rho\int_0^{L_2}(\dot{P}\\&-\dot{P}_r)^2dr+\frac{1}{2}EI\int_0^{L_2}\left(w''-w_r''\right)^2dr+\frac{1}{2}K_{p1}(\theta_1-\theta_{1r})^2+\frac{1}{2}K_{p2}(\theta_2-\theta_{2r})^2\end{aligned}\tag{14}$$

with $P=\begin{bmatrix}L_1\cos(\theta_1)+r\cos(\theta_1+\theta_2)-w\sin(\theta_1+\theta_2)\\L_1\sin(\theta_1)+r\sin(\theta_1+\theta_2)+w\cos(\theta_1+\theta_2)\\0\end{bmatrix}$

and $P_r=\begin{bmatrix}L_1\cos(\theta_{1r})+r\cos(\theta_{1r}+\theta_{2r})-w_r\sin(\theta_{1r}+\theta_{2r})\\L_1\sin(\theta_{1r})+r\sin(\theta_{1r}+\theta_{2r})+w_r\cos(\theta_{1r}+\theta_{2r})\\0\end{bmatrix}$

are respectively the real and desired position vectors for the manipulator system. $K_p$ is the proportional gain of the PD controller. The Lyapunov function $V$ is derived afterwards with respect to time to obtain the following expression:

$$
\begin{gathered}
\dot{V} = \left(I_1 + M_1 \tfrac{L_1^2}{4}\right)\left(\dot\theta_1-\dot\theta_{1r}\right)\left(\ddot\theta_1-\ddot\theta_{1r}\right) + \tfrac{1}{2} I_2\left(\dot\theta_1+\dot\theta_2-\dot\theta_{1r}-\dot\theta_{2r}\right)\left(\ddot\theta_1+\ddot\theta_2 - \right. \\
\left.\ddot\theta_{1r}-\ddot\theta_{2r}\right) + \rho\int_0^{L_2}\left\{L_1^2\dot\theta_1\ddot\theta_1 + L_1^2\dot\theta_{1r}\ddot\theta_{1r} + w\dot w\left(\dot\theta_1+\dot\theta_2\right)^2 + w_r\dot w_r\left(\dot\theta_{1r}+\dot\theta_{2r}\right)^2 + w^2\right. \\
\times\left(\dot\theta_1+\dot\theta_2\right)\left(\ddot\theta_1+\ddot\theta_2\right) + w_r^2\left(\dot\theta_{1r}+\dot\theta_{2r}\right)\left(\ddot\theta_{1r}+\ddot\theta_{2r}\right) - L_1^2\dot\theta_{1r}\ddot\theta_1 - L_1^2\dot\theta_1\ddot\theta_{1r} + [\dot w + \\
r\left(\dot\theta_1+\dot\theta_2\right)]\left[r\left(\ddot\theta_1+\ddot\theta_2\right)+\ddot w\right] + \left[r\left(\dot\theta_{1r}+\dot\theta_{2r}\right)+\dot w_r\right]\left[r\left(\ddot\theta_{1r}+\ddot\theta_{2r}\right)+\ddot w_r\right] + \\
\left\{L_1\ddot\theta_1 w_r\left(\dot\theta_{1r}+\dot\theta_{2r}\right) + L_1\dot\theta_1\dot w_r\left(\dot\theta_{1r}+\dot\theta_{2r}\right) + L_1\dot\theta_1 w_r\left(\ddot\theta_{1r}+\ddot\theta_{2r}\right) - L_1\ddot\theta_1 w\left(\dot\theta_1+\right.\right. \\
\left.\left.\dot\theta_2\right) - L_1\dot\theta_1\dot w\left(\dot\theta_1+\dot\theta_2\right) - L_1\dot\theta_1 w\left(\ddot\theta_1+\ddot\theta_2\right)\right\}\sin\left(\theta_2\right) + \left\{L_1\dot\theta_1 w_r\left(\dot\theta_{1r}+\dot\theta_{2r}\right) - L_1\right. \\
\left.\times\dot\theta_1 w\left(\dot\theta_1+\dot\theta_2\right)\right\}\dot\theta_2\cos\left(\theta_2\right) + \left\{L_1\ddot\theta_{1r} w\left(\dot\theta_1+\dot\theta_2\right) + L_1\dot\theta_{1r}\dot w\left(\dot\theta_1+\dot\theta_2\right) + L_1\dot\theta_{1r}\times\right. \\
\left. w\left(\ddot\theta_1+\ddot\theta_2\right) - L_1\ddot\theta_{1r} w_r\left(\dot\theta_{1r}+\dot\theta_{2r}\right) - L_1\dot\theta_{1r}\dot w_r\left(\dot\theta_{1r}+\dot\theta_{2r}\right) - L_1\dot\theta_{1r} w_r\left(\ddot\theta_{1r}+\ddot\theta_{2r}\right)\right\} \\
\times\sin\left(\theta_{2r}\right) + \dot\theta_{2r}\cos\left(\theta_{2r}\right)\left\{L_1\dot\theta_{1r} w\left(\dot\theta_1+\dot\theta_2\right) - L_1\dot\theta_{1r} w_r\left(\dot\theta_{1r}+\dot\theta_{2r}\right)\right\} + \left\{L_1\ddot\theta_1\times\right. \\
\left[r\left(\dot\theta_1+\dot\theta_2\right)+\dot w\right] + L_1\dot\theta_1\left[r\left(\ddot\theta_1+\ddot\theta_2\right)+\ddot w\right] - L_1\ddot\theta_1\left[r\left(\dot\theta_{1r}+\dot\theta_{2r}\right)+\dot w_r\right] - L_1\times \\
\left.\dot\theta_1\left[r\left(\ddot\theta_{1r}+\ddot\theta_{2r}\right)+\ddot w_r\right]\right\}\cos\left(\theta_2\right) + \left\{L_1\dot\theta_1\left[r\left(\dot\theta_{1r}+\dot\theta_{2r}\right)+\dot w_r\right] - L_1\dot\theta_1\left[r\left(\dot\theta_1+\right.\right.\right. \\
\left.\left.\left.\dot\theta_2\right)+\dot w\right]\right\}\dot\theta_2\sin\left(\theta_2\right) + \cos\left(\theta_{2r}\right)\left\{L_1\ddot\theta_{1r}\left[r\left(\dot\theta_{1r}+\dot\theta_{2r}\right)+\dot w_r\right] + L_1\dot\theta_{1r}\left[r\left(\ddot\theta_{1r}+\right.\right.\right. \\
\left.\left.\ddot\theta_{2r}\right)+\ddot w_r\right] - L_1\ddot\theta_{1r}\left[r\left(\dot\theta_1+\dot\theta_2\right)+\dot w\right] - L_1\dot\theta_{1r}\left[r\left(\ddot\theta_1+\ddot\theta_2\right)+\ddot w\right]\right\} + \left\{L_1\dot\theta_{1r}\times\right. \\
\left.\left[r\left(\dot\theta_1+\dot\theta_2\right)+\dot w\right] - L_1\dot\theta_{1r}\left[\dot w_r + r\left(\dot\theta_{1r}+\dot\theta_{2r}\right)\right]\right\}\dot\theta_{2r}\sin\left(\theta_{2r}\right) - \dot w w_r\left(\dot\theta_1+\dot\theta_2\right) \\
\times\left(\dot\theta_{1r}+\dot\theta_{2r}\right) - w\dot w_r\left(\dot\theta_1+\dot\theta_2\right)\left(\dot\theta_{1r}+\dot\theta_{2r}\right) - w w_r\left(\ddot\theta_1+\ddot\theta_2\right)\left(\dot\theta_{1r}+\dot\theta_{2r}\right) - w w_r\times \\
\left(\dot\theta_1+\dot\theta_2\right)\left(\ddot\theta_{1r}+\ddot\theta_{2r}\right) - \left[r\left(\ddot\theta_1+\ddot\theta_2\right)+\ddot w\right]\left[r\left(\dot\theta_{1r}+\dot\theta_{2r}\right)+\dot w_r\right] - \left[r\left(\dot\theta_1+\dot\theta_2\right)\right. \\
\left.\left.+\dot w\right]\left[r\left(\ddot\theta_{1r}+\ddot\theta_{2r}\right)+\ddot w_r\right]\right\}dr + EI\int_0^{L_2}\left\{\left(\ddot w-\ddot w_r\right)\left(\ddot w-\ddot w_r\right)\right\}dr + \\
K_{p1}\left(\theta_1-\theta_{1r}\right)\left(\dot\theta_1-\dot\theta_{1r}\right) + K_{p2}\left(\theta_2-\theta_{2r}\right)\left(\dot\theta_2-\dot\theta_{2r}\right)
\end{gathered}
$$

The simplification of the Eqs. (11), (12) and (13) allows to write the expression of $\dot V$ as:

$$
\begin{gathered}
\dot V = \left(\dot\theta_1-\dot\theta_{1r}\right)\left[\tau_1-\tau_{1r}+K_{p1}\left(\theta_1-\theta_{1r}\right)\right] \\
+\left(\dot\theta_2-\dot\theta_{2r}\right)\left[\tau_2-\tau_{2r}+K_{p2}\left(\theta_2-\theta_{2r}\right)\right]
\end{gathered}
\tag{15}
$$

Substituting Eq. (10) into Eq. (15) leads to obtain the derivative of the Lyapunov function as follows:

$$
\dot V = -K_{d1}\left(\dot\theta_1-\dot\theta_{1r}\right)^2 - K_{d2}\left(\dot\theta_2-\dot\theta_{2r}\right)^2 \tag{16}
$$

From Eqs. (14) and (16), the manipulator system is stable if: $K_{p1} \succ 0, K_{p2} \succ 0$, $K_{d1} \succ 0, K_{d2} \succ 0$.

The block diagram of the rigid-flexible manipulator system controlled by the PD controller is given by Fig. 2.

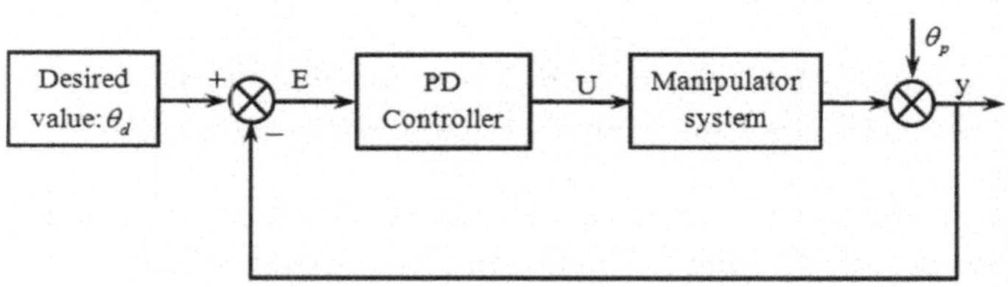


**Fig. 2** PD controller block diagram

## 4 Fuzzy Controller

The concept of fuzzy theory was introduced by Zadeh in 1965 and was used to describe complex and ill-defined dynamic models used for control synthesis. Mamdani applied the fuzzy sets theory for controlling dynamic systems, and since many researchers have developed fuzzy logic controllers (FLC) for various applications. Generally, a FLC consists of a set of linguistic conditional statements derived generally from human operators and represented the experts' knowledge about the controlled system. These statements define a set of control actions using *if −then* rules. The FLC can be considered as a fuzzy reasoning process to mimic the control actions of human operator. The conventional structure of a fuzzy logic controller is consisted of four separate blocks:

- ⋆ The fuzzification is the transformation of real variables from the outside world into fuzzy sets. An operator of fuzzification, denoted $\mu$ between 0 and 1 is used with variable measurements.
- ⋆ The rule base characterized the relations between possible events classes input and the corresponding commands. The number of subsets defining the partition of the order universe of discourse is not necessarily equal to the number of rules.
- ⋆ The inference mechanism calculates the fuzzy set concerning the control system from the basic rules and the fuzzy set corresponding to the fuzzification.
- ⋆ Finally, defuzzification is intended to transform the fuzzy set of the universe of discourse calculated by the inference mechanism, into not fuzzy value allowing the effective control of the system.

The fuzzy controller to be introduced into the forward path of the closed loop of the rigid-flexible manipulator have two inputs: the error between the desired and the measured joint angles, and its derivative, and one output that is the generated torque signal. The membership input and output functions are chosen triangular and symmetrical. The universe of discourse of inputs is divided into five fuzzy sets, and seven fuzzy sets for the output, all are between [−1, 1]. Normalizing gains are added to adjust the operation of the fuzzy controller. The basic rules of the fuzzy controller are given in Tables 1 and 2.

Figure 3 shows the structure of a fuzzy logic control of the rigid-flexible manipulator system.

**Table 1** Rigid arm rules base

| $U_1$ | | $e$ | | | | |
|---|---|---|---|---|---|---|
| | | BN | SN | Z | SP | BP |
| $\dot{e}$ | BN | BN | BN | MN | SN | Z |
| | SN | BN | MN | SN | Z | SP |
| | Z | MN | SN | Z | SP | MP |
| | SP | SN | Z | SP | MP | BP |
| | BP | Z | SP | MP | BP | BP |

**Table 2** Flexible arm rules base

| $U_2$ | | $e$ | | | | |
|---|---|---|---|---|---|---|
| | | BN | SN | Z | SP | BP |
| $\dot{e}$ | BN | BN | MN | MN | SN | Z |
| | SN | MN | MN | SN | Z | SP |
| | Z | MN | SN | Z | SP | MP |
| | SP | SN | Z | SP | MP | MP |
| | BP | Z | SP | MP | MP | BP |

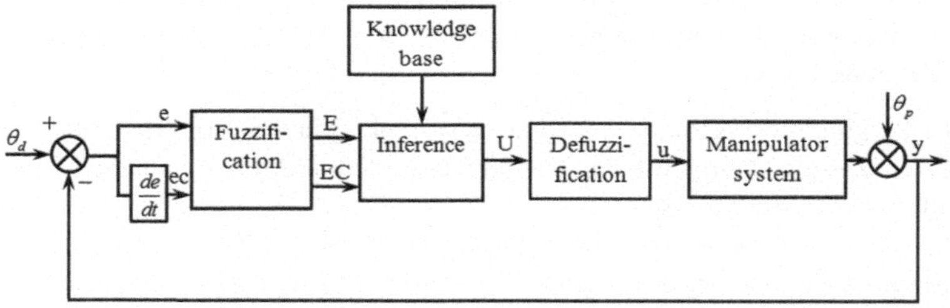


**Fig. 3** Diagram of the fuzzy controller

## 5 Gain Scheduling PD Fuzzy Controller

The PD control used with fuzzyfied gains of the rigid-flexible manipulator allows to make the regulation with online gain normalization according to error and error variation of joint variables. $K_p$, $K_d$ are assumed within prescribed ranges $\left[K_{p\min}, K_{p\max}\right]$ and $\left[K_{d\min}, K_{d\max}\right]$ respectively. So, the controller settings have the following forms:

$$\begin{cases} K_p = \left(K_{p\max} - K_{p\min}\right) \Delta K_p + K_{p\min} \\ K_d = \left(K_{d\max} - K_{d\min}\right) \Delta K_d + K_{d\min} \end{cases} \tag{17}$$

where $\Delta K_p$ and $\Delta K_d$ are expressed as:

$$\Delta K_p = \frac{K_p - K_{p\min}}{K_{p\max} - K_{p\min}}, \quad \Delta K_d = \frac{K_d - K_{d\min}}{K_{d\max} - K_{d\min}}$$

These parameters are obtained by a set of *if* − *then* fuzzy rules. The controller inputs (errors and their derivatives) have five triangular membership functions, while the outputs ($\Delta K_p$ and $\Delta K_d$) have Gaussian membership functions. An example of a knowledge rule base for the proportional gain adjustment is presented in Table 3. It is the same for the others parameters. The block diagram of the rigid-flexible manipulator system with the fuzzy PD controller is shown in Fig. 4.

**Table 3** Rule base of proportional gain for the rigid arm

| $K_{p1}$ | | $e$ | | | | |
|---|---|---|---|---|---|---|
| | | BN | SN | Z | SP | BP |
| $\dot{e}$ | BN | Big | Small | Small | Small | Big |
| | SN | Big | Big | Small | Big | Big |
| | Z | Big | Big | Big | Big | Big |
| | SP | Big | Big | Small | Big | Big |
| | BP | Big | Small | Small | Small | Big |

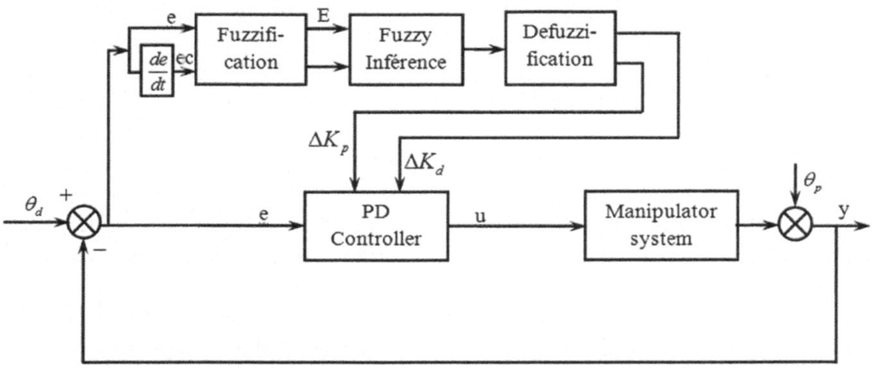


**Fig. 4** Gain scheduling PD fuzzy controller block diagram

# 6 Fractional Order Fuzzy PD Controller (FOFPD)

## 6.1 Design of the FOFPD

The structure of the conventional PD controller in the time domain is:

$$u(t) = K_p e(t) + K_d \frac{de(t)}{dt} \tag{18}$$

and in the Laplace domain is:

$$u(s) = K_p e(s) + K_d s e(s) \tag{19}$$

The discretization with the z-transform gives the following form:

$$\begin{aligned} u(z) &= K_p e(z) + K_d \frac{1 - z^{-1}}{T_e} e(z) \\ => u(k) &= K_p e(k) + K_d r(k) \end{aligned} \tag{20}$$

where $r(k) = \dfrac{e(k) - e(k-1)}{T_e}$ is the rate of change of the error signal, $e(k)$ is the error signal and $T_e$ is the sampling period. The structure of fuzzy PD controller is based upon Eq. (12). The fuzzy PD controller has $K_p e(k)$ and $K_d r(k)$ as inputs and $K_u u(k)$ as output. Hence, the equation of the fuzzy PD controller becomes:

$$U_{FPD}(k) = K_u u(k) = K_u \left(K_p e(k) + K_d r(k)\right) = K_u \left(K_p e(k) + K_d \left(D^1 e(k)\right)\right) \tag{21}$$

The fractional order controller is performed using non-integer operators for the integrator and the differentiator. In this work, only a non-integer order differentiator is used for the design of a FOFPD controller. Consequently, the equation of FOFPD controller is written as:

$$U_{FOFPD}(k) = K_u \left(K_p e(k) + K_d \left(D^{\mu} e(k)\right)\right) \tag{22}$$

where $D^{\mu}$ is the non-integer order differentiator. If $\mu = 1$, it becomes an integer order controller. The fractional order calculus is a very interesting field in mathematical calculus and has grown significantly over the last century. It has been widely used in various fields such as modeling systems described by non-integer states equations and in control engineering.

In order to improve the performance of the fuzzy PD controller and increase robustness, a fractional order differentiator is used instead of the integer order derivative at the input of the fuzzy controller. Several methods of fractional order integration and differentiation have been stated in the literature such as the definition of Riemann–Liouville (R–L), Grunwald–Letnikov (G–L), the Caputo definition and the Oustaloup approximation. In the present study, fraction order differentiator $\dfrac{d^{\mu}}{dt^{\mu}}$ has been implemented using Grunwald–Letnikov (G–L) fractional derivative definition which is briefly given as follows:

$${}_aD_t^{\alpha} f(t) = \lim_{h \to 0} \frac{1}{h^{\alpha}} \sum_{i=0}^{(t-a)/h} (-1)^i \binom{\alpha}{i} f(t - ih) \tag{23}$$

where $D$ is the mathematical operator (differentiator), $\alpha$ is the order of the operation, $t$ and $a$ are the limits and $h$ is the step size which is assumed to be very small.

$$\binom{\alpha}{i} = \frac{\alpha(\alpha-1)(\alpha-2)\ldots(\alpha-i+1)}{\Gamma(i-1)} \quad \text{where } \Gamma(i) = \int_0^{+\infty} e^{-x} e^{i-1}\, \mathrm{d}x$$

In this case, the implemented differentiator operator must be in discrete form. So, the application of the z-transform gives:

$$s^{\alpha} = \left(\frac{1 - z^{-1}}{T_e}\right)^{\alpha} = T_e^{-\alpha} \sum_{i=0}^{\infty} (-1)^i \frac{\alpha(\alpha-1)(\alpha-2)\ldots(\alpha-i+1)}{\Gamma(i-1)} z^{-i} \tag{24}$$

Therefore, the differential operator $D^{\alpha}$ in discrete form is written as follows:

$$D^{\alpha} = T_e^{-\alpha} \sum_{i=0}^{\infty} (-1)^i \binom{\alpha}{i} z^{-i} = T_e^{-\alpha} \sum_{i=0}^{\infty} b_i z^{-i}$$

where the binomial coefficient $b_i$ can be simplified by the recursive algorithm as:

$$b_i = \left(1 - \frac{1+\alpha}{i}\right) b_{i-1};\ b_0 = 1;\ \ => D^{\alpha}(f(x)) = T_e^{-\alpha} \sum_{i=0}^{\infty} b_i f(x-i)$$

It is clear from the above equation that in order to implement a fractional order operator, one must calculate and store an infinite number of coefficients. In order to realize fractional order operators, a short memory concept has been introduced. So only the last values should be calculated and stored in memory. In this work, a short memory of 100 values was considered. The above equation can be written as follows:

$$D^{\alpha}(f(x)) = T_e^{-\alpha} \sum_{i=0}^{100} b_i f(x-i)$$

The structure and implementation of the FOFPD controller of the rigid-flexible manipulator robot system is represented by Fig. 5.

For the design of the FOFPD controller, two triangular Membership Functions (MFs) namely Positive (P) and Negative (N) are used for each input of the controller ($e(k)$ and $r(k) = \Delta^{\mu} e(k)$) and three singleton membership functions are considered for the output of the controller which are positive (p), zero (z) and negative (n). These MFs are characterized by an adjustable parameter "$H$", where the MFs of the controller inputs ($e(k)$ and $r(k)$) have the following expressions:

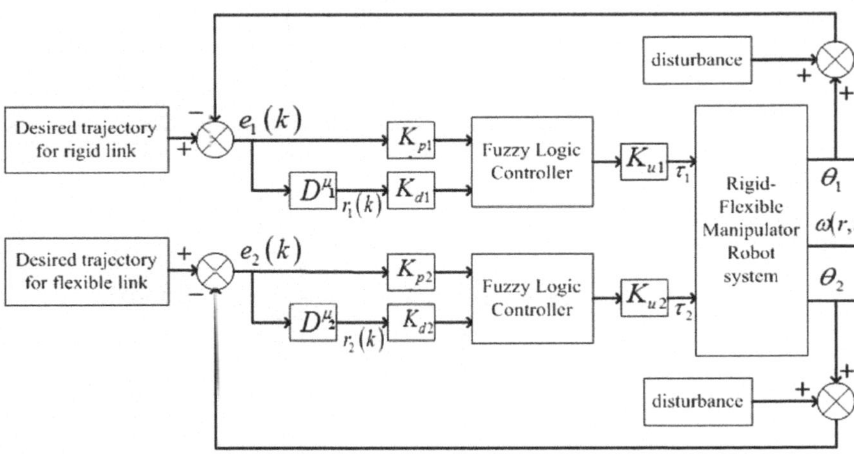


**Fig. 5** Block diagram of FOFPD controller applied to a rigid-flexible manipulator

$$\mu_p(e) = \begin{cases} 0 & \text{if } K_p e(k) < -H \\ \dfrac{K_p e(k) + H}{2H} & \text{if } -H < K_p e(k) < H \\ 1 & \text{if } K_p e(k) > H \end{cases}$$

$$\mu_n(e) = \begin{cases} 1 & \text{if } K_p e(k) < -H \\ \dfrac{-K_p e(k) + H}{2H} & \text{if } -H < K_p e(k) < H \\ 0 & \text{if } K_p e(k) > H \end{cases}$$

$$\mu_p(r) = \begin{cases} 0 & \text{if } K_d r(k) < -H \\ \dfrac{K_d r(k) + H}{2H} & \text{if } -H < K_d r(k) < H \\ 1 & \text{if } K_d r(k) > H \end{cases}$$

$$\mu_n(r) = \begin{cases} 1 & \text{if } K_d r(k) < -H \\ \dfrac{-K_d r(k) + H}{2H} & \text{if } -H < K_d r(k) < H \\ 0 & \text{if } K_d r(k) > H \end{cases}$$

The MFs of the input and output variables are represented by Figs. 6 and 7.

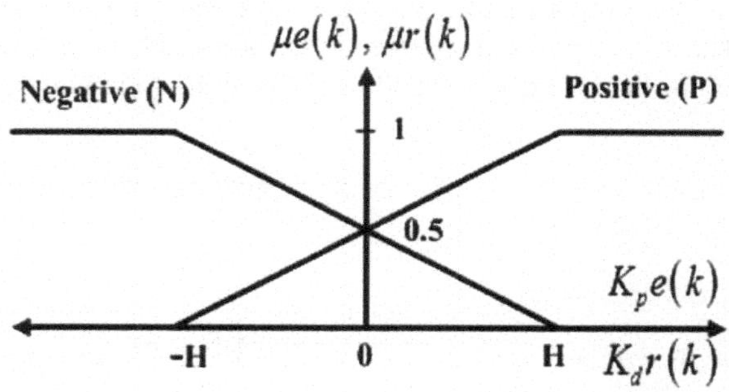


**Fig. 6** Input membership functions

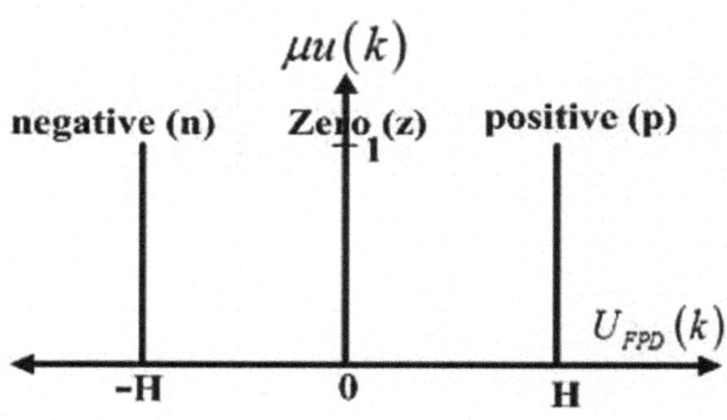


**Fig. 7** Output membership functions

To describe the rule base of the FOFPD controller using the MFs, four rules are presented as:

- R1: If $K_p e(k)$ is "$P$" and $K_d r(k)$ is "$P$" Then $U_{FPD}(k)$ is "$p$"
- R2: If $K_p e(k)$ is "$P$" and $K_d r(k)$ is "$N$" Then $U_{FPD}(k)$ is "$z$"
- R3: If $K_p e(k)$ is "$N$" and $K_d r(k)$ is "$P$" Then $U_{FPD}(k)$ is "$z$"
- R4: If $K_p e(k)$ is "$N$" and $K_d r(k)$ is "$N$" Then $U_{FPD}(k)$ is "$n$"

where $\mu_p(e) + \mu_n(e) = 1$ and $\mu_p(r) + \mu_n(r) = 1$.

The inference mechanism used for this fuzzy controller is Mamdani concept. In order to calculate the control action of the FOFPD controller, the center of gravity method is used as a defuzzification algorithm which allows the following control action:

$$U_{FOFPD} = \frac{\sum_{i=1}^{4} \mu_{FOFPDi}\, u_{FOFPDi}}{\sum_{i=1}^{4} \mu_{FOFPDi}} \tag{25}$$

where $\mu_{FOFPDi}$ is the result of the inference mechanism of rule $i$ and $u_{FOFPDi}$ is the output of the controller for the same rule. In order to obtain all the possible combinations of the inputs of the fuzzy controller, the following plane is drawn such that $K_p e(k)$ is the horizontal axis and $K_d r(k)$ is the vertical axis and with these MFs the space can be divided into 12 regions as shown in Fig. 8.

Using the method of the Mamdani min–max inference mechanism, a unique MF of the output can be obtained in each region and to each rule (Table 4).

From Eq. (25) of defuzzification and from the previous table which gives the output membership functions of each region and each rule, the analytical expression of the controller $U_{FOFPD}$ can be generated for each region. For the first region, the analytical expression of the output of the controller is determined as follows:

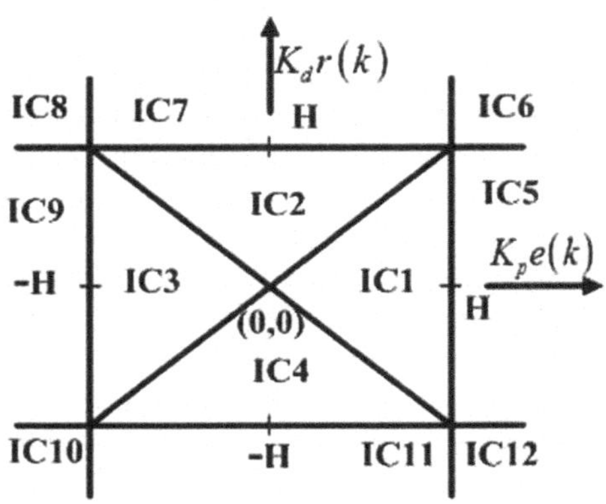


**Fig. 8** Input combinations regions of the fuzzy PD controller

**Table 4** The membership function of the output for each region and each rule

| IC region | rule 1 | rule 2 | rule 3 | rule 4 |
|---|---|---|---|---|
| IC 1 | $\mu_p(r)$ | $\mu_n(r)$ | $\mu_n(e)$ | $\mu_n(e)$ |
| IC 2 | $\mu_p(e)$ | $\mu_n(r)$ | $\mu_n(e)$ | $\mu_n(r)$ |
| IC 3 | $\mu_p(e)$ | $\mu_p(e)$ | $\mu_p(r)$ | $\mu_n(r)$ |
| IC 4 | $\mu_p(r)$ | $\mu_p(e)$ | $\mu_p(r)$ | $\mu_n(e)$ |
| IC 5 | $\mu_p(r)$ | $\mu_n(r)$ | 0 | 0 |
| IC 6 | 1 | 0 | 0 | 0 |
| IC 7 | $\mu_p(e)$ | 0 | $\mu_n(e)$ | 0 |
| IC 8 | 0 | 0 | 1 | 0 |
| IC 9 | 0 | 0 | $\mu_p(r)$ | $\mu_n(r)$ |
| IC 10 | 0 | 0 | 0 | 1 |
| IC 11 | 0 | $\mu_p(e)$ | 0 | $\mu_n(e)$ |
| IC 12 | 0 | 1 | 0 | 0 |

$$U_{FOFPD}(k) = \frac{\sum_{i=1}^{4} \mu_i u_i}{\sum_{i=1}^{4} \mu_i} = \frac{\mu_p(r) \times H + \mu_n(r) \times 0 + \mu_n(e) \times 0 + \mu_n(e) \times (-H)}{\mu_p(r) + \mu_n(r) + \mu_n(e) + \mu_n(e)}$$

$$U(k) = \frac{H\left[\dfrac{K_d r(k) + H}{2H}\right] - H\left[\dfrac{-K_p e(k) + H}{2H}\right]}{1 + 2\left[\dfrac{-K_p e(k) + H}{2H}\right]} = \frac{H\left[K_p e(k) + K_d r(k)\right]}{2\left[2H - K_p e(k)\right]}$$

For other regions, the analytical expression of the output of the controller is calculated in the same way as for the first region. Hence the analytical formula of $U_{FOFPD}$ for all regions are shown in Table 5.

Here, if the input signals $(K_p e(k), K_d r(k))$ belong to a boundary line between two neighboring regions, one of these two regions may be used, since all these control functions are continuous on the boundaries.

The tuning of controller gains in closed loop is a very important step to achieve the desired performances. There are no available specified methods in the literature for intelligent controllers tuning such as fuzzy logic. Hence, various optimization techniques are used to optimize the gains of these controllers with customized performance index such as Genetic Algorithm (GA), Particle Swarm Optimization (PSO), Backtracking Search Algorithm (BSA), etc. In this work, PSO has been used to optimize the gains of FOFPD and FPD controllers in closed loop. The PSO is one of Evolutionary Algorithms (EA) based on the population that has been developed from research into the swarm, such as fish schooling and the flow of birds developed by [10]. It becomes one of the most powerful methods to solve optimization problems.

**Table 5** The analytical expression of the controller $U_{FOFPD}(k)$ for all 12 regions

| IC region | The analytical expression of $U_{FOFPD}$ |
|---|---|
| IC 1 and IC 3 | $\frac{H\left[K_p e(k) + K_d r(k)\right]}{2\left[2H - \left\|K_p e(k)\right\|\right]}$ |
| IC 2 and IC 4 | $\frac{H\left[K_p e(k) + K_d r(k)\right]}{2\left[2H - \left\|K_d r(k)\right\|\right]}$ |
| IC 5 | $\frac{K_d r(k) + H}{2}$ |
| IC 6 | $H$ |
| IC 7 | $\frac{K_p e(k) + H}{2}$ |
| IC 8 and IC 12 | 0 |
| IC 9 | $\frac{K_d r(k) - H}{2}$ |
| IC 10 | $-H$ |
| IC 11 | $\frac{K_p e(k) - H}{2}$ |

The PSO consists of a swarm of particles moving in a n-dimensional search space to find an optimal solution. Each particle has a position and a velocity represented respectively by a position vector $X_i$ and a velocity vector $V_i$. Each particle remembers its own best position so far in a $P_{besti}$ vector where i is the index of this particle. The best position among all the neighbors of a particle is then stored in the particle $G_{best}$. At each iteration of the algorithm, the $i$th position of the particle $X_i$ evolves according to the following update rules:

$$\begin{aligned} V_i^{k+1} &= wV_i^k + c_1 r_1 \left(P_{best_i} - X_i^k\right) + c_2 r_2 \left(G_{best} - X_i^k\right) \\ X_i^{k+1} &= X_i^k + V_i^{k+1} \end{aligned} \tag{26}$$

where $w$ is the inertia weight and often decreases linearly from $w_{\max} = 0.9$ to $w_{\min} = 0.4$ during a run. In general, the inertia weight $w$ is adjusted according to the following expression:

$$w = w_{\max} - \frac{w_{\max} - w_{\min}}{ite_{\max}} ite$$

$ite$ represents the number of current iteration and $ite_{\max}$ the maximum number of iterations. Also $c_1$ and $c_2$ are the acceleration constants which influence the convergence speed of each particle and are often set to 2, $r_1$ and $r_2$ are the random numbers uniformly distributed in the interval [0,1]. For optimization, the population size was chosen 10, the maximum number of iterations was maintained at 100 and the Integral of the Absolute Error (IAE) was chosen as the objective function of the optimization algorithm to minimize the error of manipulator system. The objective function is defined as:

$$J = \int_{0}^{10} \{\xi_1 |e_1| + \xi_2 |e_2| + \xi_3 |e_3|\} \, dt$$

where $e_1$, $e_2$ and $e_3$ are the tracking error of rigid link, flexible link and the flexibility of the manipulator respectively. Also $\xi_1$, $\xi_2$ and $\xi_3$ are respectively the weights corresponding to these errors.

## 6.2 *Stability Study*

For the stability study of the rigid-flexible manipulator system with a FOFPD controller and FPD controller, the Bounded Input Bounded Output (BIBO) stability method based on the small gain theorem is used consequently.

### 6.2.1 Small Gain Theorem

Using the block diagram of the nonlinear system with a nonlinear fuzzy controller as shown in the Fig. 9, the principle of the small gain theorem is given as follows: if we consider a nonlinear dynamic system $\mathbb{S}$ and a nonlinear fuzzy controller $\mathbb{F}$ then we have:

$$\|\mathbb{F}(e(k))\| \leqslant \beta_1 \|e(k)\| + \gamma_1, \quad \beta_1, \gamma_1 \geqslant 0 \tag{27}$$

and

$$\|\mathbb{S}(u(k))\| \leqslant \beta_2 \|u(k)\| + \gamma_2, \quad \beta_2, \gamma_2 \geqslant 0 \tag{28}$$

where $\beta_1$ and $\beta_2$ are the gain of the controller $\mathbb{F}$ and the gain of the system $\mathbb{S}$ respectively, $\gamma_1$ and $\gamma_2$ are constants. From the above two equations, the system with the nonlinear fuzzy controller is BIBO stable if: $\beta_1.\beta_2 < 1$.

Moreover, for any system, the following condition is always true.

$$\|\mathbb{S}(u(k))\| \leqslant \|\mathbb{S}\| \,.\, |u(k)|$$

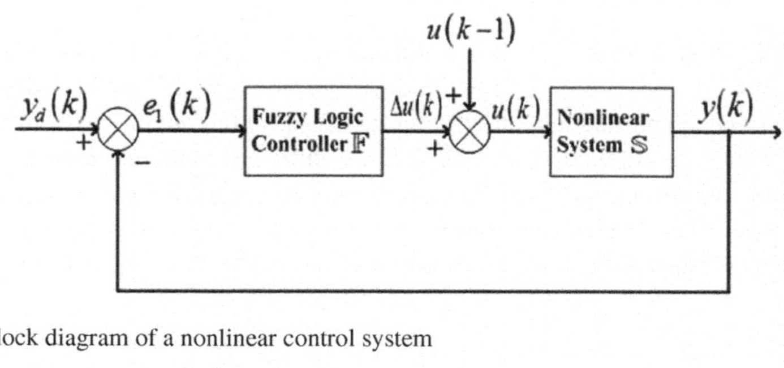


**Fig. 9** Block diagram of a nonlinear control system

where $\mathbb{S}$ is the operator norm of a nonlinear system which is defined by: $\|\mathbb{S}\| = \sup_{u_1 \neq u_2; k \geqslant 0} \frac{|\mathbb{S}(u_1(k)) - \mathbb{S}(u_2(k))|}{u_1(k) - u_2(k)}$, it means that $\beta_2 = \|\mathbb{S}\|$. Now, it is sufficient to determine $\beta_1$ for BIBO stability assessment.

### 6.2.2 BIBO Stability Condition of FOFPD and FPD Controllers

The fractional-order fuzzy PD controller differs only by the differentiator operator with the integer order fuzzy PD controller. However the internal structure of these two controllers (FOFPD and FPD) are identical, which makes possible to obtain the same BIBO stability conditions. Moreover, the input space of the controllers is divided into 12 regions, so the stability condition of each region must be developed as follows: For the nonlinear system $\mathbb{S}$ controlled by a fuzzy controller $\mathbb{F}$, the sufficient condition for the FOFPD/FPD controller should be stable are:

- The nonlinear system $\mathbb{S}$ under control has a bounded norm i.e. $\|\mathbb{S}\| < \infty$;
- The parameters of the FOFPD/FPD controller satisfy the following equation:

$$\beta_1 . \|\mathbb{S}\| < 1 \tag{29}$$

where at each region there is a corresponding expression of $\beta_1$

To obtain the value of $\beta_1$ of the first region, the calculus of the norm of a nonlinear fuzzy controller $\mathbb{F}$ is developed as:

$$\|\mathbb{F}(e(k))\| = \left\| K_u \frac{H\left[K_p e(k) + K_d r(k)\right]}{2\left[2H - \left|K_p e(k)\right|\right]} \right\|$$

$$\|\mathbb{F}(e(k))\| = \left\| K_u H \frac{\left[K_p e(k) + \frac{K_d}{T_e}(e(k) - e(k-1))\right]}{2\left[2H - K_p |e(k)|\right]} \right\|$$

$$\|\mathbb{F}(e(k))| \leq \left| \frac{K_u H\left[K_p T_e + K_d\right]}{2T_e\left[2H - K_p M_1\right]} e(k) \right| + \left| \frac{K_u H K_d e(k-1)}{2T_e\left[2H - K_p M_1\right]} \right|$$

where $M_1 = \sup_{k \geqslant 0} |e(k)| = \sup_{k \geqslant 1} |e(k-1)|$

$$=> \|\mathbb{F}(e(k))\| \leqslant \left| \frac{K_u H\left[K_p T_e + K_d\right]}{2T_e\left[2H - K_p M_1\right]} \right| |e(k)| + \left| \frac{K_u H K_d M_1}{2T_e\left[2H - K_p M_1\right]} \right|$$

Therefore $\|\mathbb{F}(e(k))\| \leqslant \beta_1 \|e(k)\| + \gamma_1$.

**Table 6** Value of $\beta_1$ for all 12 regions of FOFPD/FPD controllers

| IC region | Value of $\beta_1$ |
|---|---|
| IC 1 and IC 3 | $\left\lvert \frac{K_u H \left[K_p T_e + K_d\right]}{2T_e \left[2H - K_p M_1\right]} \right\rvert$ |
| IC 2 and IC 4 | $\left\lvert \frac{K_u H \left[K_p T_e + K_d\right]}{2T_e \left[2H - K_d M_2\right]} \right\rvert$ |
| IC 5 and IC 9 | $\left\lvert \frac{K_u K_d}{2T_e} \right\rvert$ |
| IC 7 and IC 11 | $\left\lvert \frac{K_u K_p}{2T_e} \right\rvert$ |
| IC 6, IC 8, IC 10 and IC 12 | 0 |

where $M_2 =_{k \geqslant 1}^{\sup} |r(k)| =_{k \geqslant 1}^{\sup} \left\lvert \frac{e(k) - e(k-1)}{T_e} \right\rvert \leqslant \frac{2M_1}{T_e}$

with $\beta_1 = \left\lvert \frac{K_u H \left[K_p T_e + K_d\right]}{2T_e \left[2H - K_p M_1\right]} \right\rvert$ and $\gamma_1 = \left\lvert \frac{K_u H K_d M_1}{2T_e \left[2H - K_p M_1\right]} \right\rvert$

So, to guarantee the BIBO stability in the first region, it is necessary that:

$$\beta_1.\beta_2 = \left\lvert \frac{K_u H \left[K_p T_e + K_d\right]}{2T_e \left[2H - K_p M_1\right]} \right\rvert . \|\mathbb{S}\| \leqslant 1 \tag{30}$$

For the different regions the value of $\beta_1$ is calculated in the same way as the first region, where the value of $\beta_1$ for different regions is illustrated in Table 6.

# 7 Simulation Results

Physical parameters of the rigid-flexible manipulator system used in this work are illustrated in Table 7.

## *7.1 PD and Fuzzy Controllers*

To simulate the closed loop, the empirical method of Ziegler–Nichols is used to obtain the following PD controller parameters $K_p = [0.1\ \ 0.02]$ and $K_d = [0.03\ \ 0.005]$, the fuzzy PD controller parameters vary in the following intervals $K_{p1} = [0.05\ \ 0.15]$, $K_{p2} = [0.005\ \ 0.035]$, $K_{d1} = [0.01\ \ 0.05]$, $K_{d2} = [0.001\ \ 0.009]$, and for the fuzzy controller we use for each joint two inputs and one output. The two inputs have 5 triangular membership functions, the output has 7 membership functions.

**Table 7** Parameters of the rigid-flexible manipulator

| Parameter | Symbol | Value |
|---|---|---|
| Rigid link length | $L_1$ | 0.21 m |
| Flexible link length | $L_2$ | 0.22 m |
| Mass of rigid link | $m_1$ | 0.080 kg |
| Moment of inertia of rigid link | $I_1$ | 0.0082 kg/m$^2$ |
| Moment of inertia of flexible link hub and motor | $I_2$ | $5.0536 \times 10^{-4}$ kg/m$^2$ |
| Flexural rigidity | $EI$ | 0.1143 N/m$^2$ |
| Density of flexible link | $\rho$ | 0.0182 kg/m |

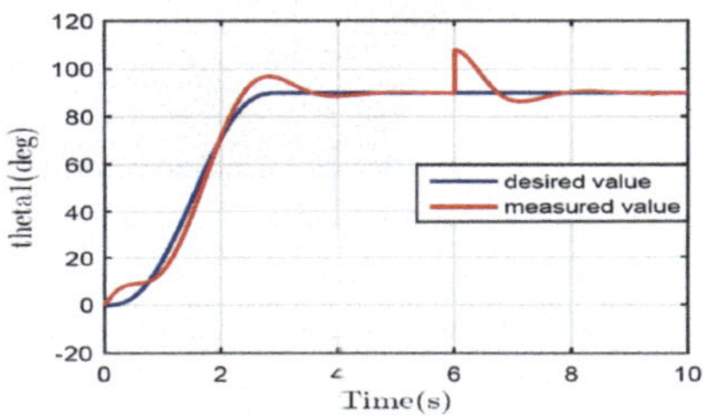


(a) Evolution of the first joint $\theta_1(t)$.

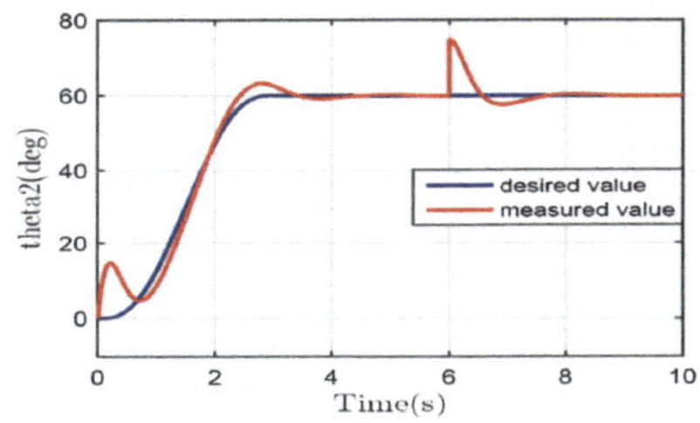


(b) Evolution of the second joint $\theta_2(t)$.

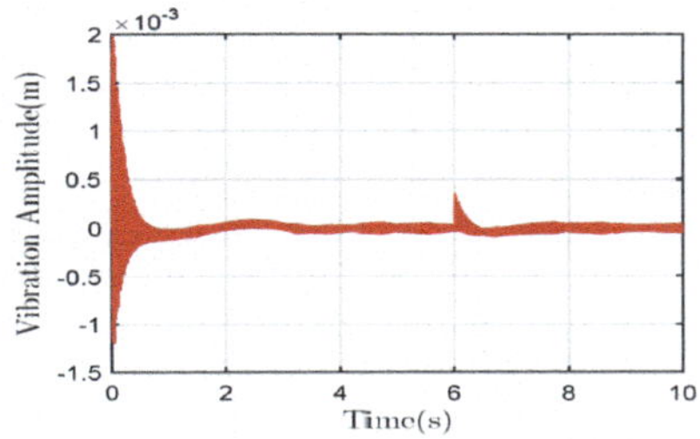


(c) Flexibility $w(r, t)$.

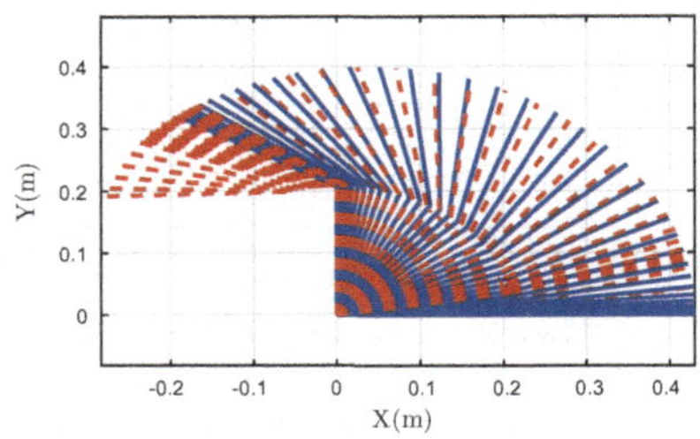


(d) XY displacement.

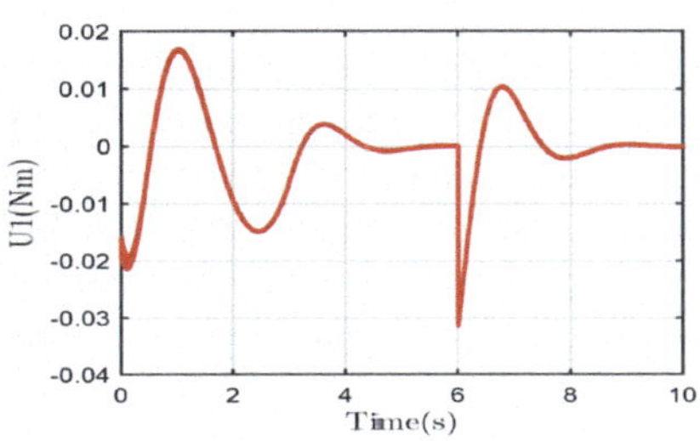


(e) Control signal of the rigid arm.

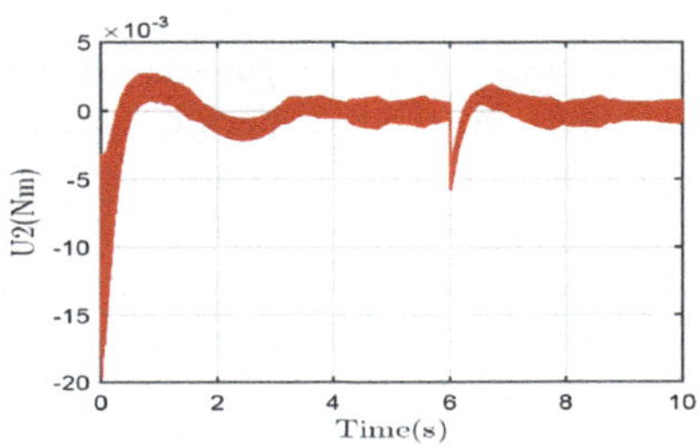


(f) Control signal of the flexible arm.

**Fig. 10** Rigid-flexible manipulator responses with PD controller

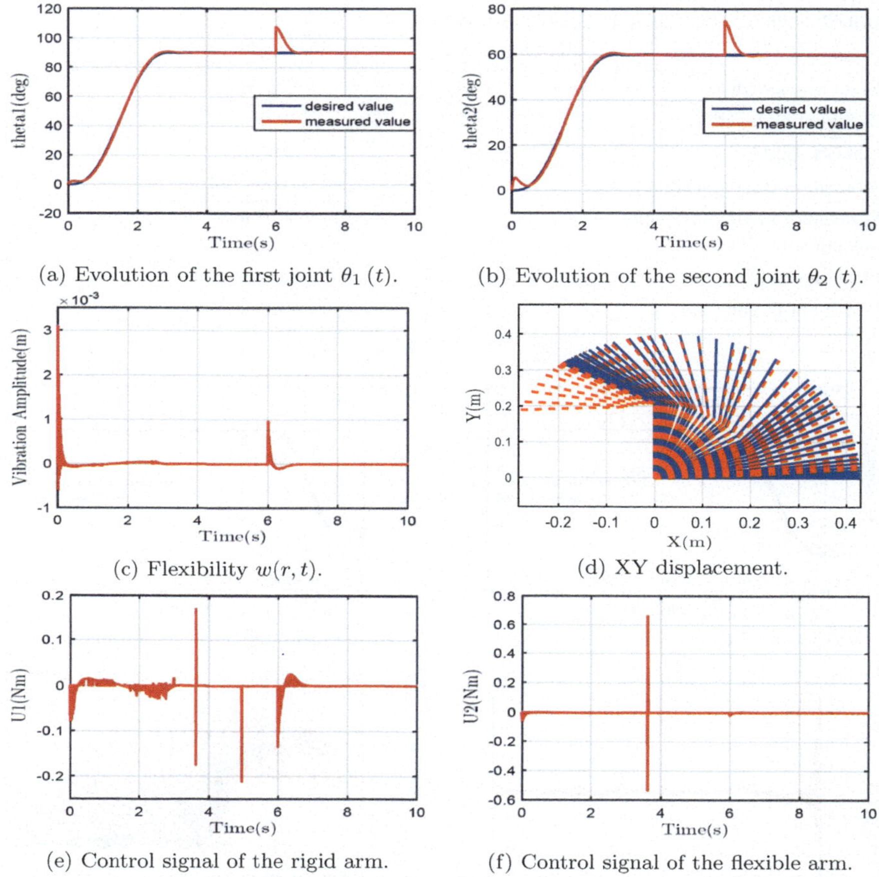


(a) Evolution of the first joint $\theta_1\,(t)$.

(b) Evolution of the second joint $\theta_2\,(t)$.

(c) Flexibility $w(r,t)$.

(d) XY displacement.

(e) Control signal of the rigid arm.

(f) Control signal of the flexible arm.

**Fig. 11** Rigid-flexible manipulator responses with fuzzy control

The assessment comparison between PD, FLC and gain scheduling PD fuzzy controllers in terms of performances shows that the FLC and gain scheduling is more efficient than PD controllers regarding path track and disturbance rejection. Moreover, the FLC satisfied vibration suppression in the end-point flexible arm. The command signal applied via PD and gain scheduling controllers has a lower magnitude compared to the FLC, which means an outstanding reduced energy consumption in this case, Figs. 10, 11, 12 and 13.

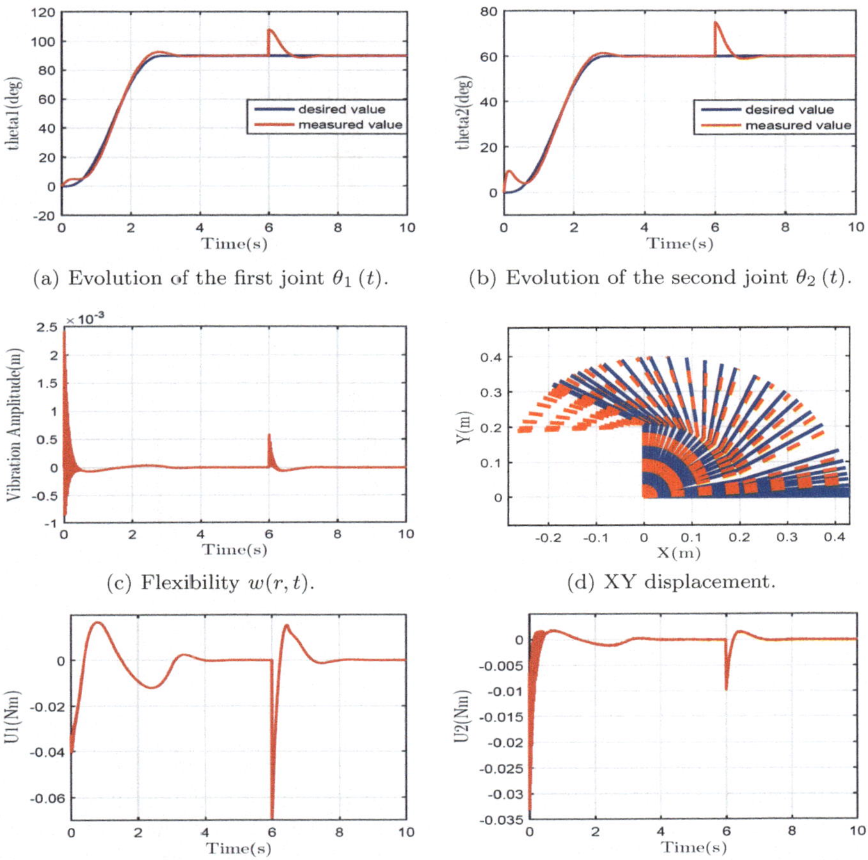


(a) Evolution of the first joint $\theta_1(t)$.

(b) Evolution of the second joint $\theta_2(t)$.

(c) Flexibility $w(r,t)$.

(d) XY displacement.

(e) Control signal of the rigid arm.

(f) Control signal of the flexible arm.

**Fig. 12** Rigid-flexible manipulator responses with Gain Scheduling PD fuzzy controller

## *7.2 FOFPD Controller*

For the simulation and to show the efficiency of the FOFPD controller compared to the FPD controller, the desired trajectory is considered a quintic polynomial between instants $t = 0\,\text{s}$ and $t = 3\,\text{s}$, then it remains constant. The responses of the rigid-flexible manipulator system with these two types of controllers using the PSO optimization algorithm are illustrated in Figs. 14 and 15.

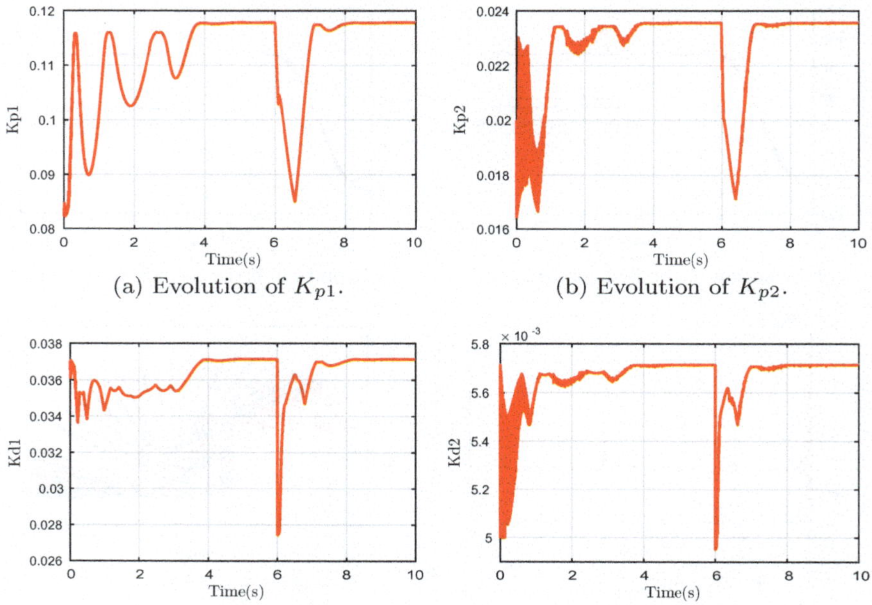


(a) Evolution of $K_{p1}$. (b) Evolution of $K_{p2}$.

(c) Evolution of $K_{d1}$. (d) Evolution of $K_{d2}$.

**Fig. 13** Evolution of the fuzzy PD controller gains

The FOFPD controller is more efficient then a FPD controller in trajectory tracking, supression of vibration in the end-point of the flexible arm and robustness against disturbances. The command signal applied to the flexible arm of the rigid-flexible manipulator by the FOFPD controller has a lower magnitude compared the FPD controller which means a reduced energy consumption in this case.

# 8 Conclusion

In this chapter, the flexibility phenomenon was dealt with as a characteristic of robotic manipulators links thanks to the major advantages required in space, medicine and industrial fields such as lightweight, diminution in useful materials and actuators, reduction in power consumption and growth in transported loads compared to robot weight. However, flexible links present significant difficulties in modelling and serious problems to develop an appropriate command law. Therefore, the work presented

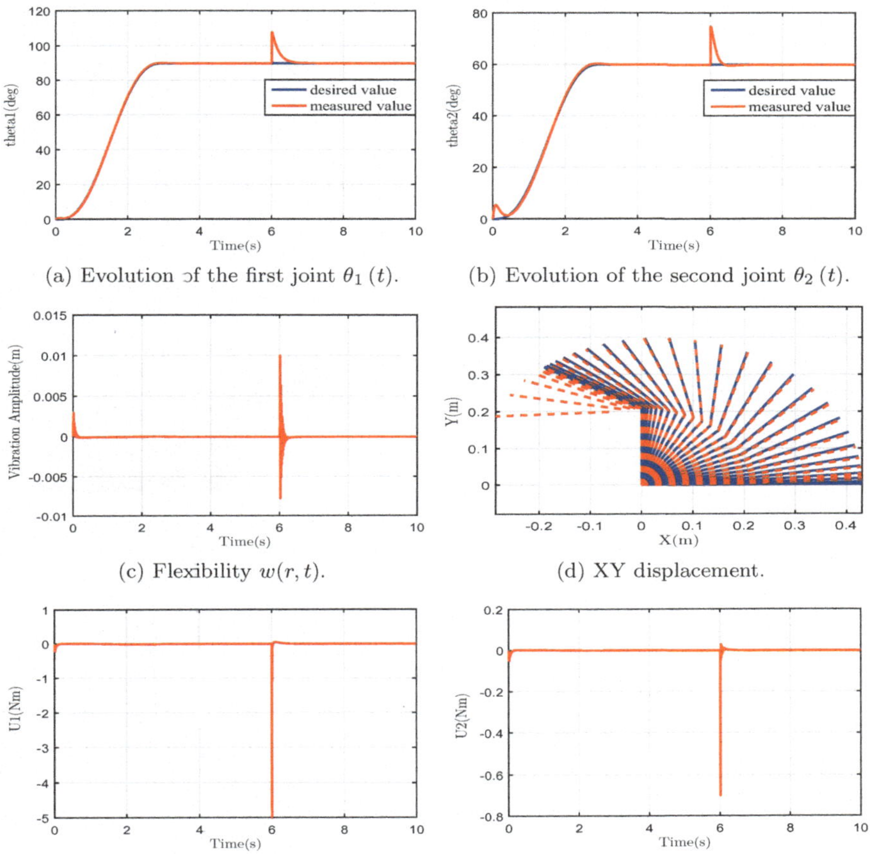


(a) Evolution of the first joint $\theta_1(t)$.

(b) Evolution of the second joint $\theta_2(t)$.

(c) Flexibility $w(r,t)$.

(d) XY displacement.

(e) Control signal of the rigid arm.

(f) Control signal of the flexible arm.

**Fig. 14** Rigid-flexible manipulator responses with fuzzy PD controller

in this chapter is placed in this context. In a first step, a dynamic model of a rigid-flexible two-links manipulator is calculated using Hamilton's principle and Lagrange equations. Then, PD and fuzzy PD controllers are applied to the manipulator to test robustness and end-point vibration suppression. Finally, a new approach of command that is the fractional order fuzzy PD (FOFPD) controller is developed to control a rigid-flexible manipulator system. This controller is implemented using the optimization algorithm Particle Swarm Optimization (PSO). The obtained simulation results show that the FOFPD controller is more efficient compared to other controllers in terms of trajectory tracking, vibration suppression, robustness against disturbances and energy consumption.

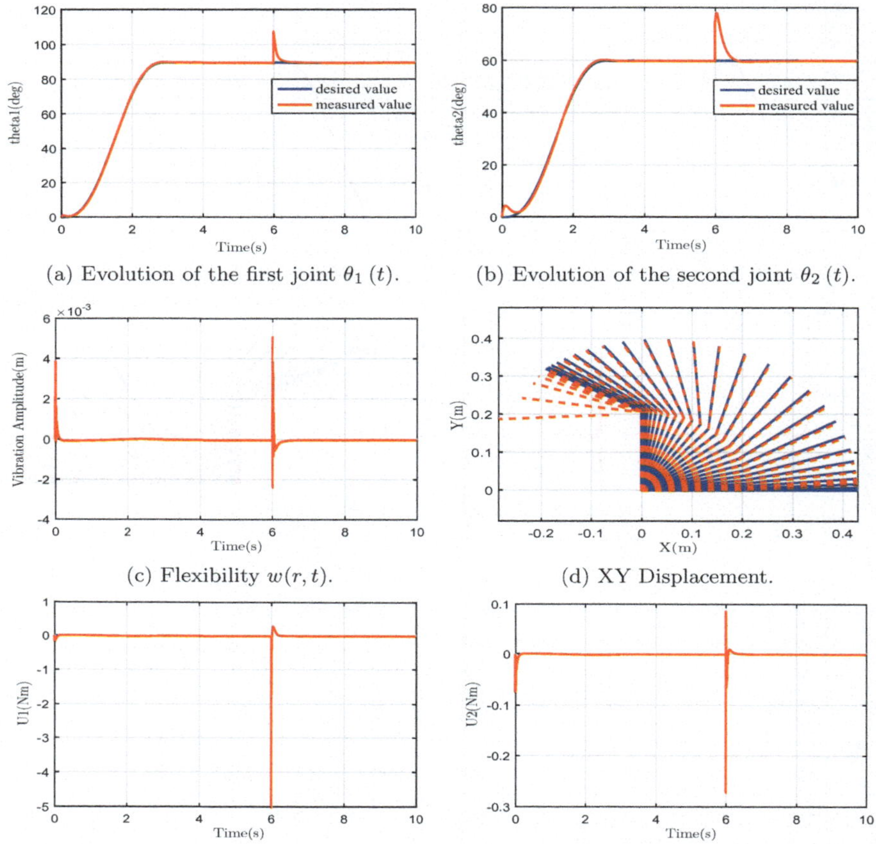


(a) Evolution of the first joint $\theta_1\ (t)$.

(b) Evolution of the second joint $\theta_2\ (t)$.

(c) Flexibility $w(r, t)$.

(d) XY Displacement.

(e) Control signal of the rigid arm.

(f) Control signal of the flexible arm.

**Fig. 15** Rigid-flexible manipulator responses with FOFPD controller

**Acknowledgements** This work was supported by the Ministry of the Higher Education and Scientific Research in Tunisia.

## References

1. Abe, A.: Trajectory planning for residual vibration suppression of a two-link rigid-flexible manipulator considering large deformation. Elsevier Mech. Mach. Theorye **44**(9), 1627–1639 (2009)
2. Alavandar, S., Jain, T., Nigam, M.J.: Bacterial foraging optimized hybrid fuzzy precompensated pd control of two link rigid-flexible manipulator. Int. J. Comput. Intell. Syst. **2**(1), 51–59 (2009)
3. Baroudi, M., Saad, M., Ghie, W.: State-feedback and linear quadratic regulator applied to a single-link flexible manipulators. IEEE International Conference on Robotics and Biomimetics.

IEEE Press, Guilin (2009)
4. Bayo, E.: A finite-element approach to control the end-point motion of a single-link flexible robot. J. Robot Syst. **4**(1), 63–75 (1987)
5. Boucetta, R., Belhadjali, S., Abdelkrim, M.N.: Global hybrid fuzzy controller for a flexible single-link manipulator. J. Eng. Appl. Sci. **6**(1), 1–5 (2011)
6. Cao, F., Liu, J.: Boundary control for a constrained two-link rigid-flexible manipulator with prescribed performance. Int. J. Control **90**(5), 1–13 (2017)
7. Chanwikrai, S., Cole, M.O.: Modeling of a rigid-flexible manipulator using hamilton's principle. Engng. J. CMU **17**(3), 19–27 (2008)
8. Chen, G., Pham, T.T.: Introduction to fuzzy systems. CRC Press, Boca Raton (2005)
9. Cost, A., Jos, S.D.A.: An introduction to fractional control. Institution of Engineering and Technology, London (2013)
10. Eberhart, R., Kennedy, J.: A new optimizer using particle swarm theory. In: IEEE Proceedings of the Sixth International Symposium on Micro Machine and Human Science, Nagoya (1995)
11. Fenili, A.: The rigid-flexible robotic manipulator: nonlinear control and state estimation considering a different mathematical model for estimation. Hindawi Publ. Corp. Shock. Vib. **20**(6), 1049–1063 (2013)
12. Fenili, A., Balthazar, J.: The rigid-flexible nonlinear robotic manipulator: modeling and control. Commun. Nonlinear Sci. Numer. Simul. **16**(5), 2332–2341 (2011)
13. Gao, H., He, W.: Fuzzy control of a single-link flexible robotic manipulator using assumed mode method. In: IEEE Youth Academic Annual Conference of Chinese Association of Automation (YAC), Wuhan (2016)
14. Gao, Y., Er, M.J., Leithead, W.E., Leith, D.J. Online adaptive control of robot manipulators using dynamic fuzzy neural networks. In: IEEE Proceedings American Control Conference (2001)
15. Hamdi, S., Boucetta, R., Belhadjali, S.: Dynamic modeling of a rigid-flexible manipulator using Hamilton's principle. In: IEEE International Conference on Sciences and Techniques of Automatic Control and Computer Engineering (STA) (2015)
16. He, W., David, A.C., Yin, Z., Sun, C.: Neural network control of a robotic manipulator with input deadzone and output constraint. IEEE Trans. Syst. Man Cybern. Syst. **46**(6), 759–770 (2016)
17. Hong, J., He, W., Le, Z., Zhang, S.: Vibration control and angular tracking of a flexible link via neural networks. In: IEEE Control and Decision Conference, Chinese (CCDC) (2015)
18. Irani, A., Talebi, H.: Tip tracking control of a rigid-flexible manipulator based on deflection estimation using neural networks: application to needle insertion. In: IEEE ISSNIP Biosignals and Biorobotics Conference (2011)
19. Kumar, V., Mittal, A.P.: Parallel fuzzy p+ fuzzy i+ fuzzy d controller: design and performance evaluation. Int. J. Autom. Comput. **7**(4), 463–471 (2010)
20. Lee, H.H., Liang, Y.: A coupled-sliding-surface approach for the robust trajectory control of a horizontal two-link rigid-flexible robot. Int. J. Control. **80**(12), 1880–1892 (2007)
21. Lochan, K., Roy, B.K., Subudhi, B.: Smc controlled chaotic trajectory tracking of two-link flexible manipulator with pid sliding surface. IFAC-PapersOnLine **49**(1), 219–224 (2016)
22. Monje, C.A., Chen, Y., Vinagre, B.M., Xue, D., Feliu-Batlle, V.: Fractional-Order Systems and Controls: Fundamentals and Applications. Springer, Heidelberg (2010)
23. Qiu, Z., Zhao, Z.: Pneumatic drive active vibration control for flexible manipulator using an adaptive interactive PD controller. In: Communications in Nonlinear Science and Numerical Simulation (2011)
24. Sabatier, J., Lanusse, P., Melchior, P., Oustaloup, A.: Fractional Order Differentiation and Robust Control Design: CRONE, H-infinity and Motion Control. Springer, Berlin (2015)
25. Spong, M., Hutchinson, S., Vidyasagar, M.: Robot Modeling and Control. Wiley, New York (2006)
26. Sun, C., He, W., Hong, J.: Neural network control of a flexible robotic manipulator using the lumped spring-mass model. IEEE Trans. Syst. Man Cybern. **47**, 1863–1874 (2016)

27. Hussein, T.M., Nemah, M.N.: Control of a two-link rigid-flexible manipulator. In: IEEE International Conference on Robotics and Mechatronics (ICROM) (2015)
28. Tarvirdizadeh, B., Alipour, K.: Trajectory optimization of two-link rigid flexible manipulators in Dynamic Object Manipulation missions. In: IEEE International Conference on Robotics and Mechatronics (ICROM) (2015)
29. Tian, L., Collins, C.: A dynamic recurrent neural network-based controller for a rigid-flexible manipulator system. Mechatronics **14**(5), 471–490 (2004)
30. Tian, L., Mao, Z.: Fuzzy neuro controller for a two-link rigid-flexible manipulator system. In: Proceedings of the 9th Internatioal Conference on IEEE Neural Information Processing. ICONIP'02 (2002)
31. Tinkir, M., Önen, Ü., Kalyoncu, M.: Modelling of neurofuzzy control of a flexible link. SAGE Publ. Proc. Inst. Mech. Eng. **224**(5), 529–543 (2010)
32. Xilun, D., Selig, J.M.: Lumped parameter dynamic modeling for the flexible manipulator. In: IEEE Intelligent Control and Automation, WCICA 2004. Fifth World Congress (2004)
33. Ying, H.: Fuzzy control and modeling: analytical foundations and applications. Wiley-IEEE Press, Hoboken (2000)
34. Zhang, S., Zhang, Y., Zhang, X., Dong, G.: Fuzzy pid control of a two-link flexible manipulator. J. Vibroengineering **18**(1), 250–266 (2016)

# Autonomous Trenchless Horizontal Directional Drilling

**Mahmoud A. K. Gomaa, Moustafa Elshafei, Abdul-Wahid A. Saif and Abdulaziz A. Al-Majed**

**Abstract** Over the past 25 years trenchless technology has flourished in many pipeline installation applications as water industry, sanitary sewer, and natural gas pipeline industry. Trenchless technology gained more popularity than open trench due to its economic and environmental factors, it facilitates the installation, replacement, or rehabilitation of underground utility systems with minimum disruption of the surface. Different trenchless installation techniques have been developed over the years but Horizontal Directional Drilling (HDD) technique is commonly used for pipelines installation beneath buildings, forests, or lakes without excavation or at least minimum digging due to its steering capability. HDD reduces environmental impact to accomplish the installation process and minimizes the drilling time and cost. Quad Motor Drilling Heads (QMDH) were recently proposed to achieve smooth and intuitive directional drilling. A QMDH uses four drilling bits, independently controlled by 4 downhole motors. In this work a real-time control and optimization of the QMDH is simulated and extensively evaluated under various drilling scenarios. The control commands include the angular velocity and torque of each of the 4 motors. The real-time control is updated at regular distance intervals. The proposed control

M. A. K. Gomaa (✉)
Intelligent Systems Lab, Memorial University of Newfoundland, St. John's, Canada

Department of Mechatronics Engineering, Helwan University Cairo, Cairo, Egypt
e-mail: makamel@mun.ca; makamel@h-eng.helwan.edu.eg

M. Elshafei
CIE Department, Zewail City for Science and Technology, City of 6 October, Giza, Egypt
e-mail: moelshafei@zewailcity.edu.eg

A.-W. A. Saif
Department of Systems and Control Engineering, King Fahd University of Petroleum and Minerals, Dhahran, Saudi Arabia
e-mail: awsaif@kfupm.edu.sa

A. A. Al-Majed
Department of Petroleum Engineering, King Fahd University of Petroleum and Minerals, Dhahran, Saudi Arabia
e-mail: aamajed@kfupm.edu.sa

 
J. Ghommam et al. (eds.), *New Trends in Robot Control*, Studies in Systems, Decision and Control 270, https://doi.org/10.1007/978-981-15-1819-5_3

algorithm selects the drilling parameters in order to minimize the time of the drilling process and the deviation from the target trajectory.

**Keywords** Automatic drilling · Drilling optimization · HDD quad-rotor system · Real-time drilling · Trenchless technology · Optimal control

## Nomenclature

| | |
|---|---|
| $A_h$ | The area of borehole ($m^2$) |
| $B, E$ | Body and inertial fixed frames, respectively |
| BHA | Bore Hole Assembly |
| $E_{rs}$ | The value of rock specific energy |
| DSS | Directional Steering System |
| HDD | Horizontal Directional Drilling |
| $P_E(k)$ | Current position of the BHA w.r.t the inertia axes |
| $\hat{P}_E(k+1)$ | Predicted position of the BHA w.r.t the inertia axes |
| $R_{op}$ | Rate of Penetration |
| $T_i$ | Motor torque (Nm) |
| $W_{ob}$ | The weight on bit (kg) |
| $w_b(k)$ | The measured depth (m): hole length up to the $k$th point |
| $\Delta\hat{w}_b(k+1)$ | Predicted value of change of measured depth (m) |
| $\dot{w}_b$ | Rate of penetration (m/s) during the $k$th interval |
| $\omega_i$ | Motor angular velocity (rpm) |
| $x_E, y_E, z_E$ | The position of the BHA in the inertial frame (m) |
| $x_B, y_B, z_B$ | The position of the BHA in the body frame (m) |
| $\phi, \theta, \psi$ | Roll, pitch, and yaw angles (rad) |

## 1 Introduction

Application of Trenchless Technologies (TT) is rapidly increasing in the underground infrastructure systems especially in urbanized area with crowded traffic and existing massive underground utilities. TT is a powerful method for installing pipelines crossing roadways, railways, rivers and mountains. It can be used for pipelines and cables installation over long distances with less time and low investment in equipment and labors. TT is commonly used for installation, inspection, replacement, maintenance, or rehabilitation of underground utilities in environmentally sensitive locations. In contrast, typical open-cut construction methods are costly and could adversely affect the community due to pollution and traffic disruptions [15].

In general, underground pipeline installation has been classified into two main categories as shown in Fig. 1;

1. Open trench which is the traditional method for installation or replacement of cables and pipelines that require a complete channel for the entire length of the pipeline through removing the workplace surface,
2. Trenchless technology.

TT can be used to install new pipelines (Trenchless Construction Methods) or to rehabilitate existing lines (Trenchless Renewal Method) [5]. Trenchless construction technologies can be classified into three groups;

1. Horizontal Earth Boring, where a mechanism is used to excavate the borehole without personnel existence inside the hole,
2. Pipe Jacking,
3. Utility Tunneling.

Both (2) and (3) require labors inside the borehole during the installation process [20]. On the other hand, there are several technologies under Horizontal Earth Boring as shown in Fig. 1. For example, Horizontal Auger Boring and Pipe Ramming are mostly used for pipelines that have limited steering requirements as road crossing. Microtunneling is used in soft soils and limited range of depth, and cannot be used for pipelines less than 25.4 cm in diameters [5].

HDD system is classified under the Horizontal Earth Boring category that does not require the human entry. HDD provides the ability of tracking the desired position of the drill bit and steering it during the drilling process. HDD system has the highest Rate of Penetration (Rop) during boring the pilot hole among all other trenchless installation techniques [13]. Unlike the other technologies for Horizontal Earth Boring, HDD is applicable to all the utilities applications as shown in Table 1 [17]. HDD is expected to be one of the most important trenchless technologies in the future for new underground construction. HDD had a great influence on the digital and communication development over the past few years through its outstanding contribution to fiber-optic and telecommunications line installation [4]. HDD process includes three steps;

1. Drilling a small diameter directional pilot hole Fig. 2,
2. Back reamer Fig. 2,
3. Pulling the pipe back through the pre-drilled pilot hole Fig. 2.

HDD can drill from 120 m up to 1,800 m in length and from 25 mm up to 900 mm in diameter [1]. HDD rig, Fig. 3, includes power unit, drilling machine, drill mud pump, mud system, drill pipe, drill tools, directional instrument. One major disadvantages of HDD is that it requires expert operators and plenty of planning before the installation process [5], so developing automatic drilling systems is important for better performance and time saving.

Before starting the drilling process, a pre-site planning should be conducted to emphasize the feasibility of HDD technology based on geologically studies [6]. If the HDD is approved to be technically and geotechnically feasible, drill entry and exit sites should be appropriately selected, in addition to the drill path between them taking into consideration a permissible allowance to avoid obstacles and unexpected

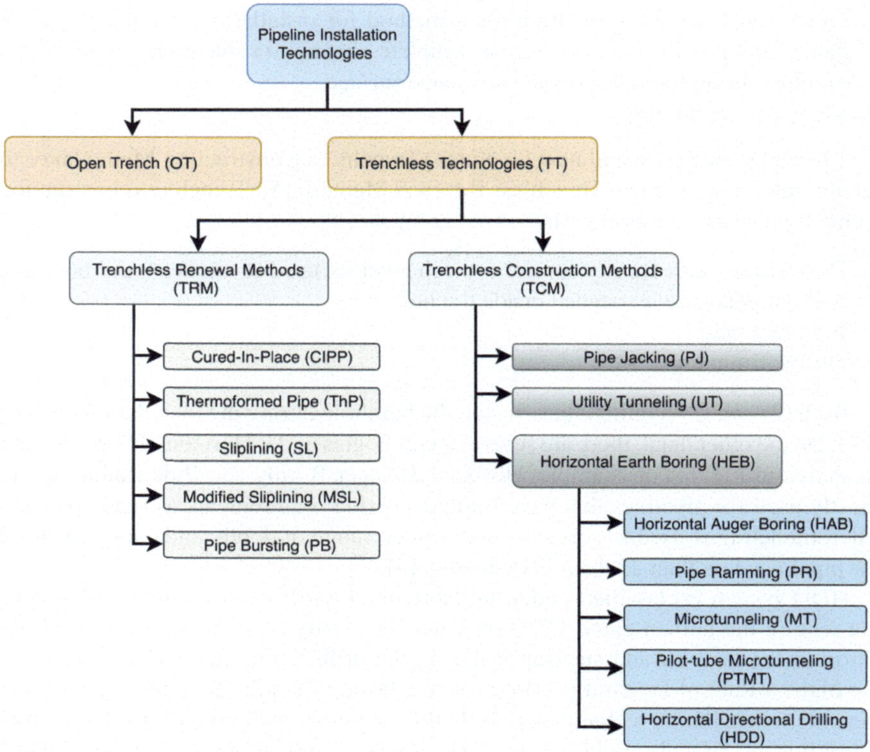


**Fig. 1** Underground pipeline installation techniques [5]

**Table 1** Appropriate techniques for trenchless new installation

| Technique | Water | Sanitary and Storm Sewers | Gas | Electricity | Tele-communications |
|---|---|---|---|---|---|
| Horizontal Auger Boring | X | | X | X | X |
| Pipe Ramming | X | X | X | X | X |
| Pipe Jacking | | X | | | |
| **Directional Drilling** | X | X | X | X | X |
| Microtunneling | | X | | | |
| Pipe Bursting | X | X | X | | |

geological concerns [6]. HDD is economically efficient because it has high installation performance, saves time, and minimizes road cuts and closed roads [6]. More details about drilling process requirements and environmental considerations can be found in [11].

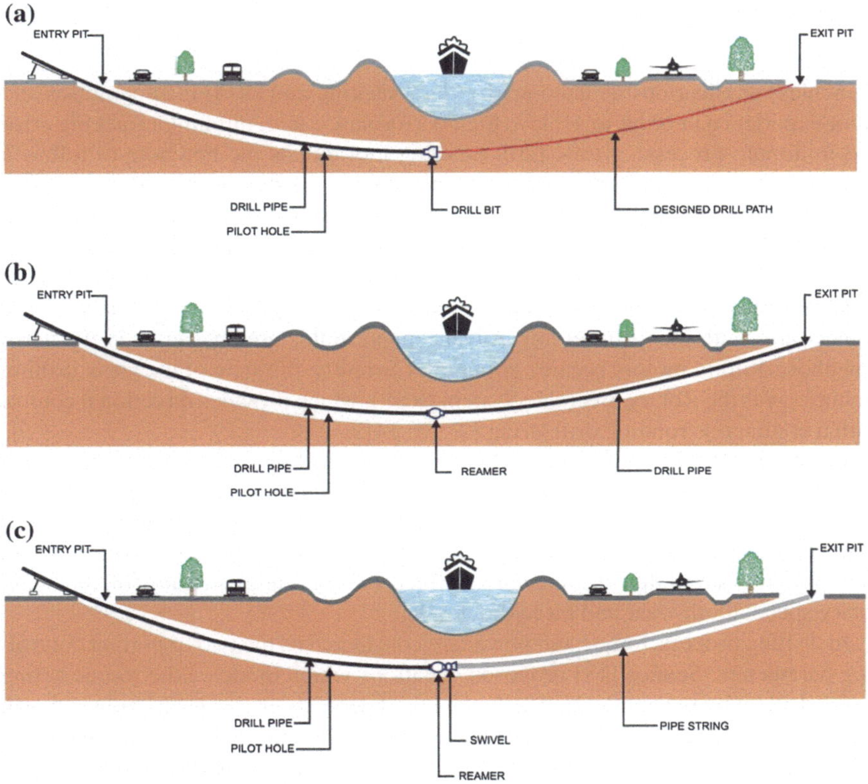


**Fig. 2** Pipeline installation using HDD: **a** Pilot hole; **b** Reamer; **c** Pull the pipe back [12]

**Fig. 3** Horizontal Directional Drilling Rig

## 2 Related Works

Trenchless applications demand appropriate tracking and steering apparatuses and labor experience in order to achieve high performance with minimum tracking error and minimum processing time [20]. Steering tool directs the borehole to follow a predefined trajectory, so selecting the proper steering method is crucial.

Directional Steering System (DSS) is important in the drilling industry in general and trenchless technologies in particular because it significantly increases drilling efficiency. It is also crucial in the oilfield industry when the oil is distributed in the form of a thin horizontal layer spreading over a large area. DSS serves to have a large contact area with the oil reservoir. Advancement in this area began in 1970s when downhole drilling motors became common. Currently, the state of the art in drilling technology is the Rotary Steerable System (RSS), which provides directional control with a continuous rotating drill string [2, 10, 14].

In traditional drilling, the main objective of the drilling engineer is to maximize Rop due to its direct impact on the drilling time and cost of drilling. Rop is affected by several factors. An earlier Rop model was proposed by Bourgoyne [3] as a function consisting of eight multiplied terms. The model includes several factors like formation strength and compaction, effects of differential pressure, bit diameter, rotary speed, tooth wear and bit hydraulic.

In drilling process, torque and power are considered as the two important controlling parameters. Spanos [18] proposed equations which modeled the torque acting on the drill bit by two methods. The first method depends on the dry friction between the borehole formation and the drill bit, while the second method is formulated as the required torque to cut the rock. Specific energy theory or log-based rock analysis is used as a standard for bit performance analysis and planning well. A model with 6 Degrees of Freedoms (DOF) was proposed by Tucker [21]. Three of the six DOF are used to pinpoint the position of drill string centroid in 3D space and the remaining three are used in expressing the shear strain, flexure, dilation, torsion and stretch, which contribute to the drill string dynamic state.

As the research flourished, a new angle in directional drilling technology was proposed by Warren [22], who claimed that Casing while drilling (CWD), which is dominant in vertical drilling in reducing drill cost and solving drilling problems could be applied to directional drilling as well. They concluded that CWD works efficiently with similar advantages for directional drilling as it did for vertical ones and removes the effort of re-running a casing after drilling which was prevalent in previous times.

A model was also proposed by Eren [9] which optimizes the drilling parameters such as weight on bit (Wob) and bit rotation speed to minimize the overall drilling cost, cost per foot and maximize Rop. Actual field data obtained via data recording systems and modern well monitoring systems was used by the developed model to predict the rate of drilling as a function of the drilling parameters. As per the study Rop of drilling could be predicted with high accuracy on the basis of the previous drill trend. To achieve minimum drill cost, the optimum bit rotation and Wob could be determined.

Koederitz [16] presented an autonomous drilling system. The performance of the drilling was evaluated by the use of a test process run by the system software on a set of target points. The results from the field test demonstrated the potential for an autonomous drilling optimization without the drilling knowledge. The procedure could be cost-effective, practical and flexible in many applications.

Geological information can be actively obtained during drilling by using Logging While Drilling (LWD) and Seismic While Drilling (SWD) tools placed at the Bottom Hole Assembly (BHA). Chen [8] utilized these tools to show an automated workflow for proactive geosteering through continuous updating of the estimates of the earth model and robust optimization of remaining well path under uncertainty. The use of the real time geological data could significantly improve drilling, and minimizes non-drilling time due to unpredictable.

Modern drilling systems include a Measurement While Drilling (MWD) unit which provides on-line measurement of the borehole path and its position in three-dimensional space. MWD transmits these monitoring parameters to the control computer for displaying, recording, printing, and providing the control parameters [7].

In this work we provide an online control scheme for real-time optimization of the drilling parameters and control of steering system to minimize the deviation from the target borehole trajectory of the pilot hole and maximize Rop. Nonlinear model for the drilling process was developed using energy balance equation. The proposed controller design problem is formulated as an optimization problem where an objective function is considered to minimize tracking error and drilling efforts.

In the next section, we present a simplified model for the DSS. Then, we provide the optimization function and procedure. The simulation results are presented at the end along with the conclusion and acknowledgements.

## 3 Directional Drilling System

### *3.1 Quad-Motors Dynamic System*

In order to determine the position of the drilling head we need to describe vectors not only in an inertial frame, but also in a body frame. The inertial frame is denoted by $\{X_E, Y_E, Z_E\}$, as shown in Fig. 4. The origin of the inertia frame is taken to be the surface starting point of drilling. By convention $X_E$ points to the geographic north, $Y_E$ points to the East, and $Z_E$ points to the earth center. The body attitude is described with respect to three virtual body axes, $\{X_B, Y_B, Z_B\}$ located at the body reference point OB. These three-reference axes are assumed to be parallel to the inertial axes.

When a vector is described with respect to the earth fixed frame it will be denoted with an $E$ subscript, and if the vector is described with respect to the body it will be

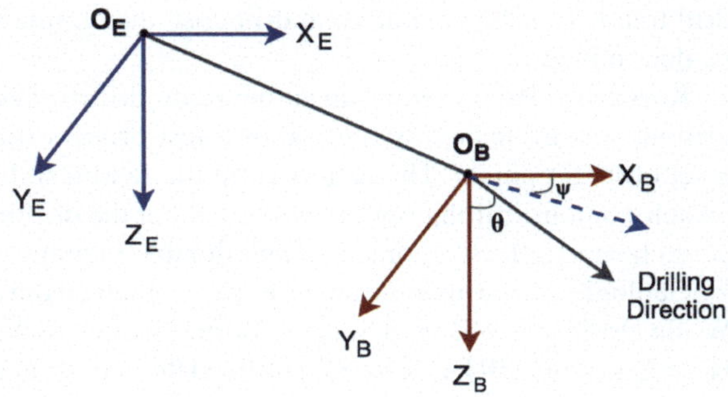


**Fig. 4** Earth and body frame

subscripted with $B$ or $b$. The position of the BHA in the inertial frame is given by the vector,

$$\zeta^T = [x_E, y_E, z_E] \tag{1}$$

We define $\{U_B, V_B, W_B\}$ to be a body fixed frame. The orientation of the BHA will be described by the rotation of the body fixed frame $\{U_B, V_B, W_B\}$, with respect to the body virtual frame $\{X_B, Y_B, Z_B\}$. The orientation of the drilling assembly is given by the three Euler angles, namely the yaw angle $\psi$ (azimuth), the pitch angle $\theta$ (inclination) and the roll angle $\phi$, and presented in the vector

$$\Omega^T = [\psi, \theta, \phi] \tag{2}$$

We assume initially, the body fixed axis $\{U_B, V_B, W_B\}$ were aligned with respect to inertial frame $\{X_B, Y_B, Z_B\}$. The instantaneous direction of motion is taken to be along $W_B$. The orientation matrix can be described by the rotational matrix $R_{BE}$ given by,

$$R_{BE} = \begin{bmatrix} c\psi s\theta & c\psi c\theta s\phi - s\psi c\phi & c\psi s\theta c\phi - s\psi s\phi \\ s\psi s\theta & s\psi s\theta s\phi + c\psi c\phi & s\psi c\theta c\phi - c\psi s\phi \\ -s\theta & c\theta s\phi & c\theta c\phi \end{bmatrix} \tag{3}$$

The rotational matrix $R_{BE}$ defines the transformation from the body axis to the inertial axes, where $cx$ denotes $\cos(x)$ and $sx$ denotes $\sin(x)$. For a point P in space, if its position w.r.t the body frame is $P_{UVW}$, then the position of the point in the inertial system is given by,

$$P_{XYZ} = \begin{bmatrix} x_E \\ y_E \\ z_E \end{bmatrix} + R_{BE}\, P_{UVW} \tag{4}$$

Let us denote the body rotational angle about $U_B$ by $\theta_1$, rotational angle about $V_B$ by $\theta_2$, and the rotational angle about $W_B$ by $\theta_3$. The Euler angles are related to the angular velocities of the BHA by the following equation

$$\begin{bmatrix} \dot{\phi} \\ \dot{\theta} \\ \dot{\psi} \end{bmatrix} = \begin{bmatrix} 1 & \sin\psi\tan\theta & \cos\psi\tan\theta \\ 0 & \cos\psi & -\sin\psi \\ 0 & \sin\psi\sec\theta & \cos\psi\sec\theta \end{bmatrix} \begin{bmatrix} \dot{\theta}_1 \\ \dot{\theta}_2 \\ \dot{\theta}_3 \end{bmatrix} \tag{5}$$

If we are interested in only the direction of the borehole, we may then ignore the roll rotation of the BHA. In this case $R_{BE}$ is given by,

$$R_{BE} = \begin{bmatrix} c\psi s\theta & -s\psi & c\psi s\theta \\ s\psi s\theta & c\psi & s\psi c\theta \\ -s\theta & 0 & c\theta \end{bmatrix} \tag{6}$$

The planned borehole trajectory can be described by several methods. However, we assume without loss of generality the target trajectory is given by a table of points indexed by the measured depth (MD), the measured distance along the borehole. The points need not be uniformly spaced. We will denote the MD by $w_b(k)$. The $k$th target point is given by the vector,

$$P_T(k) = [x_E^t(k), y_E^t(k), z_E^t(k), w_b^t, \theta^t, \psi^t] \tag{7}$$

A Directional Steering System (DSS) equipped with four motors is shown in Fig. 5 [19]. A drill bit is attached to each motor. The speed of each motor can be independently controlled, causing the rate of removal of rocks by each bit and the direction of advancement of the drilling head to be precisely controlled. The drill head assembly is attached to the end of the drill string, which includes an inner pipe for carrying the drilling fluid. The use of four motors in coordination with other traditional drilling variables allow precise control of the drilling direction and optimization of Rop. The top and bottom motors rotation (rotation about $V_B$) is obtained by increasing or decreasing the speed of the top motor while decreasing or increasing the speed of the lower motor. The body yaw movement (rotation about $U_B$) is obtained similarly using the right and left motors. The BHA roll movements is the rotation about $W_B$. The control of the four motors allows better management of the drilling operation in various drilling environment and under various operational constraints.

The drill bit position at point $k + 1$ along the trajectory in terms of the body axis is given by,

$$\Delta u_b(k+1) = \delta\theta_1[w_b(k+1) - w_b(k)] \tag{8}$$

$$\Delta v_b(k+1) = \delta\theta_2[w_b(k+1) - w_b(k)] \tag{9}$$

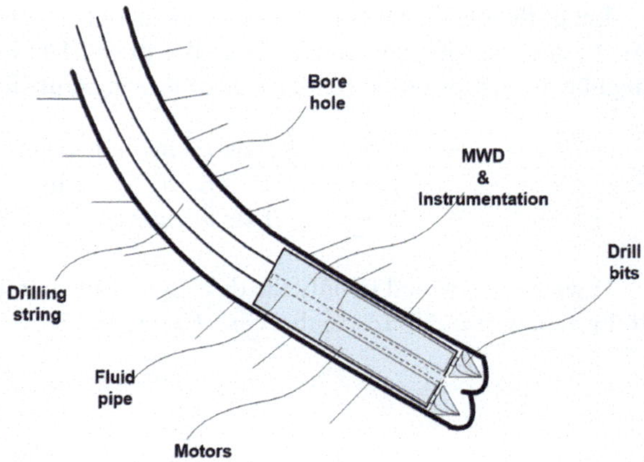


**Fig. 5** Overview of the drilling assembly

$$\Delta w_b(k+1) = w_b(k+1) - w_b(k) \cong R_{op} * t_k \tag{10}$$

where $t_k$ is the drilling time from the $k$th point to the $(k+1)$th point. The BHA predicted location with respect to the inertial axes will be given by,

$$\hat{P}_E(k+1) = P_E(k) + R_{BE} \begin{bmatrix} \Delta u_b(k+1) \\ \Delta v_b(k+1) \\ \Delta w_b(k+1) \end{bmatrix} \tag{11}$$

The motor torque is resolved by the drill bit into two components; a drag torque on a plane perpendicular to the bit axis ($T_D$), and a lift force ($F_L$). The lift force of the $i$th motor can be approximated by the relation,

$$F_{L_i} = T_i \left( \frac{2\pi}{\Delta_i} \right) \tag{12}$$

where, $\Delta_i$ is the feed rate per revolution and $T_i$ is the motor torque. The bit drag torque is the torque used to crush the rocks, $T_{Di}$ may be approximated by,

$$T_{D_i} \approx T_i \tag{13}$$

For crushing a volume of rock, a minimum quantity of energy will be required, this amount depends on the nature of the rock. Rock Specific Energy, $E_{rs}$ in $J/m3$, is defined as the energy required to crush a unit volume of the rock. From the energy balance equations, equating the mechanical power by the four motors to the rate of rock crushing, we get,

$$\sum_{i=1}^{4} T_i \omega_i + W_{ob} \dot{w}_b = \dot{w}_b A_h E_{rs} \tag{14}$$

where $\dot{w}_b$ is the rate of advance of the drilling head in m/s, also known as the rate of penetration, and $W_{ob}$ is the weight on the bit. Finally, $A_h = (\frac{\pi D_h^2}{4})$ is the cross-sectional area of borehole, and $D_h$ is hole diameter. The Rop is given by,

$$R_{op} = \frac{\bar{\Delta}\bar{\omega}}{2\pi} \tag{15}$$

where $\bar{\Delta}(t)$ is the average rate of penetration per revolution. $\bar{\Delta}(t)$ is a function of many factors including; the $E_{rs}$, $Wob$, the total motor torques, bit geometry, and drilling fluid.

From Eq. 14, the instantaneous rate of penetration is given by

$$\dot{w}_b(t) = \frac{\bar{\Delta}\,\bar{\omega}}{2\pi} = \frac{\sum_{i=1}^{4} T_i\,\bar{\omega}}{A_h \hat{E}_{rs} - W_{ob}} \tag{16}$$

The average penetration per revolution is given by

$$\bar{\Delta}(t) \cong \frac{2\pi \sum_{i=1}^{4} T_i}{A_h E_{rs} - W_{ob}} \tag{17}$$

For better description of the forces and torques acting on BHA, we introduce the following auxiliary inputs:

$$u_1 = F_{L_1} + F_{L_2} + F_{L_3} + F_{L_4} \tag{18}$$

$$u_2 = (F_{L_2} - F_{L_4})\,\mu_b \tag{19}$$

$$u_3 = (F_{L_1} - F_{L_3})\,\mu_b \tag{20}$$

$$u_4 = T_{D_1} + T_{D_3} - T_{D_2} - T_{D_4} \tag{21}$$

where $u_1$ is related to the total thrust of the four motors, while $u_2$ generates moment responsible for changing the azimuth direction. Similarly, $u_3$ generates the moment for changing the inclination, while $u_4$ generates the torque for the roll motion. $\mu_b$ is the distance between motor axes and the BHA central axes $W_B$.

The lateral rates of rotation can be approximated by

$$\begin{aligned}\dot{\theta}_1 &= K_{f_1} u_2 \\ \dot{\theta}_2 &= K_{f_2} u_3 \\ \dot{\theta}_3 &= K_{f_3} u_4\end{aligned} \tag{22}$$

where $K_{f1}, K_{f2}, K_{f3}$ are functions of the formation properties. To simplify analysis, we will assume that the BHA roll is stabilized by using differential torque control, or simply using appropriate roll stabilizers installed on the BHA body.

Now, referring to Eqs. 8 and 9

$$\begin{aligned}\delta\theta_1 &= \dot{\theta}_1 \times t_k \\ \delta\theta_2 &= \dot{\theta}_2 \times t_k\end{aligned} \tag{23}$$

In summary, the control actions $u_1$, $u_2$, and $u_3$ affect the angular velocities of the BHA in Eq. 22. The changes in the orientation of the BHA at the end of $t_k$ are then given by Eq. 23, and the position of the BHA is determined by Eq. 11.

### 3.2 Rock Parameters Identification

Since all the parameters, except rock parameters $R_p = \{E_{rs}, K_{f1}, K_{f2}, K_{f3}\}$, are directly measurable online, the $R_p$ can be inferred. The estimated value of $R_p$ is then used during optimization to predict the new BHA position during the next control step. The rock specific energy is estimated from the previous control parameters, and the measured rate of penetration using Eq. 14.

$$\hat{E}_{rs}(k) \cong \frac{\sum_{i=1}^{4} T_i(k-1)\,\omega_i(k-1) + W_{ob}(k-1)\,\dot{w}_b(k)}{A_h \dot{w}_b(k)} \tag{24}$$

To estimate the parameters $K_{f1}$, $K_{f2}$, $K_{f3}$, we go backward to Eqs. 9, 8, and 22, as follows. Let $\Delta w_b(k)$ be the last change in the measured depth during the last adjustment interval, i.e. going from step $k-1$ to step $k$. and using Eq. 11 we get,

$$\begin{bmatrix}\Delta u_b(k) \\ \Delta v_b(k) \\ \Delta w_b(k)\end{bmatrix} = R_{EB}(k)(P_E(k) - P_E(k-1)) \tag{25}$$

where $R_{EB} = R_{BE}^{-1} = R_{BE}^{T}$.

Then the formation parameters can be estimated using the current and previous state values and previous control vector.

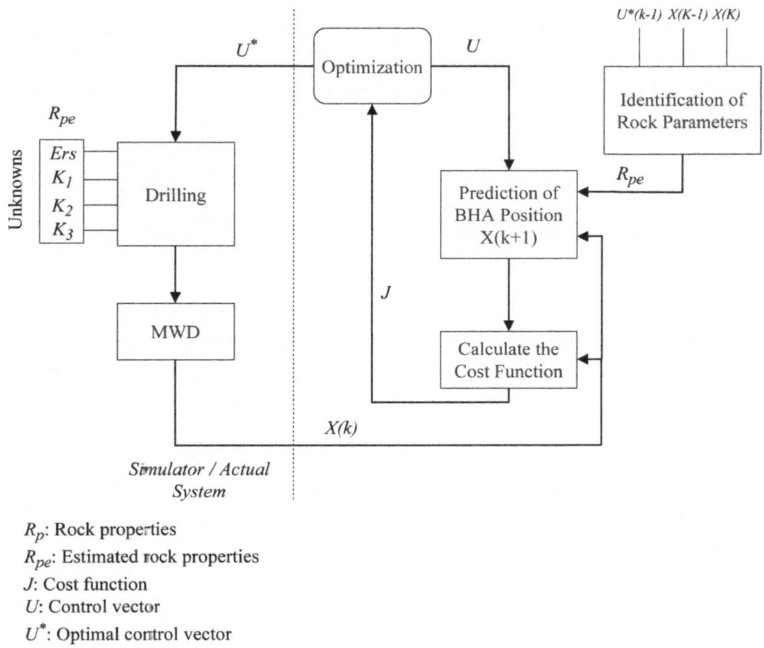


**Fig. 6** Overview of the automatic control algorithm

# 4 Drilling Optimization

An overview of the automatic control algorithm is shown in Fig. 6. At the start of each control period $k$, the MWD provides a state vector, representing the current BHA position, inclination, azimuth, MD, Rop, and the total time. We assume also we have measurements of the body angular velocities, motors angular velocity and torque. The MWD includes, among other things, the following;

1. Three perpendicular set of accelerometers for gravity measurements to determine the vertical axis.
2. Three magnetometers for determining the magnetic north. The accelerometers and magnetometers are aligned with the BHA body axis.
3. Inertial measurement system (IMU) to determine the body angular velocities, rotation, and body translational motion (acceleration, velocity, and travelled distance "MD").
4. Embedded system to perform the transformation from body measurements to the inertial frame.

The automatic control algorithm consists of 4 parts; the optimization algorithm, the position prediction step, the cost function, and rock parameters identification. The system to be controlled is the BHA dynamics subject to effect of unknown rock characteristics.

### 4.1 Optimization Algorithm

At each point $k$, we assume we have a state vector $X(k)$, and a target position $w_{bt}(k+1)$, and it is required to find the optimal control parameters $U(k)$, so that the BHA reaches the target position at the end of the adjustment period. The state vector includes;

$$X(k) = [t_{total}, w_b, x_E, y_E, z_E, \theta, \psi, \dot{w}_b, \hat{E}_{rs}] \tag{26}$$

The control input represents the angular velocities and torques, weight on bit, and the action time.

$$U(k) = [\omega_1, \omega_2, \omega_3, \omega_4, T_1, T_2, T_3, T_4, W_{ob}, t_k] \tag{27}$$

It is to be noticed that $t_k$, is the time to reach the next target position, is to be determined.

The control algorithm has been implemented as follows,

1. Identify $R_{pe}$ using the current state vector, previous state vector, and last control action $U(k-1)$.
2. The optimization algorithm proposes a control vector $U(k)$.
3. Use Eqs. 8, 11 to predict the position $\hat{P}_E(k+1)$.
4. Evaluate the objective function given the desired position $P_E^d(k+1)$.
5. Send the value of the objective function to the optimization algorithm.
6. The optimization algorithm checks if the cost function is minimized. If not go to step 2.
7. If the optimal value is reached, send the control to the BHA, and go to step 1.

### 4.2 Optimization of Cost Function

Clearly the objective is to reach the target position $P_E^d(k+1)$ at the end of each adjustment step. The predicted value of $w_b(k+1)$ is obtained as follows,

$$\hat{w}_b(k+1) = f(T, \omega, W_{ob}) = \frac{\bar{\Delta}\,\bar{\omega}}{2\pi} = \frac{\sum_1^4 T_i\,\bar{\omega}}{A_h \hat{E}_{rs} - W_{ob}} \tag{28}$$

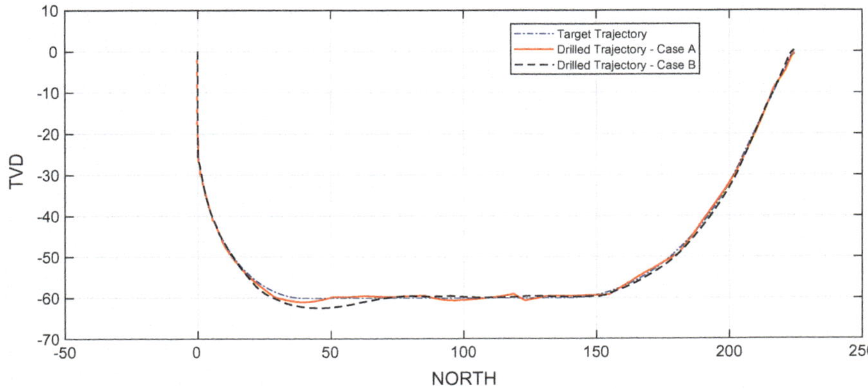


**Fig. 7** Trajectory tracking, Case A and Case B

$$\hat{w}_b(k+1) = w_b(k) + t_k f(\omega, W_{ob}) \tag{29}$$

The optimal values of control parameters are obtained by minimizing the following cost function;

$$J = (\hat{P}_E(k+1) - P_E^d(k+1))^T \Gamma_1 (\hat{P}_E(k+1) - P_E^d(k+1)) + \nabla U(k)^T \Gamma_2 \nabla U(k) \tag{30}$$

where $\nabla U(k) = U(k) - U(k-1)$ is the change in the input parameters vector, $\Gamma_1$ and $\Gamma_2$ are positive semi-definite weighting matrices, $\hat{P}_E(k+1)$ is the predicted position if $U(k)$ is used, and $P_E^d(k+1)$ is the target position.

## 5 Simulation and Discussion

The proposed optimization method is applied to follow a drilling trajectory of a real borehole section of total measured length of 292 m, and true vertical depth (TVD) = 60 m. The target trajectory is interpolated to give the target XYZ position, the MD, the inclination angle and azimuth angle every two meters of the MD, a total of 164 points, including the origin. The optimization procedure requires evaluation of the cost function Eq. 30, for any given $U(k)$. The model described in Eq. 11 is used for prediction of $\hat{P}_E(k+1)$ for the given control vector. Rock Specific Energy is assumed to be unknown but is estimated in the optimization step based on the achieved Rop during the previous control step, see Eq. 24.

Two control strategies were simulated. In case A the weighting matrices $\Gamma_1$ and $\Gamma_2$ were chosen for good compromise between the drilling time, drilling effort, and trajectory tracking accuracy. In case B, drilling time is strongly optimized while keeping the same weight for the tracking accuracy.

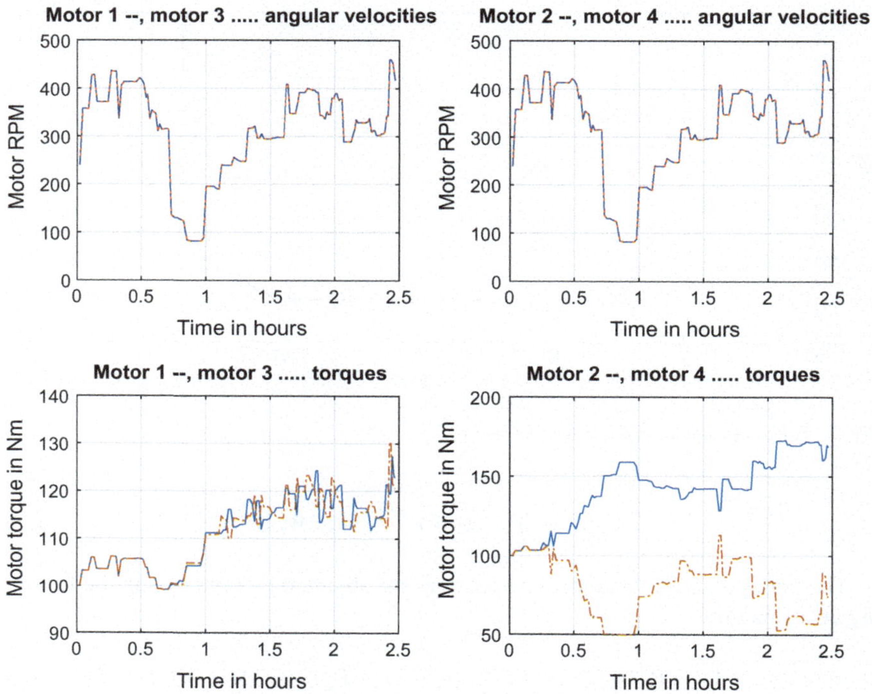


**Fig. 8** Case A: the RPM and torques for the 4 motors

The drilling trajectory of case A is shown in Fig. 7. The angular velocities and the torques of the four motors are shown in Fig. 8. The drilling time came to 2:47 h. The Rop is shown in Fig. 9. The average Rop came to 2.05 m/min. The maximum tracking error is 1.22 m.

In case B, the weighting matrices of the cost function were chosen to heavily penalize the drilling time, while maintaining the same target tracking accuracy of case A. Figure 7, shows the drilled hole trajectory of case B. The drilling time of this case came to only 1:41 h. Moreover, Fig. 10. Shows that the average Rop was 3.73 m/min., greater than the mean Rop of case A. However, the tracking error came to 2.47 m which is higher than case A. This gain in time was achieved by more demanding requirements on the drilling motors as shown in Fig. 11. The mean RPM of the four motors came to 486.6 RPM, compared to 320.7 RPM in case A. The average torque was 173.1 Nm compared to 132.85 Nm for case A. The peak power of motors came to 17.53 kW, compared to 7.52 kW in case A.

The minimization of the cost function in Eq. 30 is performed using Matlab constraint minimization function “fmincon”, by the interior-point algorithm. The lower and upper bounds on the control values are given in Table 2.

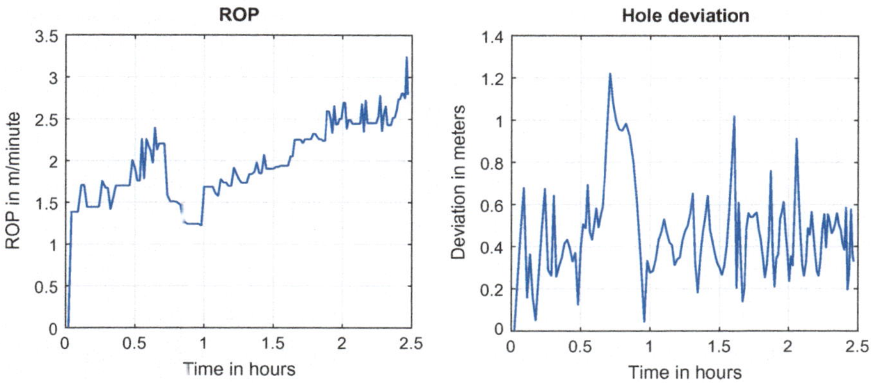


**Fig. 9** Case A: Rop and Hole deviation

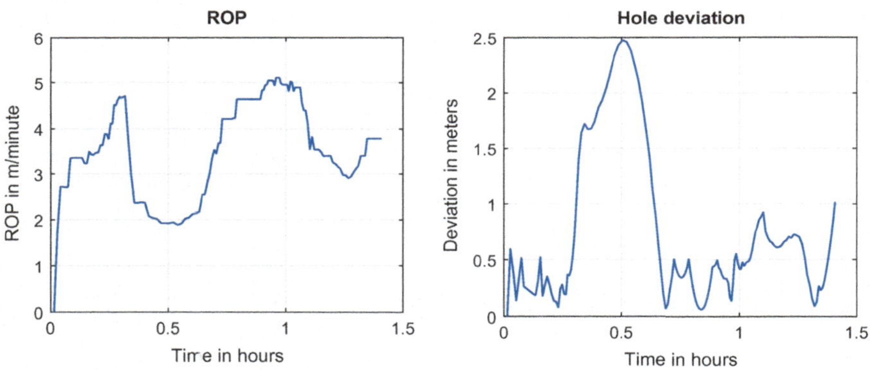


**Fig. 10** Case B: Rop and Hole deviation

## 6 Conclusion

This work presents simulation study and evaluation of real-time automatic trenchless drilling using a quad motor drilling system. The control system employs an optimization approach which combines the conventional drilling parameters as well as the directional steering control. The system assumes availability of real-time MWD information to determine the control actions to direct the BHA to the next target position. The objective function compromises between trajectory tracking accuracy, drilling effort, and drill time. The automatic control algorithm accounts for the practical operational limits using standard constraint optimization techniques. The proposed control approach has been evaluated using a real borehole section of total measured length of 292 m, two simulation scenarios with different weighting matrices were considered to illustrate the superiority of the proposed approach for the trade off between system performance parameters. The results showed how the weighting

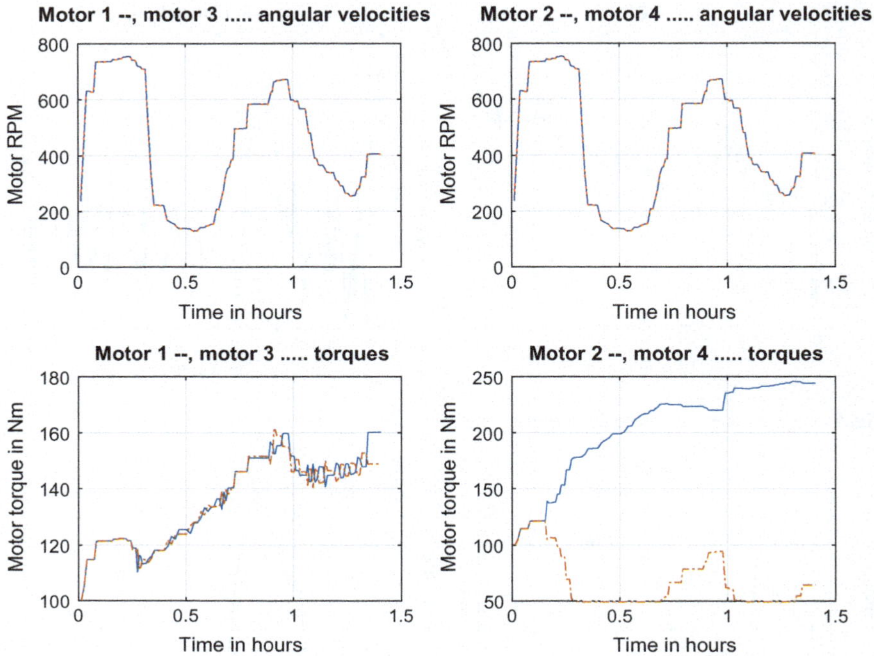


**Fig. 11** Case B: the RPM and torques for the 4 motors

**Table 2** Upper and the lower bound for control parameters

| Parameter | Min. | Max. |
|---|---|---|
| Torque ($T_i$) | 50 Nm | 500 Nm |
| Angular velocity ($\omega_i$) | 30 rpm | 900 rpm |
| $W_{ob}$ | 500 kg | 5000 kg |
| $t_k$ | 20 s | 300 s |

matrices of the objective function can be effectively tuned to increase or decrease the rate of penetration while tracking the predetermined trajectory with minimum time and control efforts. The proposed drilling system will also facilitate drilling of 3D hole trajectories.

**Acknowledgements** This work has been supported by Saudi Arabia National Science, technology, and Innovation Projects under contract number NSTIP 12-OIL3033-04.

## References


1. Ariaratnam, S.T., Lueke, J.S., Allouche, E.N.: Utilization of trenchless construction methods by canadian municipalities. J. Constr. Eng. Manag. **125**, 76–86 (1999)
2. Baker, R.: A Primer of Oil-Well Drilling. Petroleum extension Service at Austin, Texas, USA (2001)
3. Bourgoyne Jr A.T., Millheim, K.K., Chenevert, M., Young Jr, F.S.: Applied Drilling Engineering. Society of Petroleum Engineers, USA (1986)
4. Bueno, S.M.: Horizontal Directional Drilling Guide. Technical report, Trenchless Technology Magazine (2015)
5. Bugbee, K.P., Patton, B.D., Jubin, R.T. Robinson, S.M., Sullivan, N.M.: Evaluation of trenchless technologies for installation of pipelines in radioactive environments. In: Waste Management Conference (2010)
6. CAPP: Planning Horizontal Directional Drilling for Pipeline Construction. Technical report, Canadian Association of Petroleum Producers (2004)
7. Chen, C., Yanshun, Z., Chunyu, L.: Surveying method of measurement while drilling based on the inertial sensor. In: First International Conference on Pervasive Computing Signal Processing and Applications, pp. 1192–1195 (2010)
8. Chen, Y., Lorentzen, R.J., Vefring, E.H.: Optimization of well trajectory under uncertainty for proactive geosteering. SPE J. **20**, 1–16 (2015)
9. Eren, T., Ozbayoglu, M.E.E.: Real time optimization of drilling parameters during drilling operations. In: SPE Oil and Gas India Conference and Exhibition, pp. 1–14 (2010)
10. Eustes, A.W.: The evolution of automation in drilling. In: SPE Annual Technical Conference and Exhibition, pp. 1–5 (2007)
11. Hashash, Y., Javier, J., Petersen, T., Osborne, E.: Evaluation of Horizontal Directional Drilling. Technical Report ICT-R27-SP16, Illinois Center for Transportation (2011)
12. Hydrotech, E.C.: Thrust boring and horizontal directional drilling. http://hydrotechndrc.com (2015)
13. Iseley, T., Gokhale, S.B.: Trenchless Installation of Conduits Beneath Roadways. National Academy Press, Washington (1997)
14. Joshi, S., Ding, W.: The cost benefits of horizontal drilling. In: Proceedings of the American Gas Association, pp. 679–684 (1991)
15. Jung, Y.J., Sinha, S.K.: Evaluation of trenchless technology methods for municipal infrastructure system. J. Infrastruct. Syst. **13**, 144–156 (2007)
16. Koederitz, W.L., Johnson, W.E.: Real-time optimization of drilling parameters by autonomous empirical methods. In: SPE/IADC Drilling Conference and Exhibition (2011)
17. MDOT: Overview and Introduction of Trenchless Installation within MDOT ROW . Technical Report 3701, Michigan Department of Transportation (2006)
18. Spanos, P.D., Sengupta, A.K., Cunningham, R.A., Paslay, P.R.: Modeling of roller cone bit lift-off dynamics in rotary drilling. J. Energy Resour. Technol. **117**, 197–207 (1995)
19. Talib, M., Elshafei, M., Khoukhi, A., Saif, A.W.A., Abdulazeez, A.: Modeling and control of quad-rotor directional steering system. In: Proceedings of IEEE International Symposium on Innovations in Intelligent Systems and Application, pp. 76–83 (2014)
20. Tanwani, R., Iseley, T.: Tracking and steering systems in trenchless construction. J. Constr. Eng. Manag. **120**, 65–76 (1994)
21. Tucker, W.R., Wang, C.: An integrated model for drill-string dynamics. J. Sound Vib. **224**, 123–165 (1999)
22. Warren, T.M., Houtchens, B.D., Madell, G.: Directional drilling with casing. SPE Drill. Complet. **20**, 1–7 (2005)

# Stabilization of Second Order Underactuated System Using Fast Terminal Synergetic Control

**D. Zehar, A. Chérif, K. Behih, K. Benmahammed and N. Derbel**

**Abstract** The control of underactuated mechanical systems (UMS) has been the subject of active scientific research, provided by the wide applications of such systems in different disciplines. The main purpose of studying these systems is to control certain variables in parallel, knowing that underactuated systems admit more degrees of freedom than the actuators. The lack of actuator complicates the task of controlling such systems. In this chapter, our objective is to find a control law which guarantees the finite time convergence and the stability of a class of second order underactuated system, for this reason, we have chosen to use the fast terminal synergetic control (FTSC), this latter is considered among robust controls, and has already proven its effectiveness in several applications of nonlinear systems. The UMS is divided into two sub-systems, and each one has its macro variable, which is constructed in hierarchical manner. To prove the efficiency of the proposed strategy of control, we show simulation results of a second order underactuated system example.

**Keywords** Underactuated system · Terminal synergetic control · Stability · Finite time convergence

D. Zehar (✉) · A. Chérif
Department of Electromechanical, University of Bordj Bou Arréridj, El Anasser, Algeria
e-mail: djamila.zehar@univ-bba.dz

A. Chérif
e-mail: aida.cherif@univ-bba.dz

K. Behih
Department of Electrical Engineering, LSI Laboratory, University of Sétif, Setif, Algeria
e-mail: khalissabehih@yahoo.fr

K. Benmahammed
Department of Electronic, LSI Laboratory, University of Sétif, Setif, Algeria
e-mail: Khierben@gmail.com

N. Derbel
ENIS, CEMLab Laboratory, University of Sfax, Sfax, Tunisia
e-mail: nabil.derbel@gmail.com

 
J. Ghommam et al. (eds.), *New Trends in Robot Control*, Studies in Systems, Decision and Control 270, https://doi.org/10.1007/978-981-15-1819-5_4

# 1 Introduction

Applications in robotics, vehicle and automatic, which serve the needs of the human being, are essential sources that motivate the analysis and control of a category of non-linear mechanical systems called underactutaed mechanical systems UMS. An USM is a nonlinear system that has fewer control actuators than degrees of freedom to control. In recent years, there has been a major interest in the development of stabilizing control strategies for these systems, and underactuated algorithms, for mechanical systems that may be underactuated by design, or by failure of one of the actuators. The underactuted character comes from several reasons; maybe the dynamics of these systems are underactuated, such as aircrafts, or this character is due by design, since the actuators are expensive and sometimes very heavy, therefore they are avoided in the design of systems, as in robots with flexible joints, or due to an actuator failure. Another reason is that underactuation is artificially imposed, to create UMS of low order, which are academic benchmarks, in order to simplify the task of researchers in their studies in laboratories.

Researchers have come to classify UMS according to structural and dynamic properties, where some classes have only one UMS, which requires them to be studied on a case-by-case. Then the interest of simplifying and unifying the structure on which the application of an appropriate command can achieve the objective remains a difficult task to accomplish.

Control of UMS has been constantly evolving for several years, but because of their complexity (nonlinearity and underactuation) makes the study of developing a control for all of these systems more difficult, and is performed more slowly, in addition the results that are established for nonlinear systems are not directly applicable to the case of UMSs. The theorem of Brockett is not applicable, since these systems are not triangularisable through a smooth feedback. Control strategies which are based on passivity and energy are first applied to stabilize UMS, among the known works; we can mention the work in [5, 9], where they applied this strategy of control to stabilize an inverted pendulum. Then partial linearization method is studied for UMS, as in [11, 12]. Unfortunately in [4] it has been mentioned that these techniques of control which are divided in two phases and which commutate to linear controller have some deficiency in control, because they are slow, and can't respond to precision performance.

The notion of robust control has been introduced to meet the control requirements in the presence of external matched and mismatched disturbances and parametric uncertainties. For this purpose, several tools are proposed in the literature; one of the recently answered is the study based on the sliding mode control (SMC), which is a special case of the theory of systems with variable structure. Among the works using this strategy of control we can mention [13, 15], where authors have chosen different constructions of sliding surfaces, adapted to the proposed system structure. Unfortunately, this control suffers from problem of chattering, which causes significant mechanical oscillations in the actuators and eventually, lead to their rapid damage. To improve performances of the sliding mode control and in order to resolve the problem of chattering, many techniques have been combined with this latter, as in the work of [2], where they used the fuzzy approximation of the switching control, the work found in [1], which is based on backstepping procedure, and the work of [10] where they introduced neural networks.

Another approach called terminal sliding mode control (TSMC) is used to improve the asymptotic convergence to be in finite time, as in [3, 8, 16]. Using this control, time may be not fast enough, to solve this issue, researchers have developed a new strategy named fast terminal sliding mode control (FTSMC), as in [14]. Despite the fact that these controls have proved their effectiveness in the stabilization and tracking of nonlinear and underactuated processes, the phenomenon of chattering remains as a big problem, and the studies in this context are developed more and more, until the appearance of the synergetic control theory (SCT), which attempts to overcome the previously described problems of SMC. The SCT shares with SMC, the properties of order reduction and decoupling design procedure, and it has other several advantages: it operates at constant switching frequency and it does not have the chattering problems of sliding mode control, so it causes less power filtering problem, which is convenient for experimental implementations. This control is always continuous, and uses a macro variable that can be a function of two or more system state variables [17]. The characteristics of the macro-variable can be chosen by the designer, according to the parameters such as the objective of control, the response time, and limitations of the control.

This chapter, is decomposed on 5 parts, in the second part, we give a mathematical model of second order UMS, in the third part, generalities of SCT are set, in fourth part we introduce the fast terminal synergetic control (FTSC) for a class of second order UMS, and in the five part we show simulation results, we last this work by a conclusion.

## 2 Mathematical Model of Underactuated System

The underactuated mechanical systems have a less number of actuators than their degrees of freedom, where the underactuation property comes from several reasons, either the dynamics are underactuated or by conception in order to reduce the cost and the weight, and maybe for security reason when one of the actuator fails. The variety and the complexity of those systems lead to classify them in several classes and study them case by case. A class of second order underactuated systems has the following Lagrangian, as presented in [11]:

$$L(q, \dot{q}) = K - V = \frac{1}{2}\dot{q}^T M(q)\dot{q} - V(q) \tag{1}$$

where $V(q)$ is the potential energy. $q = [q_1, q_2]^T$ is the configuration vector, and:

$$\begin{bmatrix} m_{11}(q_2) & m_{12}(q_2) \\ m_{21}(q_2) & m_{22}(q_2) \end{bmatrix}$$

is the inertia matrix. The Euler-Lagrange equations are given as:

$$\frac{d}{dt}\left(\frac{\partial L}{\partial \dot{q}_1}\right) - \frac{\partial L}{\partial q_1} = U \tag{2}$$

$$\frac{d}{dt}\left(\frac{\partial L}{\partial \dot{q}_2}\right) - \frac{\partial L}{\partial q_2} = 0 \tag{3}$$

From some mathematical developments, we can obtain the following matrix representation:

$$\begin{bmatrix} m_{11}(q_2) & m_{12}(q_2) \\ m_{21}(q_2) & m_{22}(q_2) \end{bmatrix} \begin{bmatrix} \ddot{q}_1 \\ \ddot{q}_2 \end{bmatrix} + \begin{bmatrix} G_1(q,\dot{q}) \\ G_2(q,\dot{q}) \end{bmatrix} = \begin{bmatrix} U \\ 0 \end{bmatrix} \tag{4}$$

$G_1(q, \dot{q})$ and $G_2(q, \dot{q})$ represent centrifugal, coriolis and gravity terms, where:

$$G_1(q,\dot{q}) = \frac{dm_{11}(q_2)}{dq_2}\dot{q}_1\dot{q}_2 + \frac{dm_{12}(q_2)}{dq_2}\dot{q}_2^2 + b_1(q) \tag{5}$$

$$G_2(q,\dot{q}) = \frac{dm_{21}(q_2)}{dq_2}\dot{q}_1\dot{q}_2 + \frac{dm_{22}(q_2)}{dq_2}\dot{q}_2^2 + b_2(q) \tag{6}$$

where:

$$b_1(q) = \frac{\partial V}{\partial q_1}, \quad b_2(q) = \frac{\partial V}{\partial q_2}$$

These systems can be expressed by the following state representation:

$$\begin{cases} \dot{x}_1 = x_2 \\ \dot{x}_2 = f_1(x) + g_1(x)U + d_1(t) \\ \dot{x}_3 = x_4 \\ \dot{x}_4 = f_2(x) + g_2(x)U + d_2(t) \end{cases} \tag{7}$$

where $y = [x_1, x_3]^T$ is the output vector, $x = [x_1, x_2, x_3, x_4]^T$ is the state variables vector, such that $x_1 = q_1$, $x_2 = \dot{q}_1$, $x_3 = q_2$, $x_4 = \dot{q}_2$, and $f_1(x)$, $f_2(x)$, $g_1(x)$ and $g_2(x)$ are nonlinear bounded functions, $U$ is the controller signal. are bounded external perturbations. We suppose that system in Eq. (7) is bounded input bounded output and all state variables signals are measurable.

## 3 Synergetic Control Procedure

Synergistic control theory was first introduced in a general way, in the last years by Russian researchers (Kolesnikov's team). This is a new trend in the field of control, based on the principles of oriented self-organization and the use of the dynamic properties of nonlinear systems. The synergetic control is a technique quite close to the control by sliding mode, in the sense that it forces the system to evolve with a dynamic pre-chosen by the designer. It differs from the fact that the control is always continuous and uses a macro-variable, which can be a function of two or more system state variables.

To synthesis the continuous control law, we generally use macro variables, by introducing the whole nonlinear model [6].

We introduce the basics of synergetic control synthesis for an n-order nonlinear dynamic system described by:

$$\dot{x} = f(x, U, t) \tag{8}$$

where $x$ is the state space vector and $U$ is the control input. For simplicity, we will consider a single input output system. Control synthesis begins by a suitable choice of a macro-variable function as:

$$\psi = \psi(x, t) \tag{9}$$

where $\psi = \psi(x, t)$ designates chosen macro-variable and a corresponding state variables and time dependent function. Then, we choose a desirable manifold as presented in (10), such that the system can remain on it even the existence of disturbances or parameter variation.

$$\psi = 0 \tag{10}$$

A large choice is available to the designer in selecting the macro-variable features accordingly with the control objectives and practical physical constraints. The macro-variable, which may be a simple linear combination, is forced to evolve accordingly to the imposed constraint of the general following form (for $T > 0$):

$$T\dot{\psi} + \psi = 0 \tag{11}$$

The convergence degree to the manifold which is expressed by (10) is regulated by the parameter $T$. The following mathematic development gives us the suitable control law:

$$\frac{d\psi}{dt} = \frac{d\psi}{dx}\frac{dx}{dt} \tag{12}$$

The following equation is calculated, by substituting (8), (9) in (11):

$$T\frac{d\psi}{dx} f(x, U, t) + \psi(x, t) = 0 \tag{13}$$

From (13), we have the following control law:

$$U = g[x, \psi(x, t), T, t] \tag{14}$$

From the last equation, we could see that the control law is a function of $T$, the macro-variables, and the system's variables, which means it permits to choose the controller structure based on the whole nonlinear system. The system can reach the predefined manifold and remains on it, by a best selection of the macro-variables and the good choice of the manifold itself [7]. Then the global stability of the closed loop is guaranteed even the presence of parameter variation and bounded perturbations.

## 4 Fast Terminal Synergetic Control of Underactuated System

Fast terminal synergetic control requires to define nonlinear macro variables. For a class of second order underactuated system, we choose the macro variables as integrated into each other in three levels, as follows and as presented in Fig. 1:

$$\psi_1 = e_2 + k_1 e_1 + \beta_1 |e_1|^{\frac{q_1}{p_1}} \operatorname{sign} e_1 \tag{15}$$

$$\psi_2 = \psi_1 + k_2 e_3 + \beta_2 |e_3|^{\frac{q_2}{p_2}} \operatorname{sign} e_3 \tag{16}$$

$$\psi_3 = \psi_2 + k_3 e_4 \tag{17}$$

where:

$$e_1 = x_1 - x_{1d}$$
$$e_2 = \dot{e}_1$$
$$e_3 = x_3 - x_{3d}$$
$$e_4 = \dot{e}_2$$

and $x_{1d}$ and $x_{3d}$ are the desired state variables. $k_1$, $k_2$, $k_3$, $\beta_1$ and $\beta_2$ are positive constants, $p_1$, $q_1$, $p_2$ and $q_2$ are odd positive constants satisfying the following conditions:

$$p_1 > q_1 \, , \quad p_2 > q_2$$

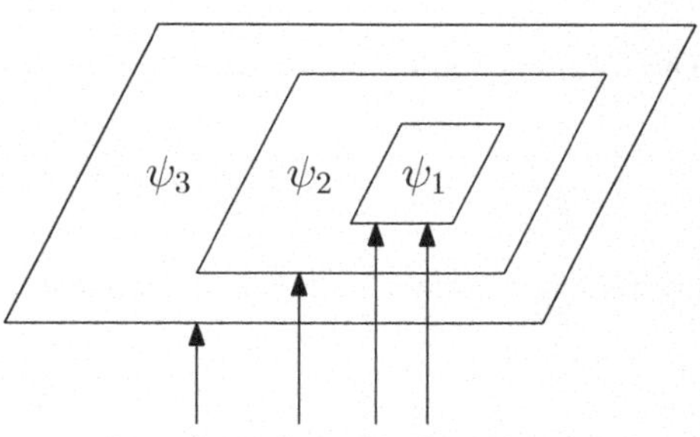


**Fig. 1** Macro-variables structure

From (11) we get:

$$T\left[\dot{e}_2+k_1e_2+\beta_1\frac{q_1}{p_1}\dot{e}_1|e_1|^{\frac{q_1}{p_1}-1}\operatorname{sign}e_1+k_2e_4+\beta_2\frac{q_2}{p_2}\dot{e}_3|e_3|^{\frac{q_2}{p_2}-1}\operatorname{sign}e_3+k_3\dot{e}_3\right]$$
$$+\psi_3=0 \tag{18}$$

Substituting the derivatives from (7) and solving the Eq. (18), we can get the following control law:

$$U=-\frac{1}{g_1+k_3g_2}\left(f_1-\ddot{x}_{1d}+k_2e_2+\beta_1\frac{q_1}{p_1}e_2|e_1|^{\frac{q_1}{p_1}-1}\mathrm{sign}e_1+k_3(f_2-\ddot{x}_{3d})\right.$$
$$\left.+k_2e_4+\beta_2\frac{q_2}{p_2}e_4|e_3|^{\frac{q_2}{p_2}-1}\mathrm{sign}e_3+\frac{1}{T}\psi_3\right) \tag{19}$$

From Eqs. (10) and (15), we have:

$$\dot{e}_1=-k_1e_1-\beta_1|e_1|^{\frac{q_1}{p_1}}\operatorname{sign}e_1$$

with an appropriate choice of $q_1$ and $p_1$, and for an initial condition $e_1(0)$, the dynamics will reach $e_1=0$ in a finite time, which means that when $e_1$ is far from zero, the dynamics became:

$$\dot{e}_1=-k_1e_1$$

with a good choice of $k_1$ we have a fast convergence, when the error is close to zero the dynamics will be approximated to:

$$\dot{e}_1=-\beta_1|e_1|^{\frac{q_1}{p_1}}\operatorname{sign}e_1$$

which is an attractor.

So the convergence time $t_{s1}$ along the manifold $\psi_1$ can be obtained as follows:

$$t_{s1}=\frac{p_1}{k_1(p_1-q_1)}\frac{\ln\left(k_1e_1(0)^{\frac{p_1-q_1}{p_1}}+\beta_1\right)}{\beta_1} \tag{20}$$

Which means that, when $e_1$ reaches $e_1=0$, the manifold $\psi_3$ will be:

$$\psi_3=k_3e_4+k_2e_3+\beta_2|e_3|^{\frac{q_2}{p_2}}\operatorname{sign}e_3$$

So for $\psi_3=0$, we have:

$$\dot{e}_3=-\frac{k_2}{k_3}e_3-\frac{\beta_2}{k_3}|e_3|^{\frac{q_2}{p_2}}\operatorname{sign}e_3$$

The convergence time $t_{s2}$ for the manifold $\psi_3$ can be given as follows:

$$t_{s2} = t_{s1} + \frac{p_2 k_3}{k_2(p_2 - q_2)} \frac{\ln\left(k_3 e_3(0)^{\frac{p_2-q_2}{p_2}} + \beta_2\right) - \ln k_2}{\beta_2} > t_{s1} \tag{21}$$

The stability is studied using the Lyapunov function expressed as follows:

$$V = \frac{1}{2}\psi_3^2 \tag{22}$$

which gives:

$$\dot{V} = \psi_3 \dot{\psi}_3 \tag{23}$$

From (11), we get:

$$\dot{V} = -\frac{1}{T}\psi_3^2 \tag{24}$$

From this equation, we see that:

$$\dot{V} = -\frac{2}{T}V < 0$$

So the asymptotic exponential stability of the system can be achieved.

## 5 Simulation Results

The overhead crane is a second order underactuated system; its structure is given in Fig. 2, it consists of a trolley of mass $M$ and a load mass $m$ suspended to the trolley by a rope of length $L$. The swing angle of the load with respect to the vertical line

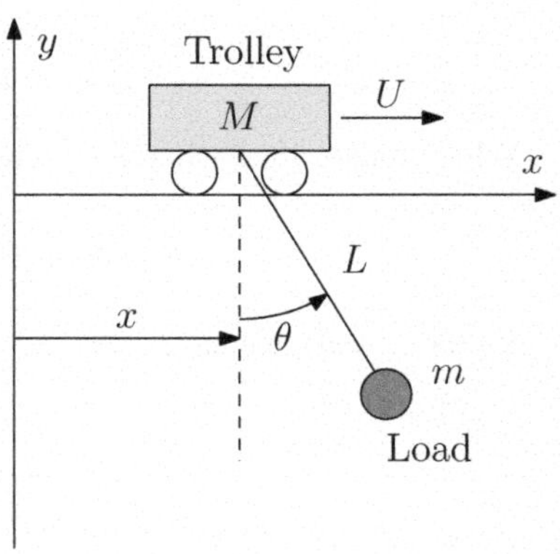


**Fig. 2** The Crane system

is $q$, the trolley position with respect to the origin is $x$, and the force applied to the trolley is $U$. The dynamics of the crane system are given by the following equations:

$$\begin{cases} (M+m)\ddot{x} + mL(\ddot{\theta}\cos\theta - \dot{\theta}^2\sin\theta) = U \\ \ddot{x}\cos\theta + L\ddot{\theta} + g\sin\theta = 0 \end{cases} \tag{25}$$

where: $m$, $M$ are respectively the masses of the load and the trolly, and $L$ is the length of the load. $U$ is the control input. The state variables are chosen as: $x_1 = x$ and $x_3 = \theta$. The constants are given by: $m = 0.8\,\text{kg}$, $M = 1\,\text{kg}$, $L = 0.3\,\text{m}$,

## 5.1 Simulation Using Synergetic Control

Our objective is to stabilize the crane system at the position $x_1 = x_{1d} = 1.5$, $x_3 = x_{3d} = 0$. This means that the trolley moves along the rail to the desired position, while maintaining the load in the upright position. The initial condition vector is $x(0) = [0, 0, -\frac{\pi}{12}, 0]^T$, Parameters of the controller are: $T = 0.1$, $k_1 = 0.25$, $k_2 = 8$ and $k_3 = 2$.

We can see from Figs. 3 and 4 that all signals track their references, and from Fig. 5 that all macro-variables converge to their manifolds. Figure 6 shows that the control signal is smooth.

To improve these results, we will use the terminal synergetic control in the second section.

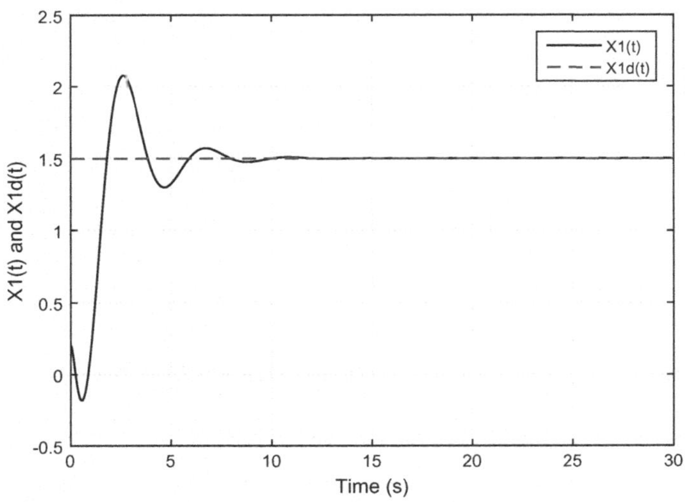


**Fig. 3** The trolley position and the desired position

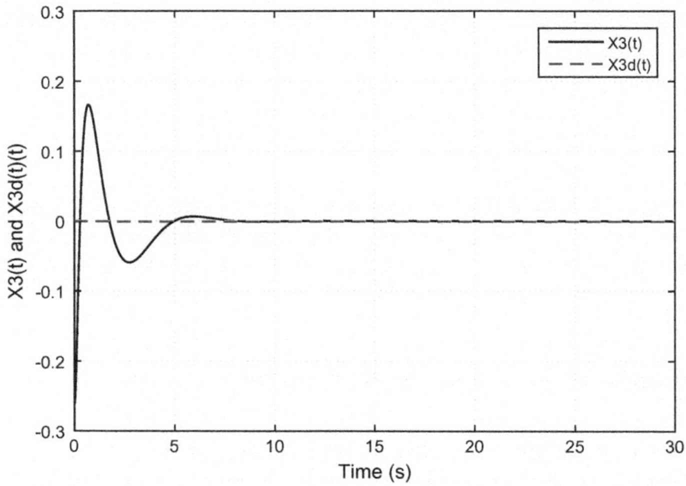


**Fig. 4** The angle of the load and the desired angle

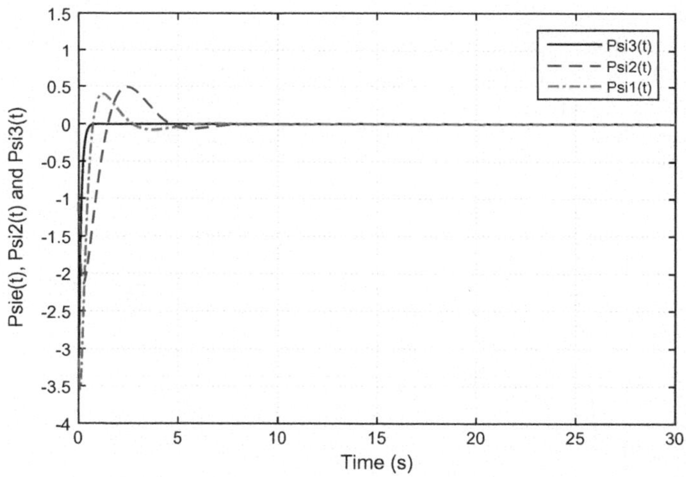


**Fig. 5** Macro variables

## 5.2 *Simulation Using Fast Terminal Synergetic Control*

Using the same initial conditions and the following parameters:

$$k_1 = 0.25\ ,\quad k_2 = 8\ ,\quad k_3 = 2\ ,\quad \beta_1 = 0.15\ ,\quad \beta_2 = 0.15$$
$$q_1 = 5\ ,\quad p_1 = 7\ ,\quad q_2 = 9\ ,\quad p_2 = 11$$

We get the following results presented in Fig. 7 and in Fig. 8.

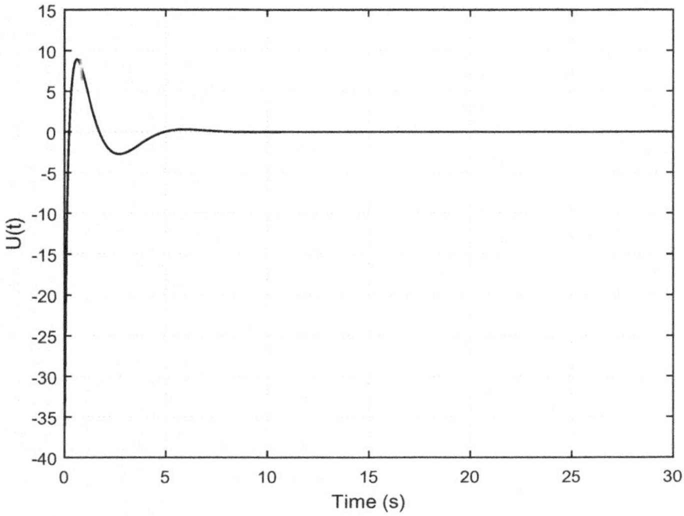


**Fig. 6** Control signal

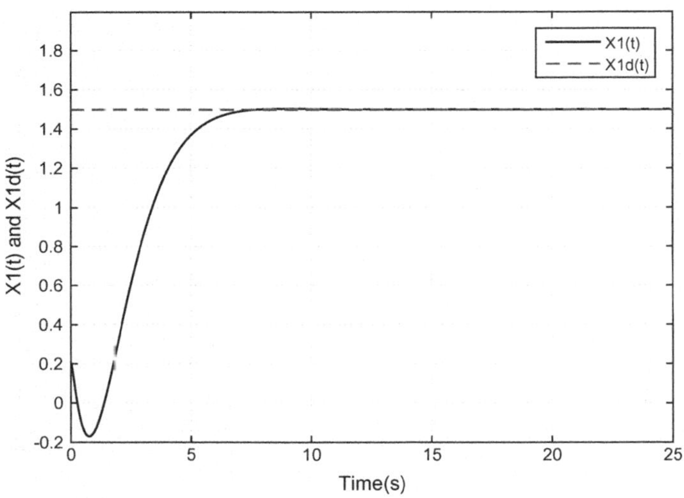


**Fig. 7** The trolley position and the desired position

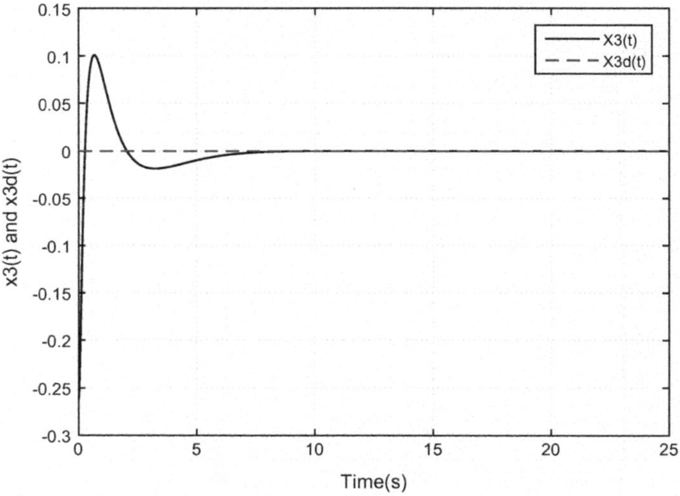


**Fig. 8** The angle of the load and the desired angle

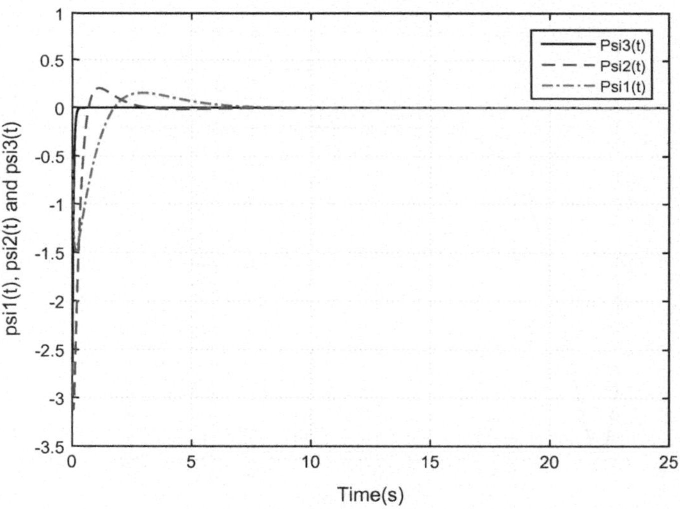


**Fig. 9** Macro variables

We can see from Figs. 7 and 8 that the position of the trolley and the angle of the load can reach their equilibrium in finite time and rapidly. The good precision and the stability performance are achieved.

Figure 9 and in Fig. 10 show respectively the macro variables and the control input, we can see that the macro variables can reach their manifold and the signal of control is very smooth.

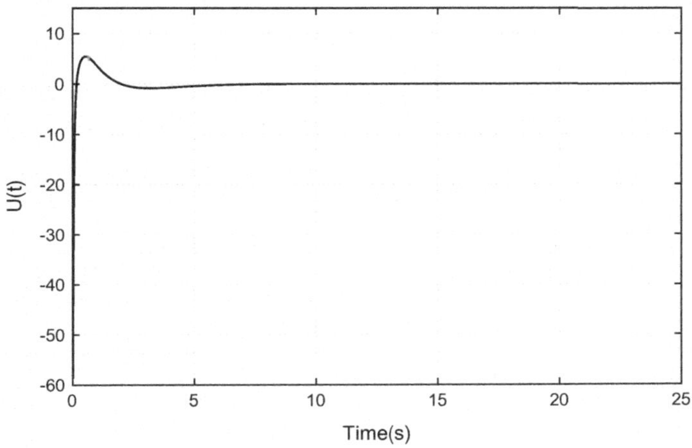


**Fig. 10** Control signal

Comparing to results in the first part using synergetic control, we see that the response of the system is faster and in finite time. Therefore the fast terminal synergetic controller is able to stabilize states with good precision, faster and in finite time and without chattering.

## 6 Conclusions

A fast terminal synergetic control for a category of second order underactuated systems has been presented in this paper. The proposed control law based on the synergetic theory is shown to be able to achieve the objective and maintain stabilization performance, this theory achieves identical performance as sliding mode control, without carrying the deficiency of chattering. Our results using synergetic control indicate that this one is effective, and the system can attain references while maintaining its stability in finite time, however these results are less significant comparing with the ones obtained using the fast terminal synergetic controller, because this later helped to improve the performances of the system, including the precision, the stability, the fast response, the convergence in finite time, the robustness and the easy implementation. Nevertheless, this strategy of control, is guaranteed in presence of bounded disturbances and parametric variations, this later will be more delicate when we have nonparametric uncertainties of the system model, so we can mention this problem to be solved as future works, and take into account all type of matched and mismatched uncertainties.

## References

1. Adhikary, N., Mahanta, C.: Integral backstepping sliding mode control for underactuated systems: swing-up and stabilisation of the cart-pendulum system. ISA Trans. **52**(6), 870–880 (2013)
2. Alaoui, S., Pages, O., Elhajjaji, A., Chaari, A., Koubaa, Y.: Robust adaptive fuzzy sliding mode control design for a class of MIMO underactuated system. IFAC Proc. **44**(1), 11127–11132 (2011). Elsevier
3. Bhave, M., Janardhanan, S., Dewan, L.: A finite time convergent sliding mode control for rigid underactuated robotic manipulator. Syst. Sci. Control Eng. **2**(1), 493–499 (2014)
4. Choukchou Braham, A.: Contribution à la stabilisation des systèmes mécaniques sous actionnés. Ph.D. thesis, Université Aboubekr Belkaid Tlemcen (2011)
5. Fantoni, I., Lozano, R., Spong, M.W.: Energy control of the pendubot. IEEE Trans. Autom. Control **45**(4), 656–661 (2000)
6. Harmas, M.N.: Synergetic control and synchronisation of chaotic systems. Int. J. Control, Autom. Commun. Syst. (IJCACS) **1**(2), 31–39 (2016)
7. Liu, C.H., Hsiao, M.Y.: Synchronization on unified chaotic systems via PI-type terminal synergetic control. In: CACS'2013, Sun Moon Lake, Taiwan. https://doi.org/10.1109/CACS.2013.6734101 (2013)
8. Liu, J., Wang, X.: Advanced Sliding Mode Control for Mechanical Systems, Springer, Berlin (2011)
9. Lozano, R., Fantoni, I.: Perspectives in Control. Passivity based control of the inverted pendulum, pp. 83–95. Springer, London (1998)
10. Mahjoub, S., Mnif, F., Derbel, N., Hamerlain, M.: Radial-basis functions neural network sliding mode control for underactuated mechanical systems. Int. J. Dyn. Control **2**, 533–541 (2014)
11. Olfati Saber, R.: Nonlinear control of underactuated mechanical systems with application to robotics and aerospace vehicles. Ph.D. thesis, Massachusettes Institute of Technology, Department Electrical Engineering and Computer Science (2001)
12. Seto, D., Baillieul, J.: Control problems in super articulated mechanical systems with application to robotics and aerospace vehicles. IEEE Trans. Autom. Control **39**(12), 2442–2453 (1994)
13. Shyu, K.K., Jen, C.L., Shang, L.J.: Design of sliding-mode controller for anti-swing control of overhead cranes. IEEE Proc. ACIIES'31 **1**, 147–152 (2005)
14. Singh, A.M., Hoang, V.T., Ha, Q.P.: Fast terminal sliding mode control for gantry cranes. In: ISARC'33 (2016)
15. Wang, W., Liu, X., Yi, J.: Structure design of two types of sliding-mode controllers for a class of under-actuated mechanical systems. IET Control Theory Appl. **1**, 163–172 (2007)
16. Yu, X., Zhihong, M.: Multi-input uncertain linear systems with terminal sliding-mode control. Automatica **34**, 389–392 (1998)
17. Zhenhua, J.: Design of power system stabilizers using synergetic control theory. IEEE PESGM (2007). https://doi.org/10.1109/PES.2007.385651

# Sliding Mode Fault Diagnosis with Vision in the Loop for Robot Manipulators

**Antonella Ferrara, Gian Paolo Incremona and Bianca Sangiovanni**

**Abstract** This chapter is devoted to the problem of Fault Diagnosis (FD) for industrial robotic manipulators within the framework of sliding mode control theory. According to this control concept, a set of unknown input higher order sliding mode observers are designed to detect, isolate and identify multiple actuators faults and corruptions. More specifically, the whole FD architecture is based on the inverse dynamics-based feedback linearized robotic MIMO system, which is equivalent to a set of linearized decoupled SISO systems, affected by uncertain terms. The FD process includes a residual generation, followed by a decision making through the evaluation of the achieved residuals. The advantages of the sliding mode approach are the good performance in terms of stability and robustness, as well as satisfactory estimate of the occurring faults. Furthermore, in order to extend the FD strategy to multiple sensor and actuator faults, a low cost vision servoing architecture is used in the scheme, allowing one to design a fault tolerant control strategy in case of sensor faults. The effectiveness of the proposed FD architecture has been carried out in simulation on a realistic simulator as well as experimentally on a COMAU SMART3-S2 anthropomorphic robot manipulator.

**Keywords** Fault diagnosis · Robot manipulators · Sliding mode observers · Uncertain systems

---

A. Ferrara · B. Sangiovanni
Dipartimento di Ingegneria Industriale e dell'Informazione, University of Pavia, Pavia, Italy
e-mail: antonella.ferrara@unipv.it

B. Sangiovanni
e-mail: bianca.sangiovanni01@universitadipavia.it

G. P. Incremona (✉)
Dipartimento di Elettronica, Informazione e Bioingegneria, Politecnico di Milano, Milan, Italy
e-mail: gianpaolo.incremona@polimi.it

 
J. Ghommam et al. (eds.), *New Trends in Robot Control*, Studies in Systems, Decision and Control 270, https://doi.org/10.1007/978-981-15-1819-5_5

# 1 Introduction

The FD problem was extensively studied for the past three decades and it plays a crucial role in case of faults/corruptions of mechanical and electromechanical systems [19, 20]. FD typically consists of three phases: (i) "detection" is the capability to reveal the presence of a fault in the system, (ii) "isolation" is the possibility to isolate the faulty component, (iii) "identification" is the reconstruction of the time evolution of the detected fault. The efficiency of FD systems is significantly important to avoid risks, reduce maintenance costs, while consequently improving industrial production and safety [10].

In the literature, the FD problem has been widely investigated also from theoretical viewpoint in case of both linear and nonlinear coupled systems. These approaches are in general distinguished between "active" or "passive". As for the active approach, input signals are injected into the system to enable the detectability of the faults [14, 27, 30], that is to give rise to abnormal behaviors, which otherwise would be not evident. On the other hand, in the passive approach a comparison is performed between the actual process input-output measurements and the corresponding signals produced, assuming fault absence, by a nominal model of the plant, or under the form of data set [10, 33].

Having in mind the robot application, in the corresponding FD literature, coupling effects are always considered negligible, which is practically false, thus making the FD problem particularly difficult to solve [16, 22]. For robotic systems, faults typically affect actuators and sensors and they can occur simultaneously on both types of these devices. Furthermore, more than one fault at a time can affect the operations. However, the majority of the contributions in robotics make the assumption that faults are single corruptions only on the actuators or only on the sensors [7, 12, 13, 21]. Indeed, it is difficult to isolate faults when hard coupling effects among the robot joints are present, making indistinguishable actuator from sensor faults and vice versa.

A preliminary contribution where the FD system for robot manipulators is extended to more realistic fault scenarios is reported in [9]. However, the FD method in [9] was still complicated by the nonlinear couplings among the different joints of the robot arm. In the following an improvement in this sense is made by first solving the problem of the rejection of the coupling effects, through the so-called inverse dynamics control approach [29]. Specifically, the internal control loop transforms the MIMO nonlinear and coupled robotic system into a set of $n$ linear decoupled SISO systems (one for each joint) characterized by bounded uncertainty terms to account for the residuals of the feedback linearization. The use of a low cost vision sensor (an IP camera) allows one to detect possible sensor faults. The proposed logic is that when a sensor fault is revealed, then the joint variables measurements, determined via image processing and the application of the inverse kinematics, are used to close the loop, thus giving rise to a sensor fault tolerant control scheme. As for the actuator faults, a set of $n$ unknown input observers are used to perform the complete fault diagnosis. In this case, in order to design robust observers able to guarantee finite time convergence, a sliding mode methodology has been adopted [3]. More specifically, observer input laws of Suboptimal Second Order Sliding Mode (SSOSM) type [1, 5–7, 11, 31] are introduced and theoretically analyzed. Note that sliding mode and SSOSM approaches have already been efficiently applied to solve motion and force control problem in case of industrial robots [2, 4, 8, 15, 17, 18], also in a switching version or combined with deep reinforcement learning algorithms [24–26].

The main benefits of the proposal are related to the limited cost of the hardware involved in the scheme, the low computational load and to the capability of the battery of SSOSM unknown input observers to solve the fault diagnosis problem even in case of multiple faults occurring, at the same time, on both actuators and sensors. The proposed FD scheme is assessed in simulation by using Matlab Simulink, a virtual simulator realized in V-REP and experimentally on a COMAU SMART3- S2 anthropomorphic industrial robot manipulator.

The chapter is organized as follows. In Sect. 2 the model of the considered robot system is described. In Sect. 3 some preliminaries on the vision servoing system are introduced, while in Sect. 4 the considered FD problem is formulated. The proposed FD scheme is instead illustrated in Sect. 5 and analyzed in Sect. 6. A case study with simulation and experimental results is reported in Sect. 7. Some conclusions are gathered in Sect. 8.

## 2 The Robot Model

Consider the robotic system with $n$ joints illustrated in Fig. 1.

Let $\ell_i,\ i = 1, 2, \dots n$ denote the length of the $i$th link, let $q_1$ denote the orientation of the first link with respect to $y$-axis clockwise positive, and let $q_j,\ j = 2, 3, \dots n$, denote the displacement of the $j$th link with respect to the $(j-1)$th one clockwise positive. $O - \{x, y, z\}$ is the base-frame of the robotic manipulator, while $O_\mathrm{e} - \{n, s, a\}$ is the end-effector frame [28, 29].

The direct kinematics of a $n$-joints manipulator describes the relationship between the joint variables $q \in \mathbb{R}^n$ and the end-effector position and orientation in the workspace, according to the following expression

$$\begin{cases} p = p(q) \\ R_\mathrm{e}^\mathrm{b} = R_\mathrm{e}^\mathrm{b}(q) \end{cases} \tag{1}$$

where $p = [p_x,\ p_y,\ p_z]^\top$ is the position of the end-effector with respect to the base frame, and $R_\mathrm{e}^\mathrm{b}$ is the rotation matrix between the reference frame attached to the end-effector (as in Fig. 1) and the base frame. The direct kinematics can be also expressed in terms of the following homogeneous transformation matrix

$$T_\mathrm{e}^\mathrm{b} = \begin{bmatrix} R_\mathrm{e}^\mathrm{b} & p \\ 0_{3\times 1}^\top & 1 \end{bmatrix} \tag{2}$$

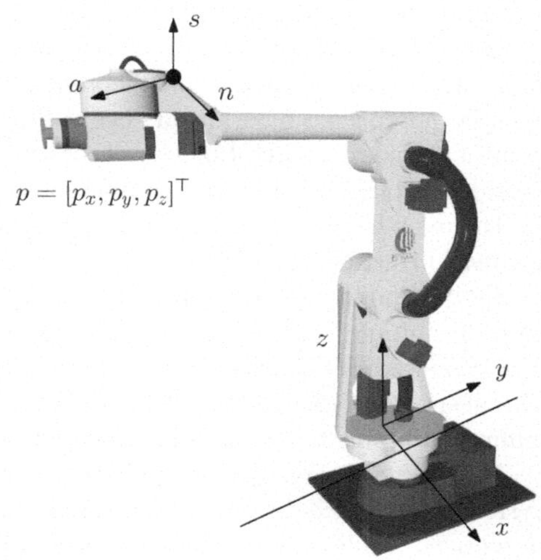


**Fig. 1** A schematic view of a spatial robot manipulator

where $0_{3\times1}$ is a column vector having null entries.

The dynamics of the robot can be described in the joint space, by using the Lagrangian approach, as

$$B(q)\ddot{q} + n(q, \dot{q}) = \tau \tag{3}$$

$$n(q, \dot{q}) = C(q, \dot{q})\dot{q} + F_{\mathrm{v}}\dot{q} + F_{\mathrm{s}}\mathrm{sign}(\dot{q}) + g(q) \tag{4}$$

where $B(q) \in \mathbb{R}^{n\times n}$ is the inertia matrix, $C(q, \dot{q}) \in \mathbb{R}^{n\times n}$ represents centripetal and Coriolis torques, $F_{\mathrm{v}} \in \mathbb{R}^{n\times n}$ is the viscous friction matrix, $F_{\mathrm{s}} \in \mathbb{R}^{n\times n}$ is the static friction matrix, $g(q) \in \mathbb{R}^n$ is the vector of gravitational torques and $\tau \in \mathbb{R}^n$ represents the motors torques.

For the sake of simplicity, the dependence of all the variables on joint variables $q$, and on time $t$ will be omitted, when it is obvious.

## 3 Some Preliminary Issues About Vision

In this section some preliminary issues about vision based control (or vision servoing loop) are introduced [28, 29].

First of all it is important to note that in the robotics framework, the use of a camera in the control scheme is an inexpensive solution to acquire data from the plant. Indeed, the camera can be located in a safer environment with respect to the robot, protected from dust and emissions, thus reducing the probability that also the vision-based measurements are corrupted by faults. However, on the other hand a vision servoing loop requires a suitable on-line data processing in order to achieve the measurements for the feedback.

In the following, for the proposal, a multi-camera system in the so-called eye-to-hand configuration (that is each camera is positioned in a fixed location) is considered. The camera field of view does not change during the task, that implies that the accuracy of the acquired measurements is kept, at least in principle, constant. Note that, the spatial vision can be achieved even by using one camera, provided that two images of the same object, taken from two different poses, are available. Moreover, in order to determine the position of the points on the end-effector of the robot manipulator with respect to a base frame, a vision control in the operative space is applied, and the transformation matrix is then computed to get position and orientation.

There exist several techniques in the framework of stereo vision, such as epipolar geometry, triangulation, absolute orientation or reconstruction from planar homography. In the following, the triangulation method combined with the inverse kinematics of the robot is analyzed (see Fig. 2). Let the planes, on which the light reflected by the object is focused and where the photosensitive sensor lies, be defined as image planes. The relationship between the coordinates of the homogeneous form of the position $\widetilde{p}$ and the corresponding coordinates in the image plane $(X_\mathrm{I},\ Y_\mathrm{I},\ 1)$ can be written as

$$\lambda[X_\mathrm{I},\ Y_\mathrm{I},\ 1]^\top = \Omega \Pi \widetilde{p} \tag{5}$$

where $\lambda > 0$, $\Omega \in \mathbb{R}^{3\times 3}$ is a matrix containing the intrinsic parameters depending on the sensor and lens characteristics, while $\Pi = [I_{3\times 3},\ 0_{3\times 1}] \in \mathbb{R}^{3\times 4}$, with $I_{3\times 3}$ being the identity matrix and $0_{3\times 1}$ being the null vector. Consider now two cameras with known intrinsic parameters, the triangulation consists of computing $p$ with respect to the base frame, starting from normalized coordinates in the image planes $s_1 = [X_1,\ Y_1]^\top$ and $s_2 = [X_2,\ Y_2]^\top$ of the projections of $p$, such that

$$\lambda_i \widetilde{s}_i = \Pi \widetilde{p}^i, \quad i = 1, 2\ . \tag{6}$$

Assuming to initially have the base frame on camera 1, the following system can be written as

$$\begin{cases} \widetilde{p}^1 = \lambda_1 \widetilde{s}_1 \\ \widetilde{p}^1 = o_2^1 + \lambda_2 R_2^1 \widetilde{s}_2 \end{cases} \tag{7}$$

where $o_2^1$ is the position vector of the frame of camera 2 with respect to the frame of camera 1, and $R_2^1$ is the corresponding rotation matrix. By solving system (7) with respect to $\widetilde{p}^1$, and considering the homogenous transformation matrix $T_1^\mathrm{b}$, finally, one has the position of the end-effector with respect to the base frame of the overall system

$$\widetilde{p} = T_1^\mathrm{b} \widetilde{p}^1\ . \tag{8}$$

What we need to show is how to find in real-time the orientation of the end-effector with respect to the base frame. This can be done by determining the homogenous transformation matrix $T_\mathrm{e}^\mathrm{b}$. Assume to acquire the vector of the projection of $N$ points

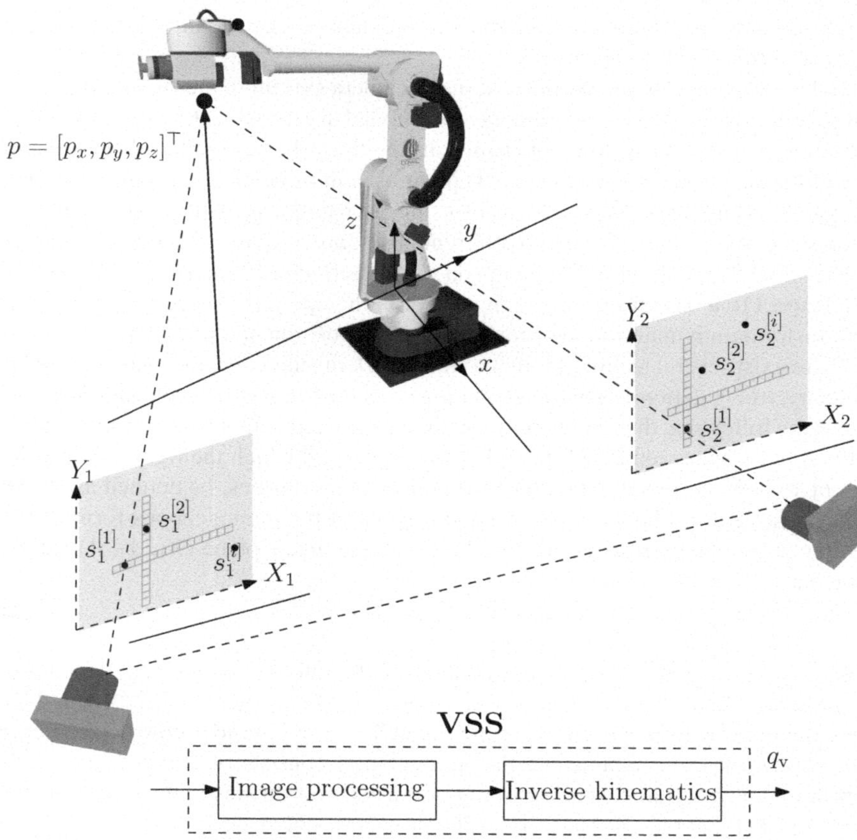


**Fig. 2** A schematic view of the considered vision servoing scheme with the inverse-kinematics, and a detail of the image processing

of the end-effector on the image plane of camera 1, the $i$th component of which is $s_1^i = [X_1^i, Y_1^i]^\top$, then, by changing the reference frame, the coordinates of these points can be expressed with respect to the reference frame of the camera itself as follows

$$\lambda_i \widetilde{s}_i^1 = \Pi T_e^1 \widetilde{p}_i^e, \quad i = 1, \ldots, N. \tag{9}$$

This is a set of $N$ equations from which it is possible to compute $T_e^1$. Then, solving system (9), in case of eye-to-hand cameras, the end-effector position and orientation with respect to the base frame results in being

$$T_e^b = T_1^b T_e^1. \tag{10}$$

**Table 1** Possible scenarios in case of fault events on a robot manipulator

| | |
|---|---|
| FE1 | Single fault on an actuator |
| FE2 | Single fault on a sensor |
| FE3 | Multiple faults on actuators |
| FE4 | Multiple faults on sensors |
| FE5 | Multiple faults on actuators and/or on sensors |

At this point, it is possible to obtain the corresponding value of the robot joint variables $q$, by inverting the kinematics of the robot manipulator [28, 29], as depicted in Fig. 2.

## 4 The Faults Model

The fault events (FEs) which are taken into account are those labeled and described in Table 1. For the sake of simplicity, it is assumed that both sensor and actuator faults can be regarded as additive terms affecting the joint variable $q$ and the torque vector $\tau$, respectively. Thus, let $\Delta\tau \in \mathbb{R}^n$ and $\Delta q \in \mathbb{R}^n$ denote the generic actuator fault and sensor fault. In case of fault events on both these components of the plant, the control system receives

$$\bar{q} = q + \Delta q \tag{11}$$

as angular displacements of the joints, while the actual torque applied to the robot is

$$\bar{\tau} = \tau + \Delta\tau \, . \tag{12}$$

Hence, the dynamical model of the robot results in being

$$B(\bar{q})\ddot{\bar{q}} + n(\bar{q}, \dot{\bar{q}}) = \bar{\tau} \, . \tag{13}$$

Note that, the considered problem is complex since, in closed loop, if multiple faults both on actuators and on sensors contemporary occur, in principle, it is not possible to discern the origin of the malfunction due to the fact that an unexpected variation of the torque causes a corresponding variation of the joint position which could be interpreted as measurement corruption, and vice versa.

### 4.1 The Multiple Faults Case: Potentiality of Vision

FD in robotics is a topic widely investigated in past years [7, 12, 13, 21], and in most cases, several assumptions have been introduced in order to solve the problem. As previously discussed, it is often assumed that fault affects a single component of the robots or multiple faults occur on a single class of components, i.e., only on the actuators or only on the sensors. The general multiple faults case is indeed not trivial to solve. As detailed in Sect. 2, the robot model is a MIMO nonlinear coupled system that implies a coupling of fault effects on the actuators and on the sensors, in such a way they become indistinguishable. So, conventional genuine observer-based fault diagnosis approaches cannot practically be used in this complex case. This is the reason why the aforementioned assumptions have to necessarily made.

A possible way to solve this problem is to design a FD system with vision in the loop. The possibility of reconstructing the joint variables by simply processing the images in which the position and orientation of the robot end-effector can be achieved represents an important ingredient also for FD. Due to the technological developments and cost reduction of the calculators, it is reasonable to devote the attention to a vision hardware, which is low-cost and easily deployable, instead of insisting on the algorithmic tools. This consideration is the motivation for the vision-based FD proposal we present in this chapter. More specifically, if a fault occurs on a sensor, the vision system allows one to detect it independently of possible actuator faults. When a sensor fault has been detected, the control loop can be closed by using the joint variables estimated by processing the images provided by cameras, which are assumed to be always well-calibrated and reliable. The actuator faults are instead detected and isolated using observers.

In the following sections the overall control strategy is presented in more details and the sliding mode based FD method is described.

## 5 The Fault Diagnosis Scheme

The proposed FD scheme for robot manipulators is illustrated in Fig. 3. The core of the scheme is the redundant sensor, that is a low cost IP camera.

Even if the use of this camera can appear trivial, it allows one to detect and isolate faults on the sensors and its measurements can be used, when a sensor faults occurs, to close the control loop, making the adopted control approach fault tolerant. For the sake of simplicity, consider the $i$th joint as

$$q_{\mathrm{c}_i}(t) = \begin{cases} q_{\mathrm{v}_i}(t) & \text{if a sensor fault occurs} \\ q_i(t) & \text{otherwise} \end{cases} \tag{14}$$

where $q_{\mathrm{c}_i}(t)$ is the $i$th component of vector $q_\mathrm{c}$, while $q_\mathrm{v}$ is the actual position of the robot provided by the redundant sensor. Since the feedback is closed using $q_\mathrm{c}$, one has

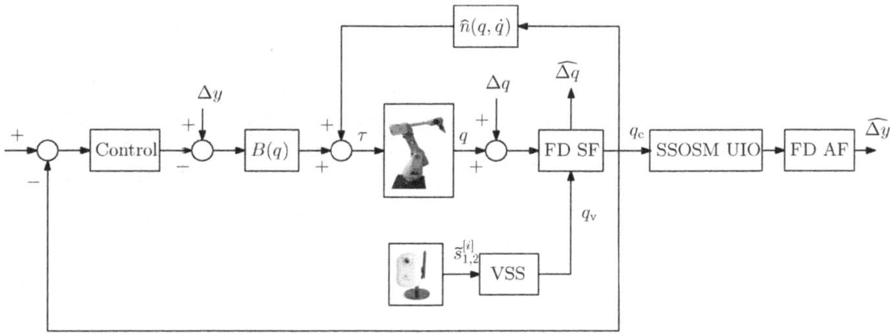


**Fig. 3** The proposed fault diagnosis scheme for the considered COMAU SMART3-S2 robot manipulator with camera Atlantis NetCamera 500

that, independently of the presence of sensor faults, the so-called "inverse dynamics" control can be adopted to transform the MIMO nonlinear robotic system into $n$ SISO perturbed double integrators. This procedure allows one to independently take into account each joint in order to perform the detection and isolation of the actuator faults. In the following, the blocks appearing in Fig. 3 are singularly described.

## *5.1 Vision Servoing and FD Logic (Sensors)*

The vision block is assumed to provide the correct position of the joints with a sampling time which can be longer than that of the measurements provided by onboard sensors. For the sake of generality, let $\tau_\text{v}$ be the vision-based data sampling time, and that $\tau_\text{s}$ be the sensor (i.e. robot) sampling time. Then, the following hold mechanism is applied

$$\begin{matrix} q_{\text{v}_i}(t) = q_{\text{v}_i}(k\tau_\text{v}),\ \forall\, k\tau_\text{v} \leq t \leq (k+1)\tau_\text{v} \\ q_{\text{s}_i}(t) = q_{\text{s}_i}(k\tau_\text{s}),\ \forall\, k\tau_\text{s} \leq t \leq (k+1)\tau_\text{s} \end{matrix},\quad \forall k \geq 0\,. \tag{15}$$

The position reconstructed from the acquired images is compared with the signal obtained from the sensors fastened on the actual robot. The resulting signals are exactly the estimate of the displacement between the sensors output and the real position of the robot, called $\widehat{\Delta q} = q_{\text{s}_i} - q_{\text{v}_i}$, with $q_{\text{s}_i}$ being the measurement of the position of the $i$th joint from the sensor and $q_{\text{v}_i}$ being the corresponding position achieved from the camera. Then, in order to detect the faults on the sensors, the following flag function is introduced:

$$R_{\text{s}_i} = \begin{cases} 0 & \text{if } \ -T_{\text{min}_{\text{s}_i}} < \widehat{\Delta q}_i(t) \leq 0 \ \text{ or } \ 0 < \widehat{\Delta q}_i(t) \leq T_{\text{max}_{\text{s}_i}} \\ 1 & \text{otherwise} \end{cases} \tag{16}$$

with $T_{\min_{s_i}}$ and $T_{\max_{s_i}}$ being the positive lower and the upper threshold values, respectively. In the following, they have been experimentally found for each joint. Even if they should be equal to zero, the noise both on the resolvers and in the vision system needs to be taken into account. If $R_{s_i}$ returns 1 on the $i$th component, then a fault is detected on the corresponding sensor. Apart from sensor fault detection and isolation, the identification of the detected fault on the $i$th sensor can be also obtained by considering the time evolution of the difference signal $\widehat{\Delta q}_i$.

## 5.2 Inverse Dynamics

The inverse dynamics of the robot manipulator [28, 29] can be written, in the joint space, relying on (3)–(4), as a nonlinear relationship between the plant inputs and the plant outputs (see Fig. 3). Introducing an auxiliary control variable $y$, one has

$$\tau = B(q_c)y + \hat{n}(q_c, \dot{q}_c) \tag{17}$$

where $q_c$ is chosen according to (14). In the following the identified matrix $B(\cdot)$ is assumed to be perfectly known, while $\hat{n}(\cdot, \cdot)$ is only an estimate of $n$. By applying the inverse dynamics control to system (3)–(4), one obtains

$$\begin{aligned} \ddot{q}_c &= y - \eta(q_c, \dot{q}_c) \\ \eta(q_c, \dot{q}_c) &= -B^{-1}(q_c)(\hat{n}(q_c, \dot{q}_c) - n(q_c, \dot{q}_c)) \end{aligned} \tag{18}$$

where $\eta(\cdot, \cdot)$ takes into account the modelling uncertainties and, possibly, external disturbances, such that $|\eta_i| < H_i$. Equation (18) represents a set of $n$ SISO decoupled systems, with state vector $x_i \in \mathbb{R}^2$, such that $x_{1_i} = q_{c_i}$ and $x_{2_i} = \dot{q}_{c_i}$, $i = 1, \ldots, n$, subject to matched uncertainty. More precisely, the dynamics of any single joint can be rewritten as

$$\begin{cases} \dot{x}_{1_i}(t) = x_{2_i}(t) \\ \dot{x}_{2_i}(t) = y_i(t) - \eta_i(t) \end{cases} . \tag{19}$$

In case of fault on the actuators, substituting (17) in (12)–(13), it is possible to describe how the fault is mirrored on the auxiliary control variable, i.e.,

$$\Delta y = B^{-1}(q_c)\Delta\tau \tag{20}$$

so that, the single joint dynamics, expressed by (19), can be rewritten as

$$\begin{cases} \dot{x}_{1_i}(t) = x_{2,i}(t) \\ \dot{x}_{2_i}(t) = y_i(t) + \Delta y_i(t) - \eta_i(t) \end{cases} . \tag{21}$$

### 5.3 Sliding Mode Observers and FD Logic (Actuators)

The FD approach for actuator faults proposed in this chapter relies on the use of a Suboptimal Second Order Sliding Mode based Unknown Input Observer (SSOSM-UIO). Assume that the inverse dynamics control, previously described, is applied so that it is now possible to design the observer on the basis of the model of the feedback linearized system. Hence, the proposed UIO is the following

$$\begin{cases} \dot{\hat{x}}_{1_i}(t) = \hat{x}_{2_i}(t) \\ \dot{\hat{x}}_{2_i}(t) = y_i(t) + u_i(t) \end{cases} \tag{22}$$

where $\hat{x}_{1_i}$ and $\hat{x}_{2_i}$ are the observer state variables, $y_i$ is the $i$th component of the auxiliary control variable, while $u_i$ is the input of the observer which relies on the SSOSM approach [5]. To this end, let $x_{1_i} = q_{c_i}$ denote the robot position acquired according to the mechanism in (14), and let $e_i \in \mathbb{R}^2$ be the estimation error, such that $e_{1_i} = x_{1_i} - \hat{x}_{1_i}$ and $e_2 = \dot{x}_{1_i} - \hat{x}_{2_i}$. Let the sliding variable be equal to a linear combination as

$$\sigma_i = \beta_i e_{1_i} + e_{2_i} \tag{23}$$

where $\beta_i$ is a positive constant. Assume $\xi_{1_i} = \sigma_i$ and $\xi_{2_i} = \dot{\sigma}_i$, then the so-called "auxiliary system" results

$$\begin{cases} \dot{\xi}_{1_i}(t) = \xi_{2_i}(t) \\ \dot{\xi}_{2_i}(t) = \beta_i \dot{e}_{2_i}(t) + \dfrac{d^{(3)} x_{1_i}(t)}{dt^3} - \dot{y}_i(t) - w_i(t) \\ w_i(t) = \dot{u}_i(t) \end{cases}, \tag{24}$$

where the drift term in the second equation of (24) is bounded, i.e., defining $F_i$ as the bound for joint $i$, one has

$$\left| \beta_i \dot{e}_{2_i}(t) + \frac{d^{(3)} x_{1_i}(t)}{dt^3} - \dot{y}_i(t) \right| < F_i\,. \tag{25}$$

Note that this is true also in practice due to the mechanical constrained nature of the robotic system.

According to the SSOSM control law in [3], the auxiliary control variable is chosen as

$$w_i(t) = \alpha_i W_{\max_i} \operatorname{sign}\left(\xi_{1_i}(t) - \frac{1}{2}\xi_{\max_i}(t)\right) \tag{26}$$

with $\alpha_i$ and $W_{\max_i}$ being positive parameters such that

$$\alpha_i \in (0,\ 1] \cap (0,\ 3) \tag{27}$$

$$W_{\max_i} > \max\left(\frac{F_i}{\alpha_i};\ \frac{4F_i}{3-\alpha_i}\right), \tag{28}$$

with $\xi_{\max_i}$ being the last extremal value of the sliding variable. The previous conditions are needed to enforce a second-order sliding mode on the sliding manifold $\sigma_i = 0$. Furthermore, the observer input is continuous, i.e.,

$$u_i(t) = \int_{t_0}^{t} w_i(\tau) d\tau \tag{29}$$

with beneficial effects in terms of chattering alleviation.

Now the concept of "equivalent control", introduced by [32], is exploited, that is the continuous time control which solves the equation $\dot{\sigma}_i = 0$ is taken into account. Let $u_{\mathrm{eq}_i}$ be the $i$th equivalent control component in our case. It can be used as a residual to perform the FD, according to the following rule

$$R_{\mathrm{a}_i} = \begin{cases} 0 & \text{if } -T_{\min_{\mathrm{a}_i}} < u_{\mathrm{eq}_i} \le 0 \quad \text{or} \quad 0 < u_{\mathrm{eq}_i} \le T_{\max_{\mathrm{a}_i}} \\ 1 & \text{otherwise} \end{cases} \tag{30}$$

with $T_{\min_{\mathrm{a}_i}}$ and $T_{\max_{\mathrm{a}_i}}$ being the activation thresholds, determined according to the known bounds of the uncertainty contributions. The accuracy of the proposed FD procedure obviously depends on the modelling uncertainty, since this latter is inevitably mixed with the faults, but this aspect is common to any FD strategy.

## 6 Analysis of the Fault Diagnosis Strategy

By applying the inverse dynamics control in (17), any single joint model is captured by a linear second order system completely decoupled from the rest of the robotic system. Then the FD rules (16) and (30) can be applied to each joint. Since by virtue of the camera sensor faults can be detected, isolated and identified, what is left to show is that the equivalent signals $u_{\mathrm{eq}_i}$, $i = 1, \ldots, n$, are actually estimates of the actuator faults. The following theorem can be proved.

**Theorem 1** *Given the nonlinear coupled robotic model* (3) *and* (4), *applying the inverse dynamics control in* (17), *and the fault detection and isolation strategy in* (16) *and* (30), *considering the SSOSM UIO in* (22)–(26), *then any component $i$ of the observer input law provides, after a transient time $t_{\mathrm{r}}$, an estimate $\widehat{\Delta y_i}$ of the $i$th fault, with an associated estimation error $\widetilde{\Delta y}_i = \Delta y_i - \widehat{\Delta y}_i$ such that $|\widetilde{\Delta y}_i| < H_i$.* □

***Proof*** In case of faults on the actuators, the system becomes that in (21). Choosing the sliding variables as in (23), a sliding mode is enforced in the finite time $t_{\mathrm{r}}$ by applying (26), according to the finite time convergence theorem proved in [5]. If $\sigma_i$ is steered to zero in $t_r$, and maintained equal to zero in the subsequent time instants, then the error $e_i$, $i = 1, 2$ turns out to be exponentially vanishing for $t \geq t_r$. When the $n$-order system, describing the joints dynamics after the application of the inverse dynamics control, exhibits a sliding mode, then the concept of equivalent control is well defined. To determine its component-wise expression one can compute the first-time derivative of the $i$th sliding variable, i.e.,

$$\dot{\sigma}_i(t) = \beta_i e_{2_i} + \ddot{q}_{c_i} - y_i - u_i \tag{31}$$

and solve the equation $\dot{\sigma}_i = 0$, that is relying on (22), one has

$$\beta_i e_{2_i} + y_i + \Delta y_i - \eta_i - y_i - u_{\text{eq}_i} = 0 \tag{32}$$

so that

$$u_{\text{eq}_i} = \beta_i e_{2_i} + \Delta y_i - \eta_i \ . \tag{33}$$

As previously observed, the second order sliding mode enforcement in the observation error state space implies that $e_{2_i}$ tends to zero. Then, it follows that

$$u^{\infty}_{\text{eq}_i}(t) = \lim_{t \to \infty} u_{\text{eq}_i}(t) = \Delta y_i(t) - \eta_i(t) \ . \tag{34}$$

Then, it is apparent that $u^{\infty}_{\text{eq}_i}$ is an estimate of the actuator fault $\Delta y_i$. As expected, it is indistinguishable from the uncertainty $\eta_i$. To conclude the proof, one can observe that

$$|\widetilde{\Delta y}_i| = |\Delta y_i - \widehat{\Delta y}_i| = |\Delta y_i - u^{\infty}_{\text{eq}_i}| = |\eta_i| < H_i \ . \tag{35}$$

Note that, according to [32], a practical way to retrieve the equivalent control signal is to filter the discontinuous control input via a low-pass filter. So, in the present case, $u_{\text{eq}_i}$ can be found at the output of the integrator indicated in (29). This means that, by virtue of the SSOSM UIO structure, no additional filters need to be included in the scheme, the residuals being directly generated.

## 7 The Case Study

In this section simulation and experimental results are reported to complement the theoretical discussion.

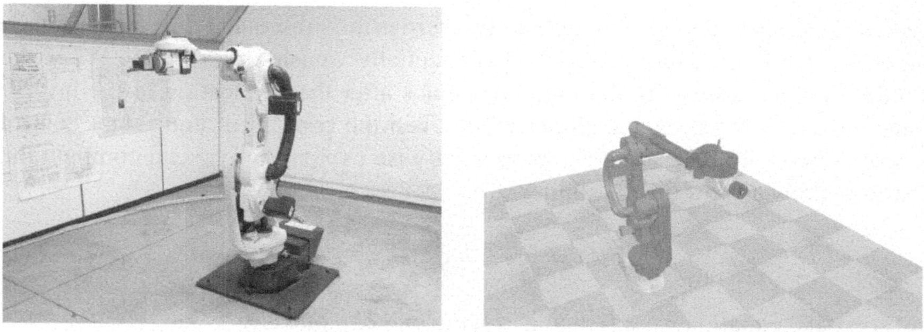

(a) COMAU SMART3-S2 (b) V-REP robot

**Fig. 4** The COMAU SMART3-S2 anthropomorphic industrial robotic manipulator. **a** Real setup. **b** Virtual robot in V-REP simulator

**Table 2** Scenarios

| (a) Scenario 1 | | | | (b) Scenario 2 | | | |
|---|---|---|---|---|---|---|---|
| $\#_{joint}$ | FE | $t_o$ [s] | $\Delta * (t)$ $(* \doteq q,\ y)$ | $\#_{joint}$ | FE | $t_o$ [s] | $\Delta * (t)$ $(* \doteq q,\ y)$ |
| 1 | AF | 11 | $10 \sin(2t)$ [rad/s$^2$] | 1 | AF | 9 | 30 [rad/s$^2$] |
| | SF | 8 | 0.2 [rad] | | SF | 8 | $0.3 \sin(2t)$ [rad] |
| 2 | AF | 7 | 30 [rad/s$^2$] | 2 | AF | 9 | $40 \sin(t)$ [rad/s$^2$] |
| | SF | 11 | $0.15 \sin(2t)$ [rad] | | SF | – | – |
| 3 | AF | – | – | 3 | AF | 11 | –60 [rad/s$^2$] |
| | SF | – | – | | SF | – | – |

## 7.1 *Experimental Setup*

The setup used for the verification and validation of the concept underlying the proposal is a COMAU SMART3-S2 industrial anthropomorphic robot manipulator, which consists of six links and six rotational joints driven by brushless electric motors. Six resolvers, fastened to the axes of the actuators, measure the joints positions, while the controller has a sampling time of 0.001 s. For the sake of simplicity, only three joints are used in the testing phase (see Fig. 4a). Since we consider a planar manipulator, the model of the robot identified in [8] can be adopted for simulation, and a single camera can be used as vision sensor. This latter is, in our lab setup, a low-cost IP camera of type Atlantis NetCamera 500, with Round Trip Time (RTT) approximately equal to 30 ms. This vision sensor is however reliable and able to guarantee good performance.

To assess the effectiveness of the proposed fault diagnosis strategy, two different scenarios have been considered. Table 2 reports the shape of the injected faults. They occur both on the actuators (AF) and on the sensors (SF), and their occurrence time is denoted with $t_o$.

Note that we do not consider faults on sensor 3, even if the proposed method can deal with faults occurring on any actuator and on any sensor, since the intrinsic noise of sensor 3 of the lab setup is so significant that the test results cannot be reliable.

## 7.2 *Matlab Simulation Results*

In this subsection the proposed FD scheme is assessed in a Matlab Simulink environment, relying on the model of the considered robot manipulator. Note that the model has been identified on the basis of real data through experimental tests, so that the simulation model is quite realistic. Furthermore, in order to recreate the real robot, the noise term $\eta = [\eta_1, \eta_2, \eta_3]^\top$, experimentally acquired, is injected to the accelerations of the joints. These matched uncertain terms (reported in Fig. 5) are bounded with bounds which can be determined by signal processing methods during the experimental tests. These are equal to $-2.3 \leq \eta_1 \leq 2.1$, $-10.3 \leq \eta_2 \leq 0.3$ and $-6.9 \leq \eta_3 \leq 59.2$.

Since the sliding mode based observers converges in a finite time $t_r$, that is after $t_r$ the observer transient phase has elapsed, we consider the time instant $t_{ob} \geq t_r$, from which the observed variables in (22) become usable. Moreover, the IP camera is simulated with a sampling time, as in the real case, of 30 ms.

The simulation tests are performed using the Euler solver and numerical integration step equal to 0.001 s. Table 3a reports the thresholds used by FD logic both for the actuators and for the sensors, while Table 3b contains the parameters of the SSOSM based observer used for each joint. A low-pass filter with a time constant of 0.1 s is finally applied to better identify the shape of the detected faults.

Figure 6 shows results obtained in case of Scenario 1. It is possible to see that all the faults have been detected, isolated and identified.

Note that the presence of both actuator and sensor faults on the same joint causes a mutual effect which could generate wrong detection. This is apparent, for instance, in the plot relevant to sensor 2 at the time instant 7 s, when a fault on actuator 2 occurs and this determines a peak also on the residuals of sensor 2. Moreover, the existence of the uncertainty terms, which are not negligible for joint 2, slightly deteriorates the identification of the fault.

In Fig. 7, it is shown the behavior of the proposed strategy in case of Scenario 2. Also in this case, all the faults have been correctly detected, isolated and identified.

Note that the matched uncertainties affecting joint 3 are particularly high, as previously mentioned, which implies a deterioration in the identification of the fault occurring on that joint.

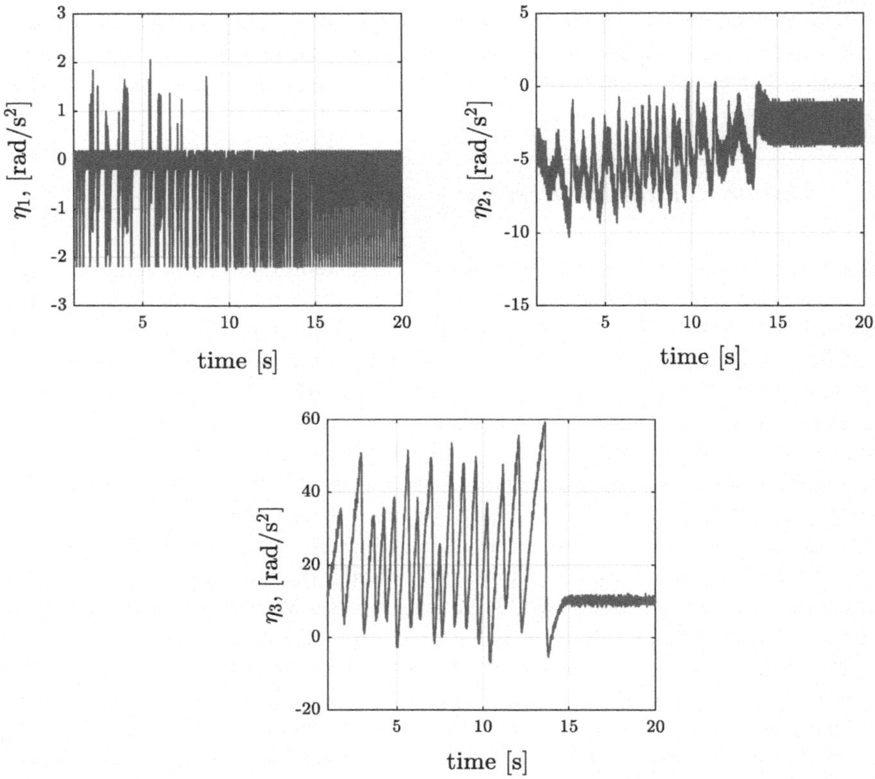


**Fig. 5** The matched uncertainties affecting the robotic system for each joint

**Table 3** FD thresholds and SSOSM based observer parameters (simulation)

| (a) Thresholds | | | | | (b) Parameters | | | |
|---|---|---|---|---|---|---|---|---|
| $\#_{\text{joint}}$ | $T_{\text{min}_{s_i}}$ | $T_{\text{max}_{s_i}}$ | $T_{\text{min}_{a_i}}$ | $T_{\text{max}_{a_i}}$ | $\#_{\text{joint}}$ | $\beta_i$ | $\alpha_i$ | $W_{\text{max}_i}$ |
| 1 | 0.02 | 0.02 | 3 | 3 | 1 | 10 | 0.9 | 30000 |
| 2 | 0.05 | 0.05 | 11 | 5 | 2 | 10 | 0.9 | 60000 |
| 3 | 0.04 | 0.02 | 8 | 65 | 3 | 10 | 0.9 | 100000 |

## *7.3 V-REP Simulation Results*

V-REP [23] is a general purpose robot simulator developed by Coppelia Robotics, that allows the design of 3D-rendered dynamic environments relying on physics engine such as the Open Dynamics Engine (ODE) and Bullet. The simulator offers a preset library of ready-to-use models of sensors, actuators and robots, and allows the user to import custom models. A virtual replica of the COMAU SMART3-S2

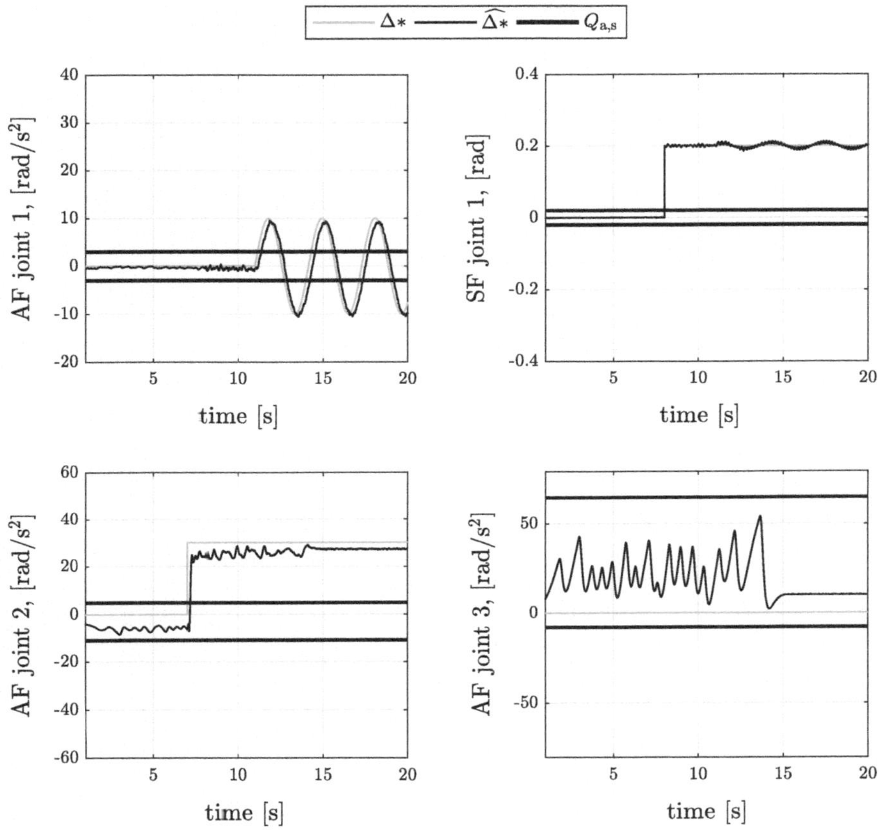


**Fig. 6** Scenario 1: multiple faults both on sensors 1 and 2 and on actuators 1 and 2 (simulation results) with $Q_{a,s}$ indicating the thresholds

anthropomorphic manipulator was built using 3D CAD softwares and assembled in V-REP for simulations. The availability of customizable vision sensors makes it possible to simulate realistic vision systems embedded on a real plant. Nevertheless, the simulator offers some basic image processing functions in order to have relevant information (i.e., feature extraction) available for computation. To this end, a camera sensor was placed in parallel to the robot and adjusted so that the entirety of the planar model would be captured in the picture frame. Moreover, filters were added to obtain three separate segmented images of the links attached to the joints taken into consideration. Thus, from the segmented image of the link it is possible to retrieve the angular position of the respective joint (Fig. 4b). The simulator features several functions for operating on robot joint objects, and is therefore possible to retrieve real-time data from the robot position encoders at each joint as well. With this information available, it is possible to reproduce the vision sensor FD Logic illustrated in Sect. 5.1

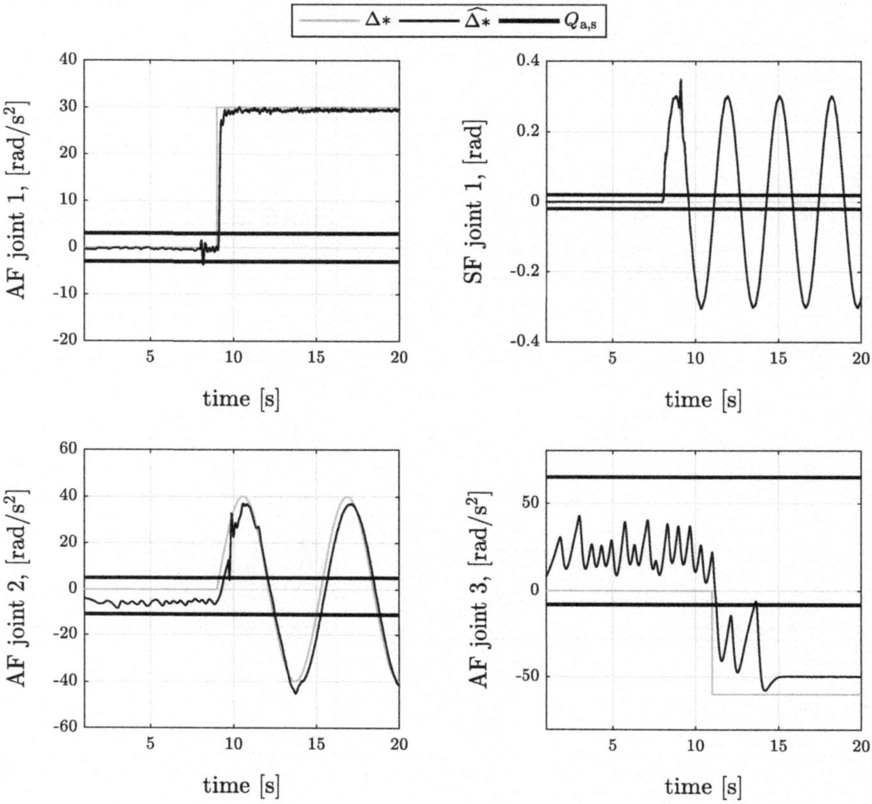


**Fig. 7** Scenario 2: multiple faults both on actuators 1, 2 and 3, and on sensor 1 (simulation results) with $Q_{a,s}$ indicating the thresholds

using the V-REP simulator, which in turn interfaces with Matlab for data analysis. Using the retrieved data, the sliding mode observer used for actuator FD illustrated in Sect. 5.3 was implemented to run in parallel to the V-REP simulation. Artificial faults reported in Table 4a were injected on both the encoder signals and the joint actuators. The simulation tests are performed using the ODE engine with a 0.005 s time-step. Figure 8 shows results obtained in case of the considered scenario. All the faults on sensors and actuators have been correctly detected, isolated and identified, showing the efficacy of the proposal even on a very realistic virtual industrial environment.

**Table 4** V-REP scenario and image from the vision sensor of joint 1

(a) V-REP Scenario

| #joint | FE | $t_o$ [s] | $\Delta * (t)$ $(* \doteq q, y)$ |
|---|---|---|---|
| 1 | AF | 11 | $0.2\sin(2t)$ [rad/s$^2$] |
| | SF | 4 | 0.2 [rad] |
| 2 | AF | 15 | $-0.1\cos(2t)$ [rad/s$^2$] |
| | SF | 6 | $0.15\sin(2t)$ [rad] |
| 3 | AF | – | – |
| | SF | – | – |

## 7.4 Experimental Results

Experimental tests on the actual COMAU SMART3-S2 industrial anthropomorphic robot manipulator have been also executed tuning the observers starting from the values adopted in simulation. The parameters are reported in Table 5a and b. Also in this case, a low-pass filter is applied to the detected signals to improve the fault identification.

Figures 9 and 10 show the effectiveness of the proposed strategy in the same scenarios considered in simulation. In all cases, the simulation results are confirmed also experimentally.

## 7.5 Performance Analysis

To assess the effectiveness and the robustness properties of the proposed fault diagnosis strategy, a large number of experimental tests have been carried out. Yet, the experiments are onerous and time consuming, due to all the preliminary practical steps necessary to put on the tests. Consider, for instance, that in our lab, it is possible to perform on average 5 test daily, not to excessively stress the mechanical part of the robot manipulator. So, to complement the experimental results, an intensive simulation testing has been performed, confident of the good accuracy of the available simulation model. The performance analysis has been accomplished considering that the vision sensor used in the diagnostic scheme is characterized by a typical frame rate.

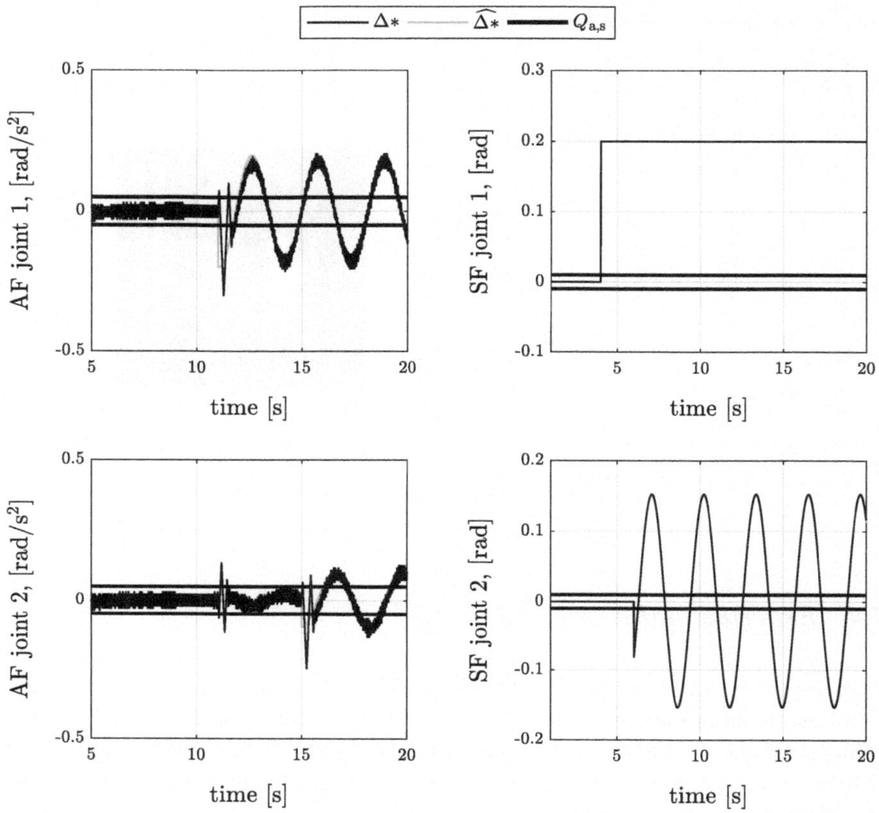


**Fig. 8** V-REP Scenario 1: faults both on sensors 1 and 2 and on actuators 1 and 2 with $Q_{a,s}$ indicating the thresholds

In particular, the IP camera of type Atlantis NetCamera 500 has a nominal interval $T_c$ between two subsequent frames of 30 ms. So, two tests have been considered: the first test (Test 1) in which the camera frame rate is equal to the nominal one, and the second (Test 2) in which, this can vary, causing a variation of the time interval between two subsequent measurements provided by the camera. More specifically, in Test 1 and Test 2, 100 randomly generated faults of step or sinusoidal type, with different amplitude and time of occurrence, have been injected to the simulation model of the COMAU SMART3-S2 robot with 1 ms as sampling time. The number of simulation runs was 100 in Test 1 (1 per fault) and 400 in Test 2 (4 per fault, to consider the 4 different $T_c$ values). The simulations have been carried out on a processor Intel Core i7-2670QM CPU @ 2.20 GHz, 8 GB RAM.

Then, the following percentages are determined: (i) percentage of false alarm ($\%_{FA}$) when a fault is detected, but no fault event has occurred, (ii) percentage of mis-detection ($\%_{MD}$) if a fault is not detected when it is present, (iii) percentage of

**Table 5** FD thresholds and SSOSM based observer parameters (experimental tests)

| (a) Thresholds | | | | | (b) Parameters | | | |
|---|---|---|---|---|---|---|---|---|
| #joint | $T_{\text{mins}_i}$ | $T_{\text{maxs}_i}$ | $T_{\text{mina}_i}$ | $T_{\text{maxa}_i}$ | #joint | $\beta_i$ | $\alpha_i$ | $W_{\text{max}_i}$ |
| 1 | 0.02 | 0.02 | 3 | 3 | 1 | 10 | 0.9 | 40000 |
| 2 | 0.05 | 0.05 | 11 | 5 | 2 | 10 | 0.9 | 150000 |
| 3 | 0.04 | 0.02 | 8 | 65 | 3 | 10 | 0.9 | 400000 |

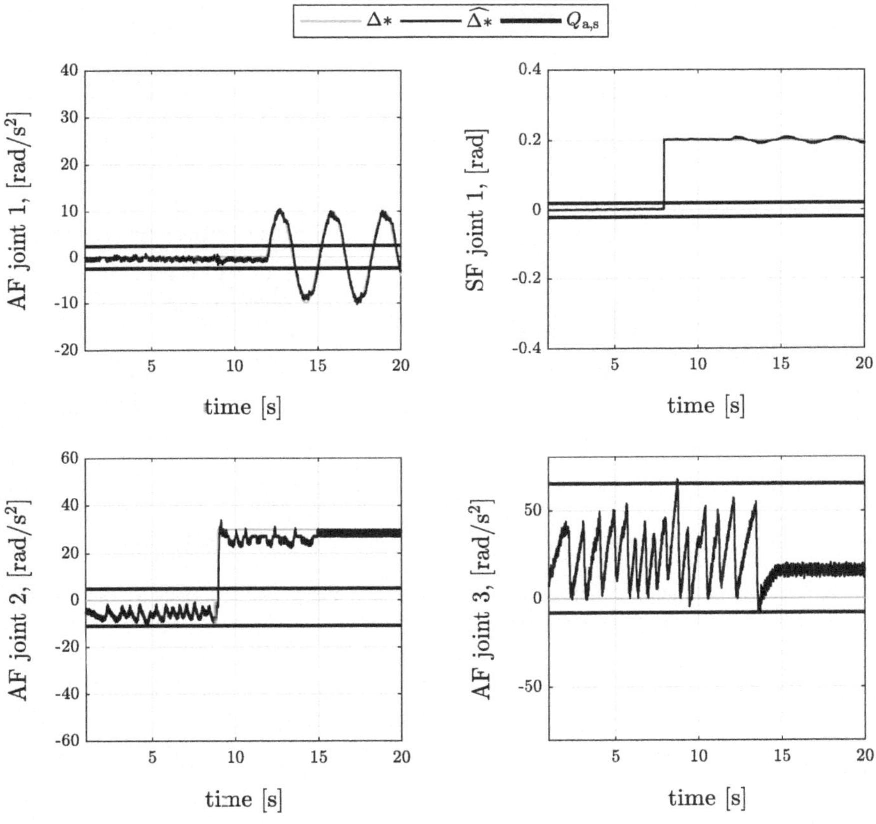


**Fig. 9** Scenario 1: faults both on sensors 1 and 2 and on actuators 1 and 2 (experimental results) with $Q_{a,s}$ indicating the thresholds

mis-isolation ($\%_{\text{MI}}$) if a fault on the $i$th joint is detected on the $j$th joint with $i \neq j$, (iv) percentage of correct FD ($\%_{\text{FD}}$) if the fault diagnosis is correctly performed. The results of Test 1 and Test 2 are reported in Table 6a and b, respectively.

The FD strategy guarantees a 100% correct diagnosis in case of single or multiple faults injected only on the actuators or only on the sensors. In case of multiple

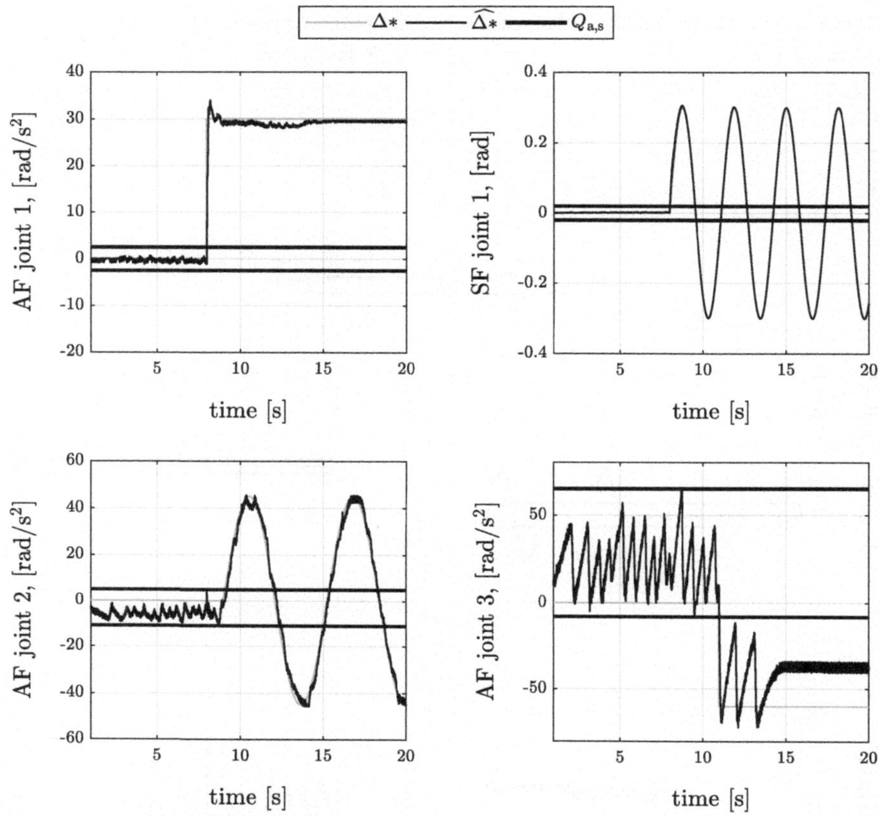


**Fig. 10** Scenario 2: multiple faults both on actuators 1, 2 and 3, and on sensor 1 (experimental results) with $Q_{a,s}$ indicating the thresholds

**Table 6** Performance of the diagnosis strategy

| Test 1 | | | | | Test 2 | | | | |
|---|---|---|---|---|---|---|---|---|---|
| Scenario | $\%_{FA}$ | $\%_{MD}$ | $\%_{MI}$ | $\%_{FD}$ | Diagnosis | $T_c$ [ms] | | | |
| FE1 | 0 | 0 | 0 | 100 | | 1 | 10 | 20 | 30 |
| FE2 | 0 | 0 | 0 | 100 | $\%_{FA}$ | 0 | 0 | 0 | 0 |
| FE3 | 0 | 0 | 0 | 100 | $\%_{MD}$ | 0 | 0 | 0 | 0 |
| FE4 | 0 | 0 | 0 | 100 | $\%_{MI}$ | 0 | 20 | 20 | 44 |
| FE5 | 0 | 0 | 44 | 56 | $\%_{FD}$ | 100 | 80 | 80 | 56 |

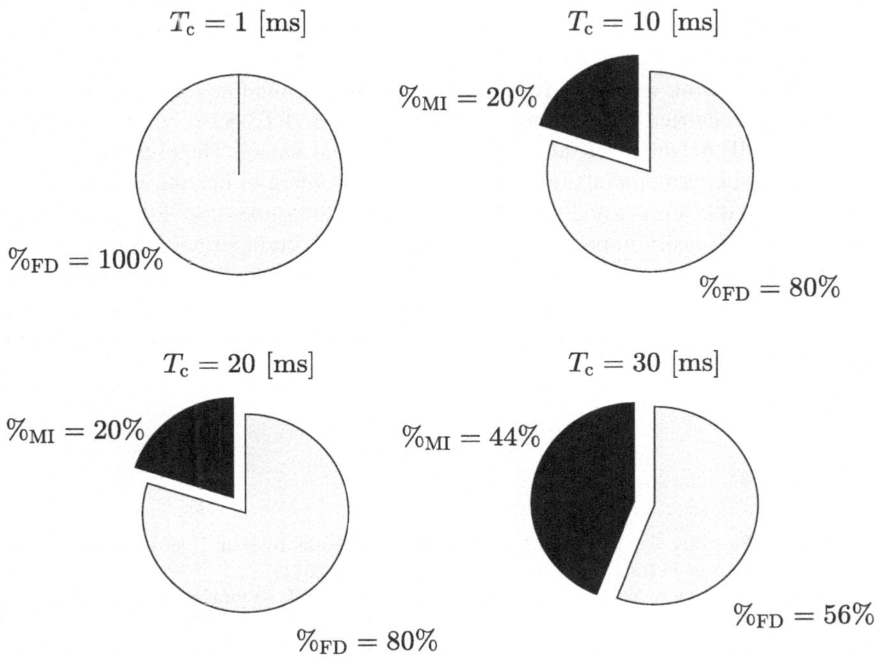


**Fig. 11** Performance of the proposed FD strategy in case of different values of $T_c$

faults both on the actuators and on the sensors, different results have been obtained, depending on the time interval $T_c$ between two acquired images. In case of $T_c = 30$ ms one has 44% of mis-isolation and 56% of correct FD, while false alarm or mis-detection are not present. But when the camera frequency increases, with $T_c$ equal to 20, 10 and 1 ms, the results improve. Note that, when the vision sensor works at the same frequency of the resolvers ($T_c = 1$ ms), a 100% of correct FD is obtained. A graphical rendering of Table 6 is reported in Fig. 11 for the reader's convenience.

## 8 Conclusions

In this chapter, a fault diagnosis strategy applied to robot manipulators has been presented. The proposed approach is based on the powerful combination of an inverse dynamics-based feedback linearizing control and of a fault detection and isolation logic. The use of a low-cost vision sensor allows one to obtain a complete sensor fault diagnosis, while providing good estimates of the controlled variables which can be used to close the control loop if a faulty behavior occurs, thus making the whole control system tolerant to sensor faults. Furthermore, the additional application of SSOSM-UIOs for the actuator fault diagnosis guarantees appreciable robustness features. Several simulation tests, considering realistic scenarios with varying multiple faults occurring in different parts of the robotic systems, have been performed both on Matlab Simulink and on a relistic V-REP robot simulator. Experimental tests have been also carried out relying on a COMAU SMART3-S2 robot manipulator, and using an IP Atlantis NetCamera 500, as redundant sensor. The present proposal successfully solves a crucial and practically open problem of dealing with multiple faults both on the sensors and the actuators, even at the same time. It is a low complexity solution, easily implementable in practice with commercial low cost vision devices. Moreover, even if the FD strategy formulation is carried out in the continuous time framework, its performances are satisfactory even experimentally, that is when a discrete-time implementation is made with sampling times which are analogous to that typical of industrial robotics.

## References

1. Alwi, H., Edwards, C., Tan, C.P.: Fault Detection and Fault-Tolerant Control Using Sliding Modes. Advances in Industrial Control. Springer, Berlin (2011)
2. Bartolini, G., Caputo, W., Cecchi, M., Ferrara, A., Fridman, L.: Vibration damping in elastic robotic structures via sliding modes. J. Robot. Syst. **14**(9), 675–696 (1998a)
3. Bartolini, G., Ferrara, A., Levant, A., Usai, E.: On Second Order Sliding Mode Controllers. Lecture Notes in Control and Information, pp. 329–350. Springer, London (1999)
4. Bartolini, G., Ferrara, A., Usai, E.: Adaptive reduction of the control effort in chattering-free sliding-mode control of uncertain nonlinear systems. Appl. Math. Comput. Sci. **8**(1), 51–71 (1998b)
5. Bartolini, G., Ferrara, A., Usai, E.: Chattering avoidance by second-order sliding mode control. IEEE Trans. Autom. Control **43**(2), 241–246 (1998c)
6. Brambilla, D., Capisani, L., Ferrara, A., Pisu, P.: Fault detection for robot manipulators via second-order sliding modes. IEEE Trans. Ind. Electron. **55**(11), 3954–3963 (2008)
7. Capisani, L.M., Ferrara, A., Ferreira de Loza, A., Fridman, L.M.: Manipulator fault diagnosis via higher order sliding-mode observers. IEEE Trans. Ind. Electron. **59**(10), 3979–3986 (2012)
8. Capisani, L.M., Ferrara, A., Magnani, L.: Design and experimental validation of a second-order sliding-mode motion controller for robot manipulators. Int. J. Control **82**(2), 365–377 (2009)
9. Capisani, L.M., Ferrara, A., Pisu, P.: Sliding mode observers for vision-based fault detection, isolation and identification in robot manipulators. In: Proceedings American Control Conference, pp. 4540–4545. Baltimore, Maryland, USA (2010)
10. Chiang, L., Braatz, R., Russell, E.: Fault Detection and Diagnosis in Industrial Systems. Advanced Textbooks in Control and Signal Processing. Springer, London (2001)
11. Davila, J., Fridman, L., Poznyak, A.: Observation and identification of mechanical systems via second order sliding modes. Int. J. Control **79**(10), 1251–1262 (2006)
12. De Luca, A., Mattone, R.: An adapt-and-detect actuator FDI scheme for robot manipulators. In: Proceedings International Conference on Robotics and Automation, vol. 5, pp. 4975–4980. Barcelona, Spain (2004)
13. De Luca, A., Mattone, R.: An identification scheme for robot actuator faults. In: Proceedings IEEE/RSJ International Conference on Intelligent Robots and Systems, pp. 1127–1131. Alberta, Canada (2005)
14. Esna Ashari, A., Nikoukhah, R., Campbell, S.: Active robust fault detection in closed-loop systems: Quadratic optimization approach. IEEE Trans. Autom. Control **57**(10), 2532–2544 (2012)

15. Ferrara, A., Incremona, G.P.: Design of an integral suboptimal second-order sliding mode controller for the robust motion control of robot manipulators. IEEE Trans. Control Syst. Technol. **23**(6), 2316–2325 (2015)
16. Halder, B.: Robust Nonlinear Fault Detection and Isolation of Robotic System: A Novel Nonlinear Analytic Redundancy Method. VDM Verlag, Germany (2009)
17. Incremona, G.P., Ferrara, A., Magni, L.: MPC for robot manipulators with integral sliding modes generation. IEEE/ASME Trans. Mechatron. **22**(3), 1299–1307 (2017)
18. Incremona, G.P., Saccon, A., Ferrara, A., Nijmeijer, H.: Trajectory tracking of mechanical systems with unilateral constraints: Experimental results of a recently introduced hybrid pd feedback controller. In: Proceedings 54th IEEE Conference on Decision and Control, pp. 920–925. Osaka, Japan (2015)
19. Isermann, R.: Fault-Diagnosis Systems: An Introduction from Fault Detection to Fault Tolerance. Springer, Berlin (2006)
20. Natke, H., Cempel, C.: Model-Aided Diagnosis of Mechanical Systems. Fundamentals, Detection, Localization, Assessment. Springer, London (2011)
21. Papadimitropoulos, A., Rovithakis, G.A., Parisini, T.: Fault detection in mechanical systems with friction phenomena: an online neural approximation approach. IEEE Trans. Neural Netw. Learn. Syst. **18**(4), 1067–1082 (2007)
22. Rigatos, G.G.: Fault Diagnosis in Robotic and Industrial Systems, 1st edn. iConcept Press Ltd., Hong Kong (2012)
23. Rohmer, E., Singh, S.P.N., Freese, M.: V-REP: a versatile and scalable robot simulation framework. In: Proceedings of the International Conference on Intelligent Robots and Systems (IROS) (2013).
24. Sangiovanni B., Rendiniello A., Incremona G. P., Ferrara A., Piastra M.: Deep reinforcement learning for collision avoidance of robotic manipulators. In: Proceedings of European Control Conference, pp. 2063–2068 (2018)
25. Ferrara, A., Incremona, G.P., Sangiovanni, B.: Integral sliding mode based switched structure control scheme for robot manipulators. In: 15th International Workshop on Variable Structure Systems, pp. 168–173 (2018)
26. Sangiovanni B., Incremona G.P., Ferrara A., Piastra M.: Deep reinforcement learning based self-configuring integral sliding mode control scheme for robot manipulators. In: IEEE Conference on Decision and Control, pp. 5969–5974 (2018)
27. Scott, J.K., Findeisen, R., Braatz, R.D., Raimondo, D.M.: Input design for guaranteed fault diagnosis using zonotopes. Automatica **50**(6), 1580–1589 (2014)
28. Siciliano, B., Khatib, O. (eds.): The Handbook of Robotics. Springer, Berlin (2008)
29. Siciliano, B., Sciavicco, L., Villani, L., Oriolo, G.: Robotics-Modelling Planning and Control, 3rd edn. Springer, London, (2009)
30. Simandl, M., Puncochar, I.: Active fault detection and control: unified formulation and optimal design. Automatica **45**(9), 2052–2059 (2009)
31. Spurgeon, S.K.: Sliding mode observers: a survey. Int. J. Syst. Sci. **39**(8), 751–764 (2008)
32. Utkin, V.I.: Sliding Modes in Optimization and Control Problems. Springer, New York (1992)
33. Venkatasubramanian, V., Rengaswamy, R., Yin, K., Kavuri, S.N.: A review of process fault detection and diagnosis: part i: quantitative model-based methods. Comput. Chem. Eng. **27**(3), 293–311 (2003)

# $L_1$ Adaptive Control of a Lower Limb Exoskeleton Dedicated to Kids' Rehabilitation

**Boutheina Maalej, Ahmed Chemori and Nabil Derbel**

**Abstract** In this chapter, four adaptive controllers have been proposed to control a 2-DOF exoskeleton dedicated to kids' rehabilitation. These control laws are implemented at the hip and the knee joints. In fact, tracking the gait scheme with an intense and a precise work may allow children to increase their brain plasticity. Through the proposed study, it is shown that the augmented $L_1$ adaptive controller is robust regards to parametric variations. Besides, to validate this controller, different scenarios and simulations were carried out to prove its effectiveness.

**Keywords** Rehabilitation · Cerebral palsy · Exoskeletons · Classical adaptive control · $L_1$ adaptive controller

## 1 Introduction

Robotic systems have recently impacted the human life. Indeed, human-robot interaction in several domains is considered as one of the most remarkable achievement all over the world. Nowadays, exoskeletons can be considered as a significant example of human-robot interaction. Initially, exoskeletons were developed for military applications [1]. Then, for industrial applications and medical uses. Since walking is a key feature of independency, restoring safe walking is one of the main goals of

B. Maalej (✉) · N. Derbel
Laboratory of Control & Energy Management, ENIS, Digital Research Center of Sfax, University of Sfax, Sfax, Tunisia
e-mail: maalej.boutheina@gmail.com

N. Derbel
e-mail: n.derbel@enis.rnu.tn

B. Maalej
Clinical Investigation Center, University of Gabes, Sfax, Tunisia

A. Chemori
LIRMM, University of Montpellier, CNRS, Montpellier, France
e-mail: ahmed.chemori@lirmm.fr

 
J. Ghommam et al. (eds.), *New Trends in Robot Control*, Studies in Systems, Decision and Control 270, https://doi.org/10.1007/978-981-15-1819-5_6

robotic rehabilitation. There exist several disabilities that may cause the movement disorders and affect the brain which is the responsible of the body functions. Any damage in the brain tissue before, during or after childbirth may affect certain areas of the brain. Besides, depending on the degree of injury, it can cause permanent disorders, characteristic of a non-progressive injury. Among the potentially determining factors of irreversible brain damage, the most frequently observed include infections of the nervous system, hypoxia (lack of oxygen), head injuries, etc. Children with cerebral motor disorders represent 3 per 1000 newborns. In this context, several medical applications appear in order to help paralyzed kids to restore their locomotion [2]. In this work, we consider the case of lower limb rehabilitation using exoskeletons in order to assist children movements [3]. In the sequel, we will be interested in kids who have between 2 and 13 years old. Hence, we propose to implement robust controllers that can be adapted to the difference of children morphologies [4]. An adaptive control is a controller with adjustable parameters. In fact, it can adapt to changes in the process dynamics and the disturbance characteristics. Moreover, adaptive techniques can also be used to provide automatic tuning of controllers. In this chapter, four control laws are proposed to solve the problem of parametric variations. The chosen controllers include two nonlinear state feedback adaptive controllers, the $L_1$ adaptive control and augmented $L_1$ adaptive control. The first approach, based on Slotine works [5], consists in a PD feedback part and a full dynamic feedforward tacking into consideration parametric variations. The second approach is also based on Slotine works [5]. It consists in adding an integral control action to the previous approach in order to eliminate undesirable steady-state position errors. The third approach consists in applying $L_1$ adaptive control proposed by Naira Hovakimyan [6]. The fourth approach is the augmented $L_1$ adaptive control [7] which combines the $L_1$ adaptive control with a Proportional-Integral control to eliminate the time lag and to improve the tracking performances. Simulation results will be presented with a comparative study which aims to show the robustness of the augmented $L_1$ adaptive control.

## 2 Description and Modeling of Exoskeletons

Exoskeletons are made in order to help kids suffering from several diseases, such as the cerebral palsy [8], to restore their walk again by a cyclical and alternative rhythmic activities [9]. The basic idea is to control the lower limb exoskeleton (Fig. 1) at the hip and knee joints.

The dynamic model includes the human limb and the exoskeleton, it is written as follows [10]:

$$M(q)\ddot{q} + C(q,\dot{q})\dot{q} + G(q) = \tau \tag{1}$$

with,

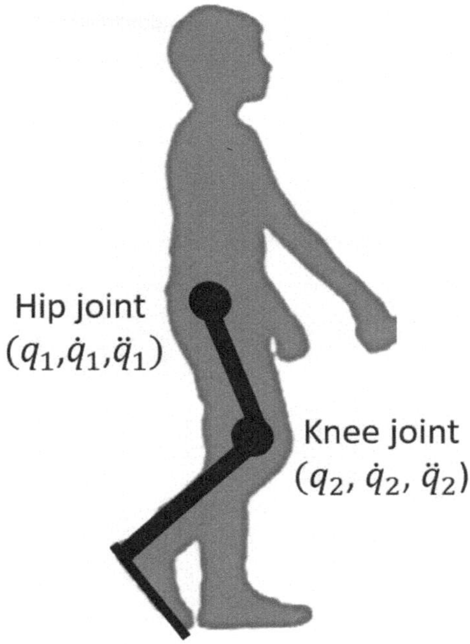


**Fig. 1** Concept of the 2-joint lower limb exoskeleton

$$M_{11} = \frac{1}{8}l_1^2 m_1 + \frac{1}{4}m_2\left(\frac{1}{4}l_2^2 + l_1^2 + l_1 l_2 \cos q_2\right) + \frac{1}{2}m_s(l_1^2 + l_2^2 k_2^2 + 2l_1 l_2 k_2 \cos q_2)$$
$$+\frac{1}{2}k_1^2 l_1^2 m_t + \frac{1}{2}(I_1 + I_2 + I_s + I_t)$$
$$M_{12} = M_{21} = \frac{1}{2}m_2\left(\frac{1}{2}l_2^2 + l_1 l_2 \cos q_2\right) + \frac{1}{2}m_s(l_2^2 k_2^2 + 2l_1 l_2 k_2 \cos q_2) + (I_2 + I_s)$$
$$M_{22} = \frac{1}{8}m_2 l_2^2 + \frac{1}{2}m_s l_1^2 + l_2^2 k_2^2 + \frac{1}{2}(I_2 + I_s)$$
$$C_{11} = \left(\frac{1}{2}m_2 + m_s k_2\right)l_1 l_2 \sin q_2$$
$$C_{12} = -\left(\frac{1}{2}m_2 + m_s k_2\right)l_1 l_2 \sin q_2$$
$$C_{21} = \left(\frac{1}{4}m_2 + 1/2 m_s k_2\right)l_1 l_2 \sin q_2$$
$$C_{22} = 0$$
$$G_1 = -\left(\frac{1}{2}m_1 + m_2 + k_1 m_t + m_s\right)g l_1 \sin q_1 - (1/2 m_1 + m_s k_2) g l_2 \sin(q_1 + q_2)$$
$$G_2 = -\left(\frac{1}{2}m_1 + m_s k_2\right)g l_2 \sin(q_1 + q_2)$$

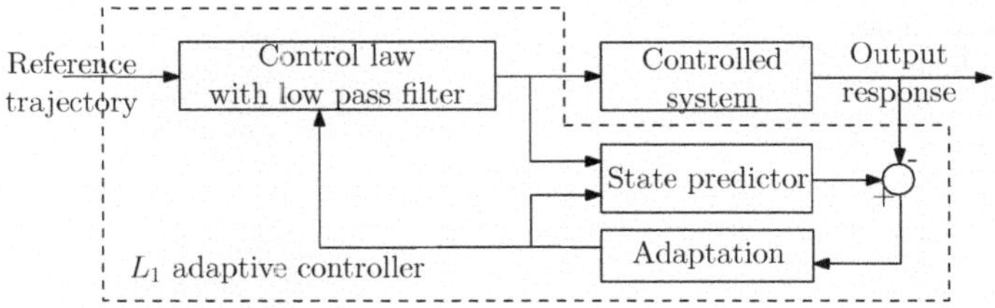


**Fig. 2** Block diagram of $L_1$ adaptive control [6]

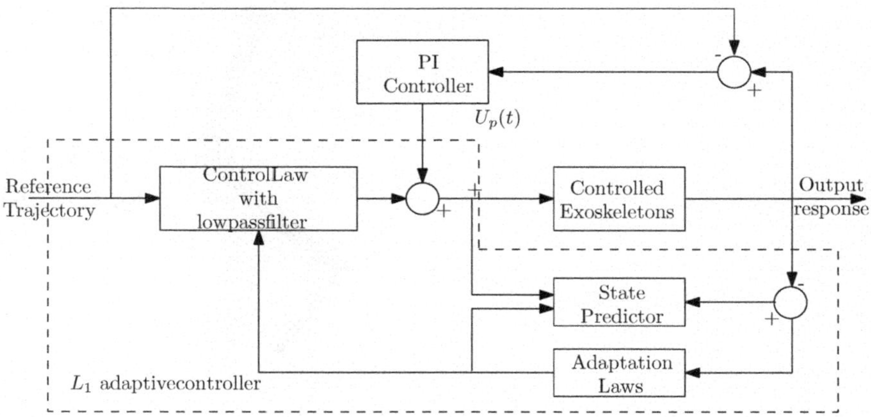


**Fig. 3** Block diagram of the augmented $L_1$ adaptive control [11]

where:

$q = [q_1\ q_2]^T \in \mathbb{R}^2$ represents the position vector of the hip and knee joints, respectively,
$\dot{q} = [\dot{q}_1\ \dot{q}_2]^T \in \mathbb{R}^2$ is the speed vector,
$\ddot{q} = [\ddot{q}_1\ \ddot{q}_2]^T \in \mathbb{R}^2$ is the acceleration vector,
$M(q) \in \mathbb{R}^2$ is the inertia matrix, which is symmetric, uniformly bounded and positive definite,
$C(q, \dot{q})\dot{q} \in \mathbb{R}^2$ represents the Coriolis, and centrifugal forces and torques,
$G(q) \in \mathbb{R}^2$ denotes the gravity torques,
$\tau \in \mathbb{R}^2$ is the vector of actuator torques,
$m_1$, $m_2$, $l_1$, $l_2$ represent the mass and length of thigh and shank segments of the exoskeleton, respectively,
$m_t$, $m_s$ denote thigh and shank masses of human limb,
$k_1, k_2$ are the position of the center of mass of thigh and shank segments, respectively,
$I_1$, $I_2$, $I_s$, $I_t$ represent the moments of inertia of thigh and shank of the exoskeleton and human limb, respectively,
$g$ is the gravity acceleration.

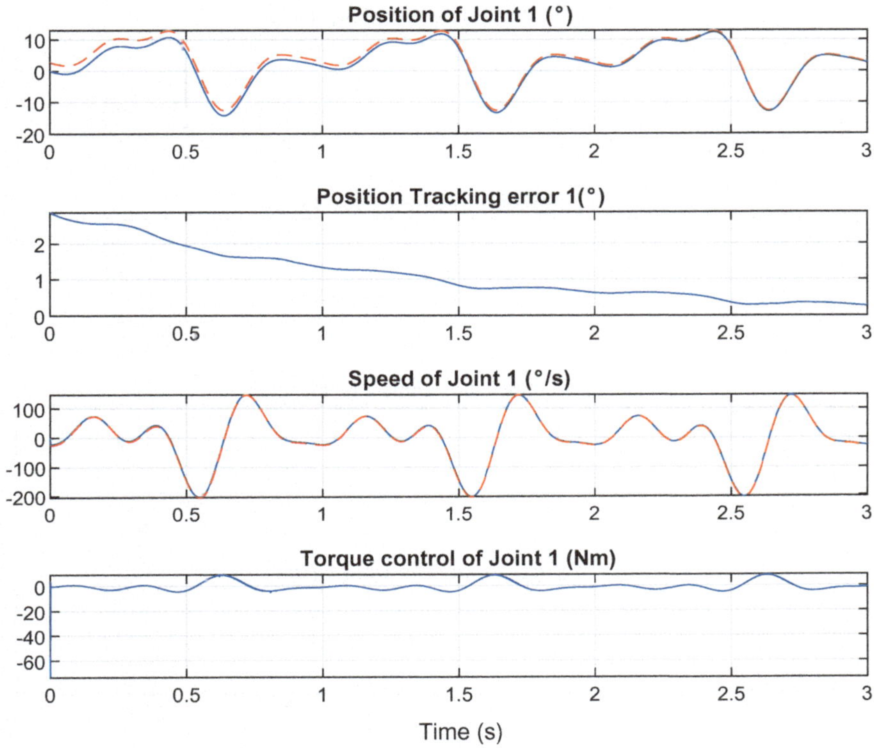


**Fig. 4** Classical adaptive control of the hip joint. For the position and the speed, dashed line: desired trajectory, solid line: actual trajectory

# 3 Proposed Control Solution

## 3.1 Adaptive Control [5]

Let consider that $q_d$ is the desired joint position and $\dot{q}_d$ is the desired velocity. The main idea is to develop a globally stable adaptive controller which is obtained from a Lyapunov stability analysis. The Lyapunov function associated to the system is:

$$V(t) = \frac{1}{2}\left[\dot{\widetilde{q}}^T M(q)\dot{\widetilde{q}} + \widetilde{a}^T \Gamma \widetilde{a} + \widetilde{q}^T K_p \widetilde{q}\right] \tag{2}$$

where

$a$: represents the vector of unknown manipulator parameters
$\widehat{a}$ : represents its estimate
$\widetilde{a} = \widehat{a} - a$ : denotes the parameter estimation error vector

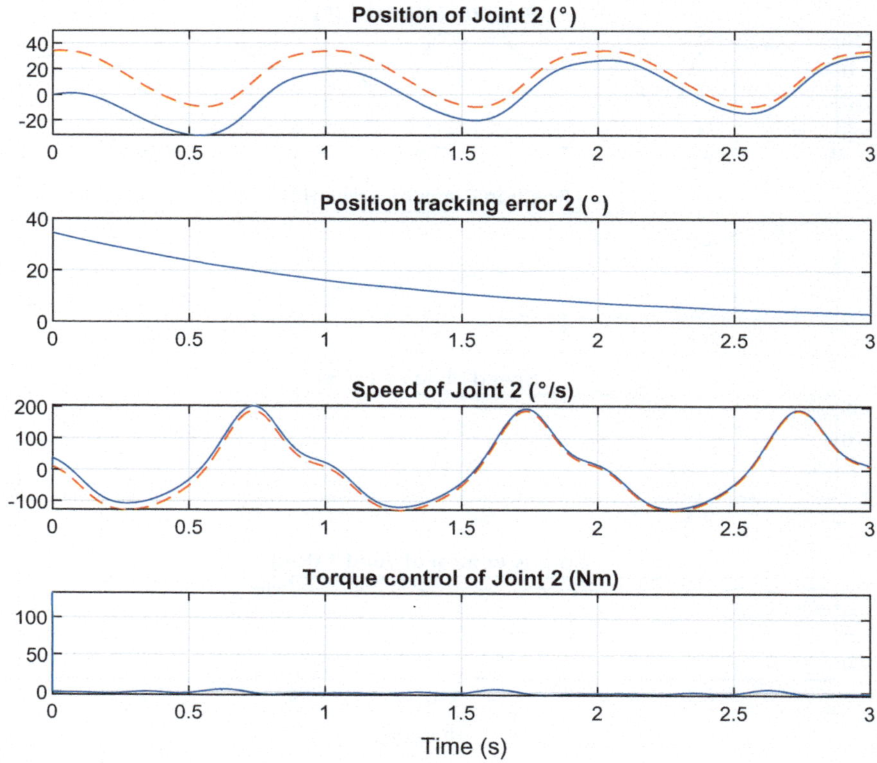


**Fig. 5** Classical adaptive control of the knee joint. For the position and the speed, dashed line: desired trajectory, solid line: actual trajectory

$K_p$, $\Gamma$: symmetric positive definite matrices
$\tilde{q} = q - q_d$: represents the position tracking error
$\dot{\tilde{q}} = \dot{q} - \dot{q}_d$: represents the velocity tracking error

The derivative of $V$ with respect to time is:

$$
\begin{aligned}
\dot{V}(t) &= \dot{\tilde{q}}^T M \ddot{\tilde{q}} + \frac{1}{2}\dot{\tilde{q}}^T \dot{M} \dot{\tilde{q}} + \tilde{a}^T \Gamma \dot{\tilde{a}} + \tilde{q}^T K_p \dot{\tilde{q}} \\
&= \dot{\tilde{q}}^T \left(\tau - C(q,\dot{q})\dot{q} - G(q) - M\ddot{q}_d\right) + \dot{\tilde{q}}^T \left(\frac{1}{2}(\dot{M} - 2C) + C\right)\dot{\tilde{q}} \\
&\quad + \tilde{a}^T \Gamma \dot{\tilde{a}} + \dot{\tilde{q}}^T K_p \tilde{q} \\
&= \dot{\tilde{q}}^T \left(\tau - M(q)\ddot{q}_d - C(q,\dot{q})\dot{q}_d - G(q) + K_p \tilde{q}\right) + \tilde{a}^T \Gamma \dot{\tilde{a}} \qquad (3)
\end{aligned}
$$

Using the property of skew symmetric matrix, we have:

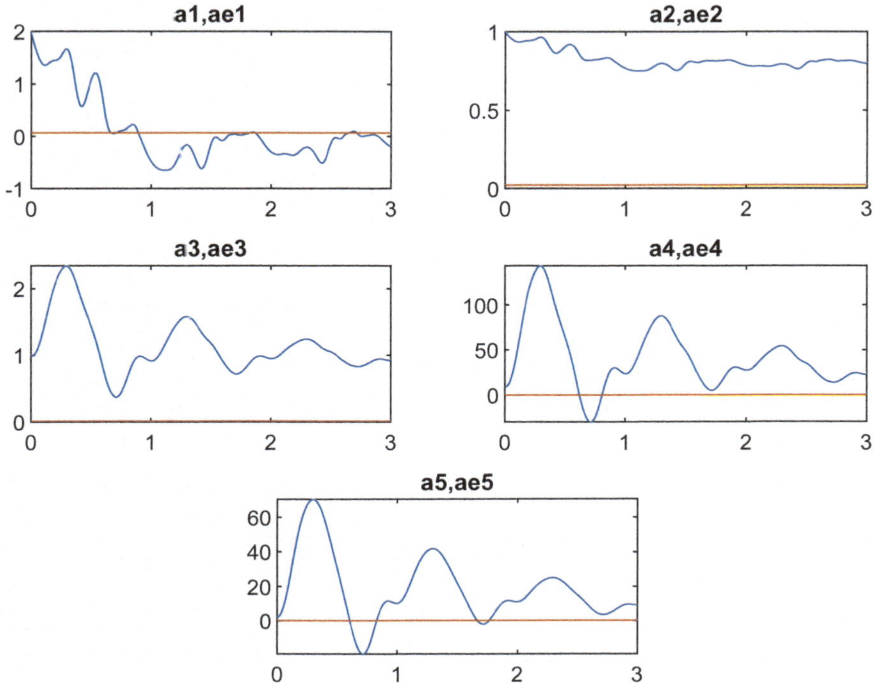


**Fig. 6** Evolution of the estimated Parameters for the first approach of Slotine

$$\frac{1}{2}\frac{d}{dt}(\dot{q}^T M\dot{q}) = \dot{q}^T\left(\tau - G(q)\right) \tag{4}$$

$$\dot{q}^T\left(\frac{1}{2}\dot{M} - C\right)\dot{q} = 0 \tag{5}$$

Considering the following control law:

$$\tau = \widehat{M}\ddot{q}_d + \widehat{C}(q,\dot{q}) + \widetilde{G}(q) - K_p\widetilde{q} - K_D\dot{\widetilde{q}} \tag{6}$$

gives

$$\dot{V} = \dot{\widetilde{q}}^T\left[\widetilde{M}(q)\ddot{q}_d + \widetilde{C}(q,\dot{q})\dot{q}_d + \widetilde{G}(q) - K_D\dot{\widetilde{q}}\right] + \widetilde{a}^T\Gamma\dot{\widetilde{a}} \tag{7}$$

where:

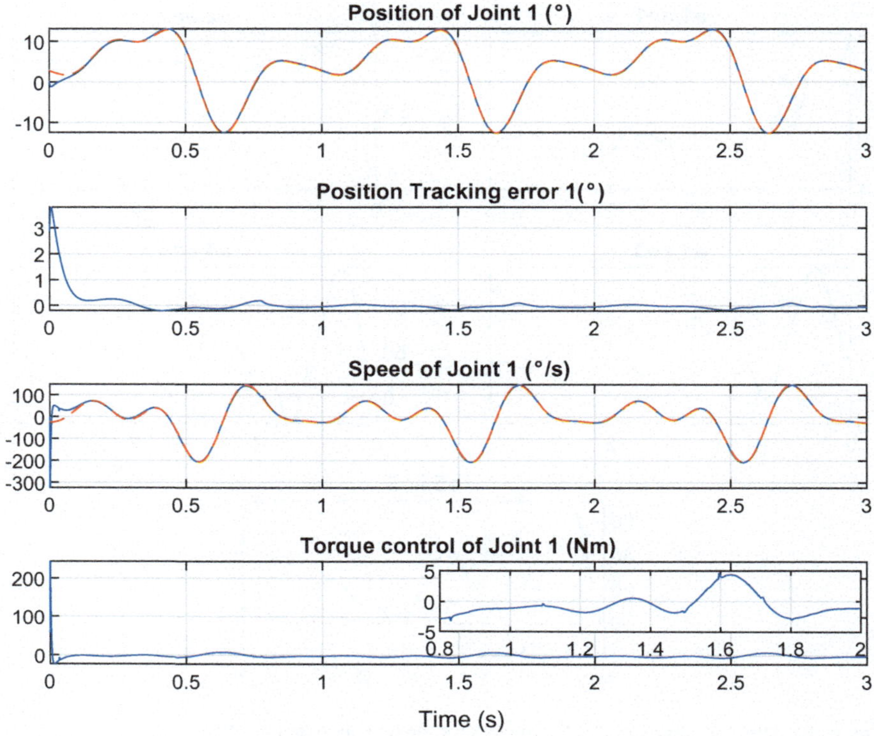


**Fig. 7** Adaptive control with integral action of the hip joint. For the position and the speed, dashed line: desired trajectory, solid line: actual trajectory

$$\widetilde{M}(q) = \widehat{M}(q) - M(q) = \sum_i M_i \widetilde{a}_i \tag{8}$$

$$\widetilde{C}(q,\dot{q}) = \widehat{C}(q,\dot{q}) - C(q,\dot{q}) = \sum_i C_i \widetilde{a}_i \tag{9}$$

$$\widetilde{G}(q) = \widehat{G}(q) - G(q) = \sum_i G_i \widetilde{a}_i \tag{10}$$

Then, let's write:

$$\widetilde{M}(q)\ddot{q_d} + \widetilde{C}(q,\dot{q})\dot{q_d} + \widetilde{G}(q) = Y\widetilde{a} \tag{11}$$

where $Y = Y(q, \dot{q}, \dot{q}_d, \ddot{q}_d)$ is an $n \times m$ matrix. Therefore:

$$\dot{V} = -\dot{\widetilde{q}}^T K_D \dot{\widetilde{q}} + \widetilde{a}^T [\Gamma \dot{\widetilde{a}} + Y^T \dot{\widetilde{q}}] \tag{12}$$

Assuming that variations of the unknown vector $a$ can be neglected, we obtain:

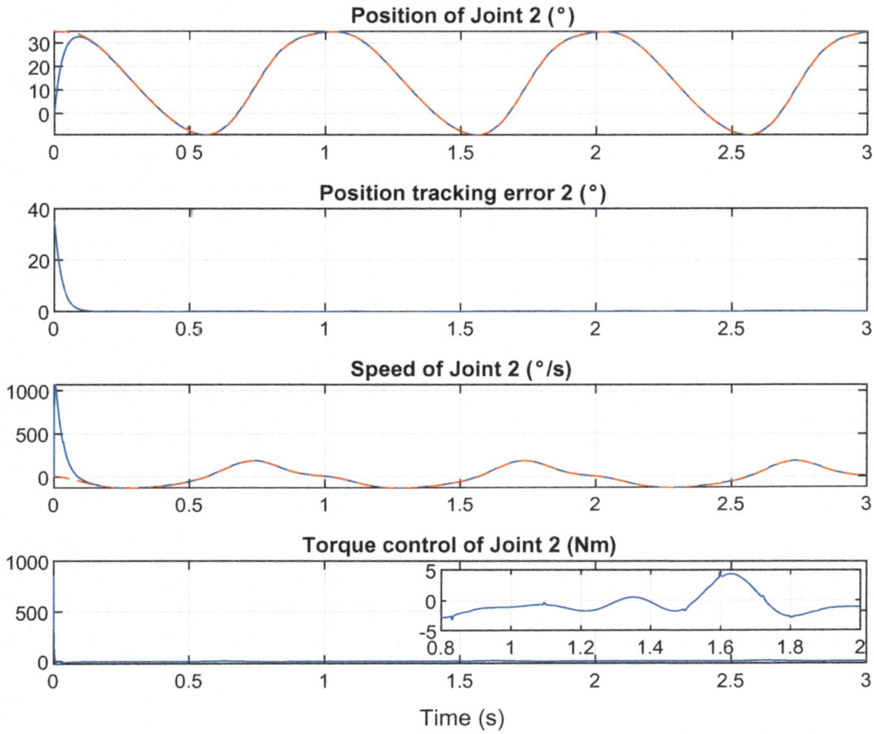


**Fig. 8** Adaptive control with integral action of the knee joint. For the position and the speed, dashed line: desired trajectory, solid line: actual trajectory

**Table 1** Tracking performance comparison

| | Classical adaptive controller | | $L_1$ adaptive controller | | Augmented $L_1$ adaptive controller | |
|---|---|---|---|---|---|---|
| | Joint 1 | Joint 2 | Joint 1 | Joint 2 | Joint 1 | Joint 2 |
| $IAE$ | 0.0037 | 0.0169 | 0.0972 | 0.1578 | 0.0054 | 0.0609 |
| $ISE$ | 0.0062 | 0.0181 | 0.0057 | 0.0279 | $1.2702\times10^{-4}$ | 0.0183 |
| $IAER$ | 28.2872 | 3.1664 | 0.4075 | 0.2513 | 0.0228 | 0.0970 |

$$\dot{V} = -\dot{\tilde{q}}^T K_D \dot{\tilde{q}} \leq 0 \tag{13}$$

using the adaptive law of the parameter vector $a$:

$$\dot{\hat{a}} = -\Gamma^{-1} Y^T \dot{\tilde{q}} \tag{14}$$

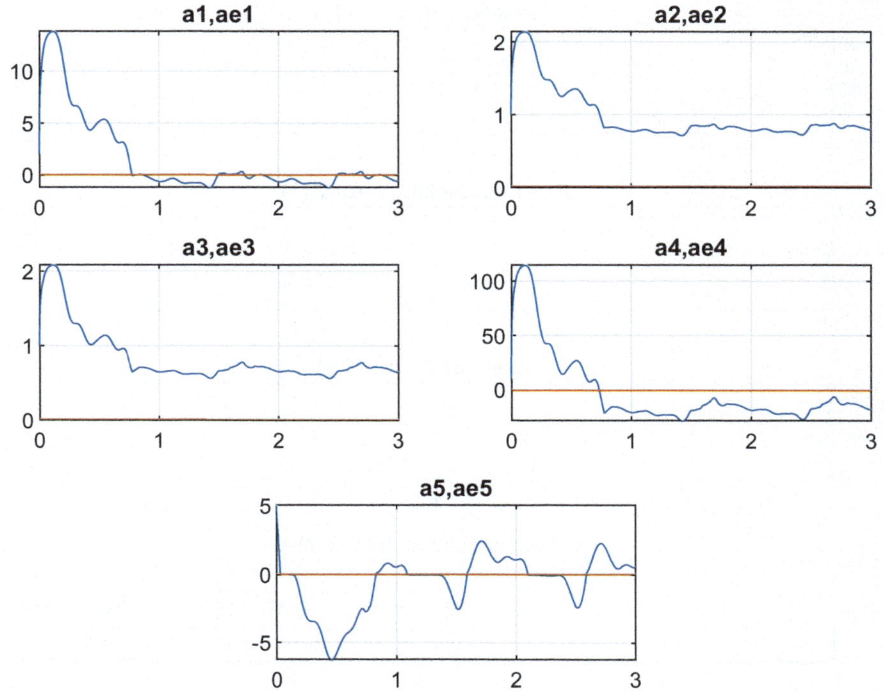


**Fig. 9** Evolution of the estimated Parameters for the second adaptive approach of Slotine

## 3.2 Adaptive Control with an Integral Action [5]

The idea is to eliminate the steady-state position errors by restrict them to lie on a sliding surface.

$$\dot{\tilde{q}} + \Lambda\tilde{q} = 0 \tag{15}$$

where $\Lambda$ is a positive definite diagonal matrix.

The virtual reference trajectory is expressed by

$$q_r = q_d + \Lambda \int \tilde{q} dt \tag{16}$$

As a consequence, $\dot{q}_d$ and $\ddot{q}_d$ are replaced by

$$\dot{q}_r = \dot{q}_d - \Lambda\tilde{q} \tag{17}$$

$$\ddot{q}_r = \ddot{q}_d - \Lambda\dot{\tilde{q}} \tag{18}$$

If we define

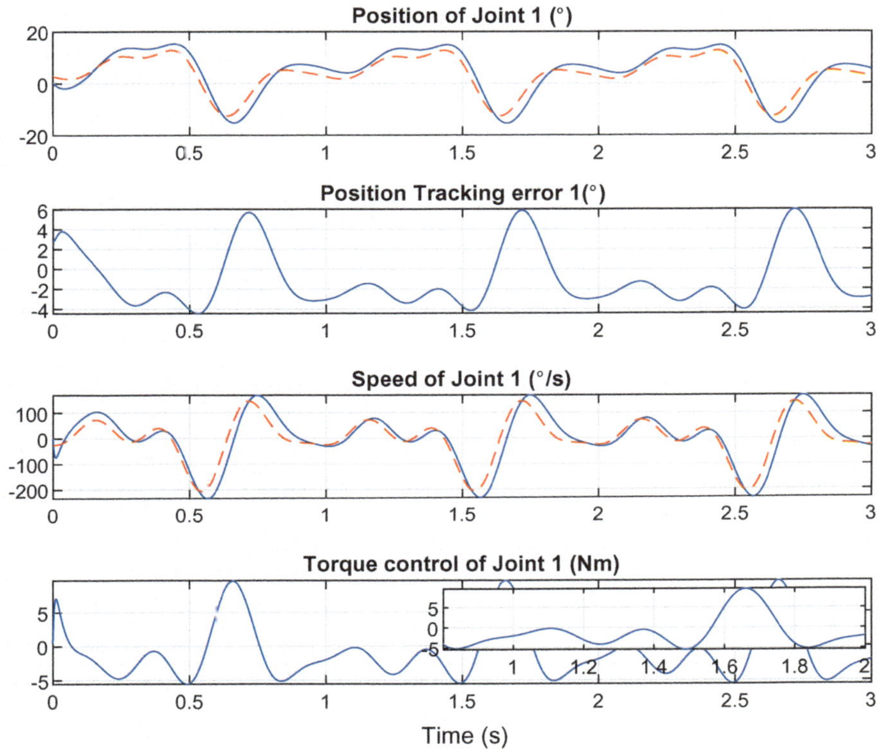


**Fig. 10** Classical $L_1$ adaptive control of the knee joint. For the position and the speed, dashed line: desired trajectory, solid line: actual trajectory

$$s = \dot{\tilde{q}}_r = \dot{q} - \dot{q}_r = \dot{\tilde{q}} + \Lambda\tilde{q} \tag{19}$$

The control and the adaptation laws become:

$$\tau = \widehat{M}\ddot{q}_r + \widehat{C}(q, \dot{q})\dot{q}_r + \widetilde{G}(q) - K_D s \tag{20}$$

$$\dot{\widehat{a}} = -\Gamma^{-1} Y^T(q, \dot{q}, \dot{q}_r, \ddot{q}_r)s \tag{21}$$

We use a Lyapunov function to demonstrate the global convergence of the tracking:

$$V = \frac{1}{2}s^T M s + \frac{1}{2}\widetilde{a}^T \Gamma \widetilde{a} \tag{22}$$

Its first time derivative leads to:

$$\dot{V} = -s^T K_D s \leq 0 \tag{23}$$

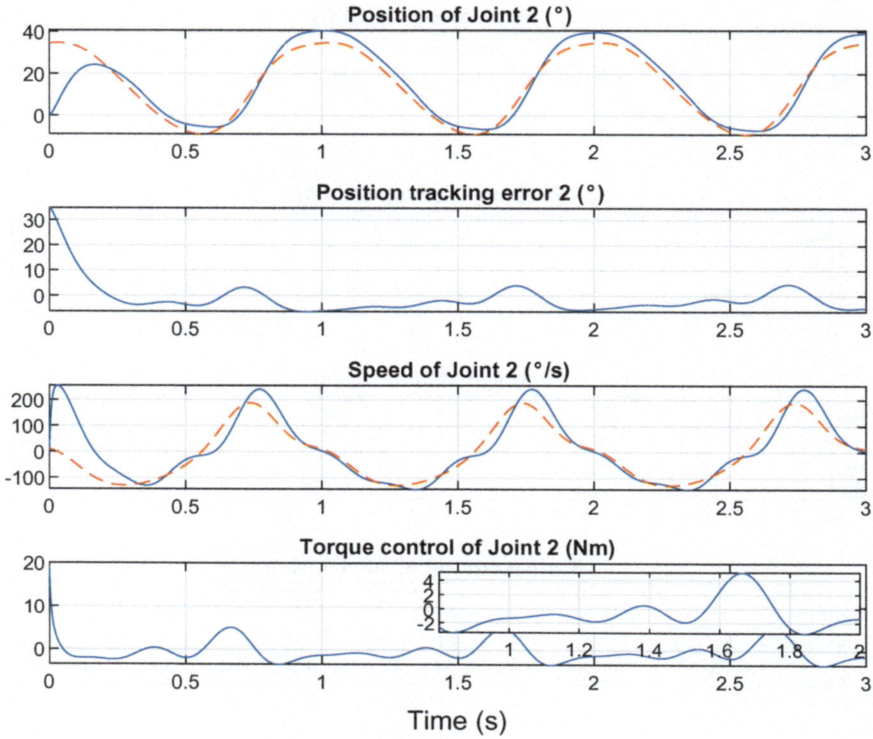


**Fig. 11** Classical $L_1$ adaptive control of the hip joint. For the position and the speed, dashed line: desired trajectory, solid line: actual trajectory

The control law does not contain $K_p$, since the position error $\widetilde{q}$ is already included in $\dot{\widetilde{q}}_r$. Moreover where $s$ becomes equal to 0, $\widetilde{q}$ converges to 0.

It is well known that the adaptive controller has several limitations such as (i) the initial value of the parameters, (ii) the persistent excitation of the estimated parameters, and (iii) the stability. Hence, $L_1$ Adaptive controller is presented in the next subsection which overcome these limitations.

### *3.3 $L_1$ Adaptive Control [11]*

$L_1$ adaptive control is inspired from the Model Reference Adaptive Control (MRAC), it is structured into four principal parts as illustrated in Fig. 2 namely the controlled system, the state predictor, the adaptation mechanism and the control law including a low pass filter. Considering that the control input vector $\tau(t)$ is a compound of two

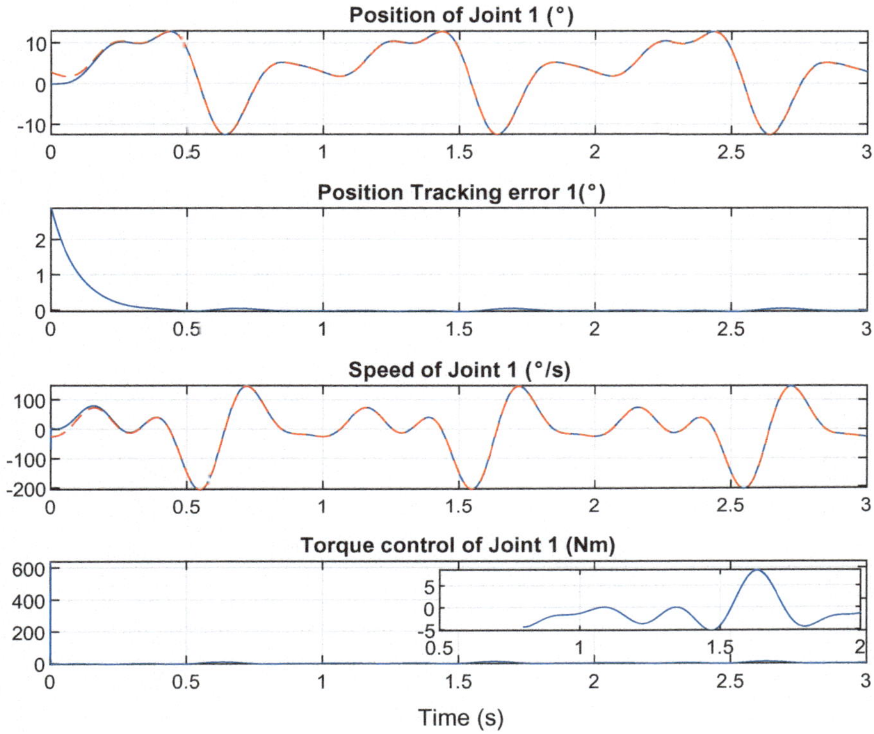


**Fig. 12** Position, speed and torque control of the hip joint, obtained with a Augmented $L_1$ adaptive controller in the nominal case. Solid line: actual trajectory, dashed line: desired trajectory

parts, a fixed state-feedback term that defines the evolution of the transient response and an adaptive term $\tau_{ad}$ that compensates the nonlinearities of the system.

The expression of the torque is:

$$\tau(t) = A_m r(t) + \tau_{ad}(t) \tag{24}$$

where:

$A_m \in \mathbb{R}^{2\times 2}$ is a Hurwitz matrix
$\tau_{ad} \in \mathbb{R}^{2\times 2}$ is the adaptive component

The tracking error is expressed as:

$$r = (\dot{q} - \dot{q}_d) + \Lambda(q - q_d) \tag{25}$$

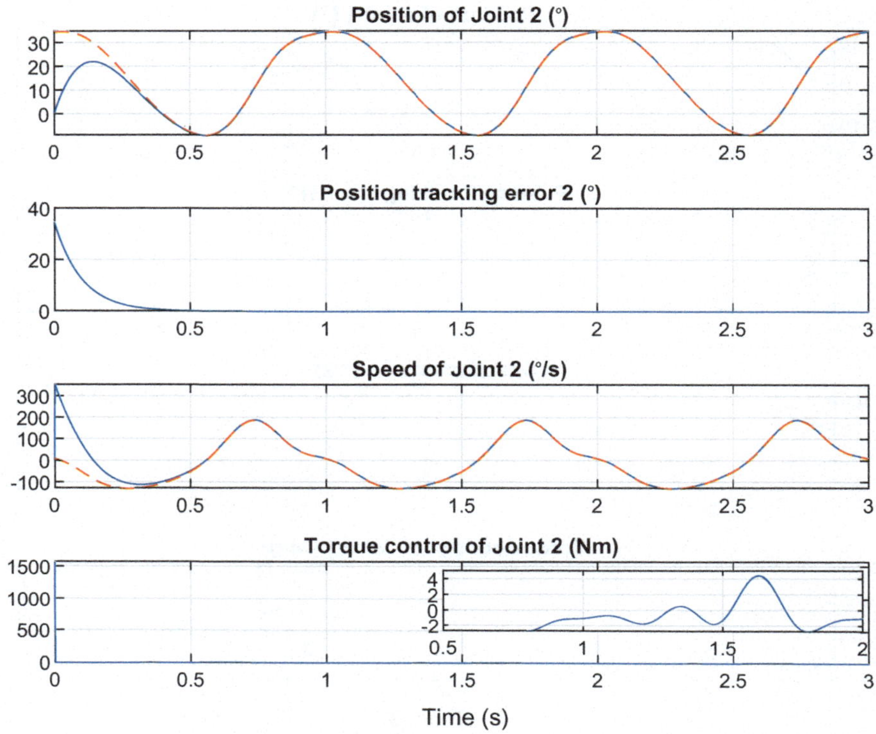


**Fig. 13** Position, speed and torque control of the knee joint, obtained with a Augmented $L_1$ adaptive controller in the nominal case. Solid line: actual trajectory, dashed line: desired trajectory

with $\Lambda \in \mathbb{R}^{2\times 2}$ is a symetric positive definite matrix. It is difficult to know exactly all nonlinearities of the system, hence the adaptive control can be defined in such a way to cancel the effect of these nonlinearities. For that, we consider the state predictor which is based on estimated parameters obtained from the adaptation mechanism:

$$\dot{\hat{r}} = A_m \hat{r}(t) + \tau_{ad}(t) - [\hat{\theta}(t)\|r_t\|_\infty + \hat{\sigma}(t)] - K\tilde{r}(t) \tag{26}$$

where:

$\tilde{r}(t) = \hat{r}(t) - r(t)$ is the prediction error
$K \in \mathbb{R}^{2\times 2}$ used to reject high frequency noises.

Using the projection method, we obtain the estimate of $\theta(t)$ and $\sigma(t)$:

$$\dot{\hat{\theta}}(t) = \Gamma Proj(\hat{\theta}(t), P\hat{r}(t)\|r_t\|_\infty) \tag{27}$$

$$\dot{\hat{\sigma}}(t) = \Gamma Proj(\hat{\sigma}(t), P\hat{r}(t)) \tag{28}$$

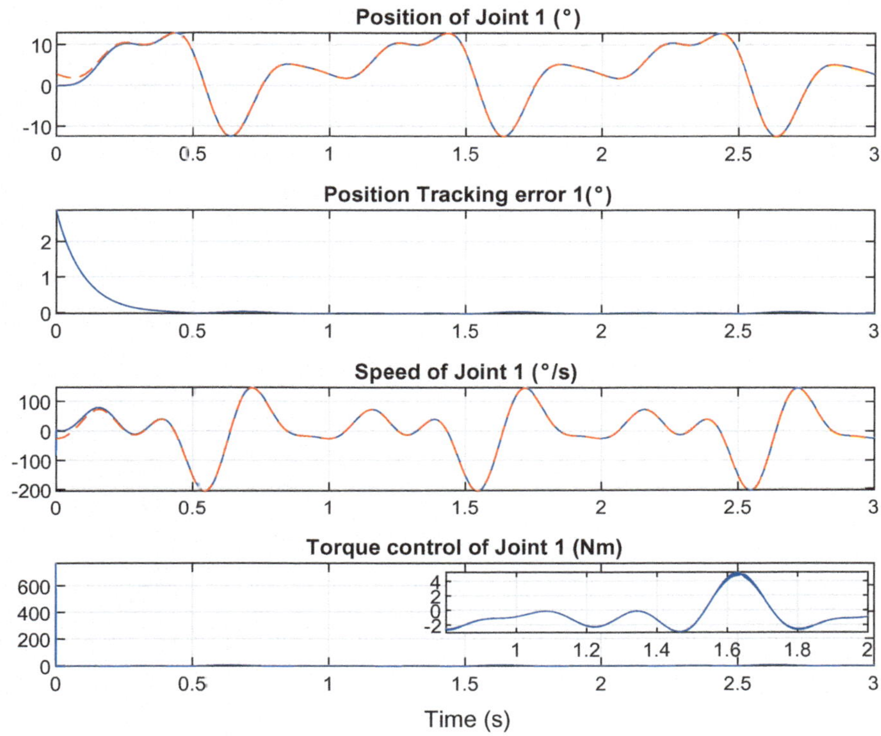


**Fig. 14** Influence of −30% of masses and −15% of length uncertainties on the hip joint tracking with augmented $L_1$ adaptive controller

where $\Gamma$ is the adaptive gain.

Let's $P$ the solution of the algebric Lyapunov equation:

$$A_m^T P + P A_m = -Q \tag{29}$$

The adaptive control is as follows:

$$\tau_{ad}(s) = C(s)\hat{\eta}(s) \tag{30}$$

where $\hat{\eta}(s)$ is the Laplace transform of $\hat{\eta}(t) = \hat{\theta}(t)\|r_t\|_\infty + \hat{\sigma}(t)$, and $C(s)$ is a bounded-input bounded-output (BIBO) stable strictly proper transfer function.

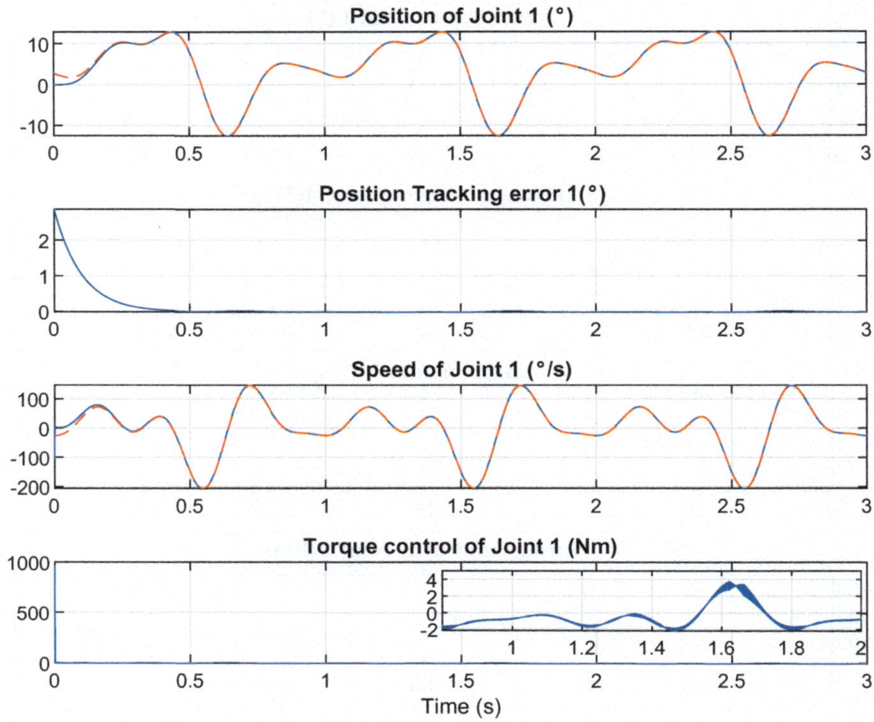


**Fig. 15** Influence of −50% of masses and −25% of length uncertainties on the hip joint tracking with augmented $L_1$ adaptive controller

## *3.4 Augmented $L_1$ Adaptive Control [11]*

The implementation of the classical $L_1$ adaptive control indicates the presence of a time lag. Hence, the idea is to develop an augmented version of the $L_1$ adaptive control. Figure 3 shows the block diagram of the augmented $L_1$ adaptive control. In fact, a linear PI controller is used as an additional part to the filtered control input.

The expression of the torque becomes:

$$\tau(t) = A_m r(t) + \tau_{ad}(t) + K_p(q - q_d) + K_i \int (q - q_d)dt \tag{31}$$

where $K_p$ and $K_i \in \mathbb{R}^{2\times2}$ are diagonal positive definite matrices.

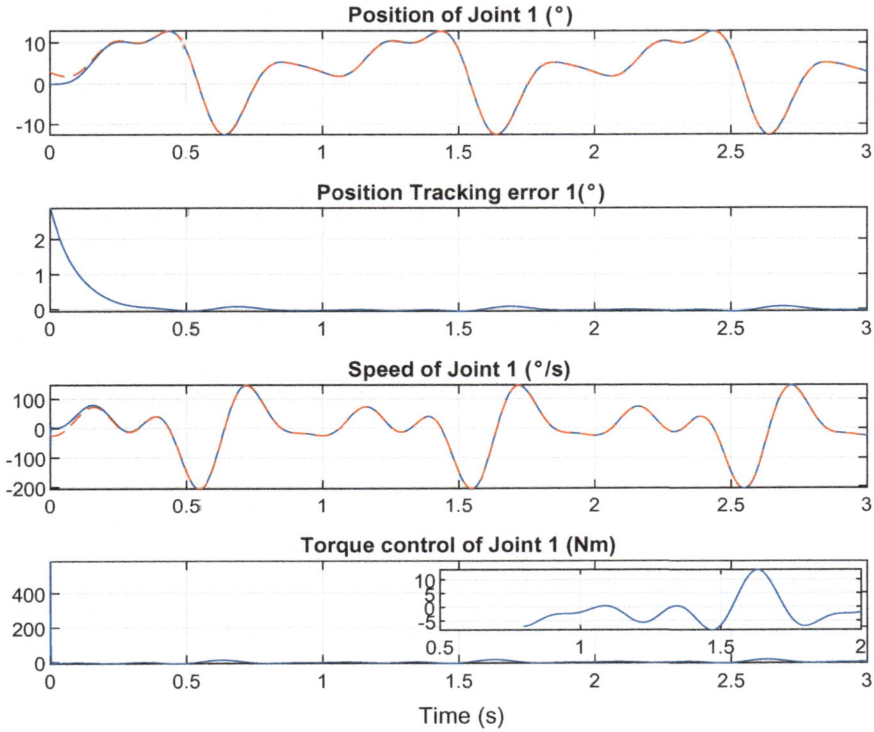


**Fig. 16** Influence of 30% of masses and 15% of length uncertainties on the hip joint tracking with augmented $L_1$ adaptive controller

## 4 Simulation Results

Lower limb exoskeletons are actuated at the knee and hip joints. To validate the different controllers proposed in the previous section, some simulations are presented. First, Figs. 4 and 5 show respectively the position evolution of the hip and the knee joints, the tracking error, their speed and their applied torques using the first approach of Slotine in the nominal case. Besides, Fig. 6 show the convergence of the estimated unknown parameter to the system parameters.

The second part consists in the adaptive control using an integral action. Results are presented in Figs. 7, 8 and 9.

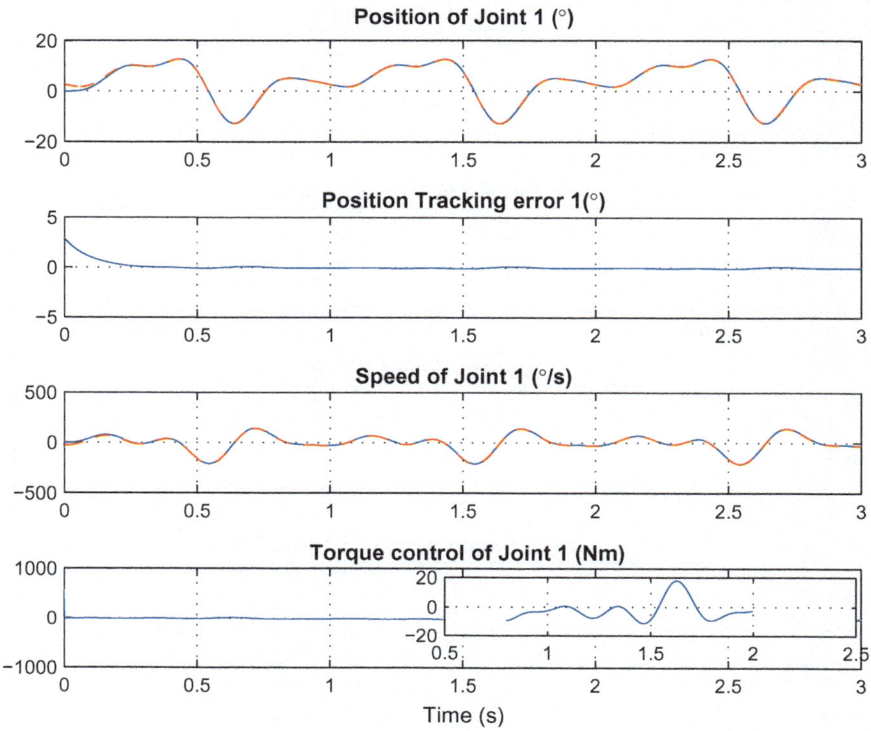


**Fig. 17** Influence of 50% of masses and 25% of length uncertainties on the hip joint tracking with augmented $L_1$ adaptive controller

These figures show a good tracking of the desired trajectories using the second approach of Slotine.

Then, a classical $L_1$ adaptive control is implemented. Figures 10 and 11 show a time lag between the actual and the desired trajectory. Hence, the idea of implementing an augmented $L_1$ adaptive control. The obtained results are illustrated in Figs. 12 and 13, and show a good tracking of the desired reference trajectories.

In order to prove the robustness of the proposed extended $L_1$ adaptive control. First, we demonstrate the performance of the proposed controllers using three criteria named as (i) the Integral of the Absolute Error (IAE), (ii) the Integral of the Squared Error (ISE) and (iii) the relative integral of absolute error (IAER):

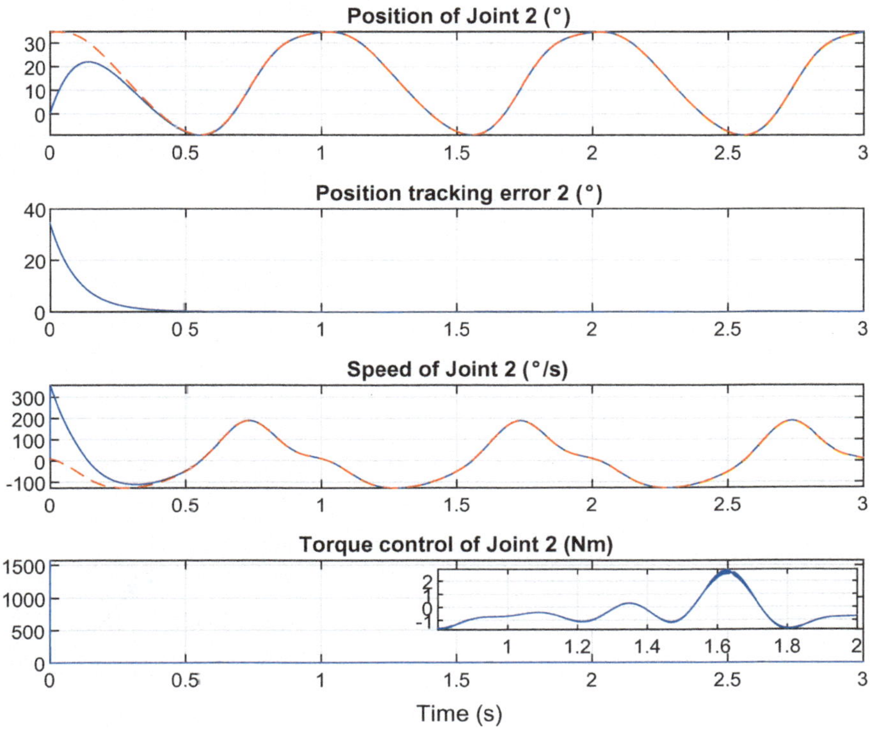


**Fig. 18** Influence of −30% of masses and −15% of length uncertainties on the knee joint tracking with augmented $L_1$ adaptive controller

$$IAE = \int |e|dt \tag{32}$$

$$ISE = \int e^2 dt \tag{33}$$

$$IAER = \frac{\int |e|dt}{\int |q_d|dt} \tag{34}$$

$e$ denotes the tracking error.

The obtained results are summarized in Table 1.

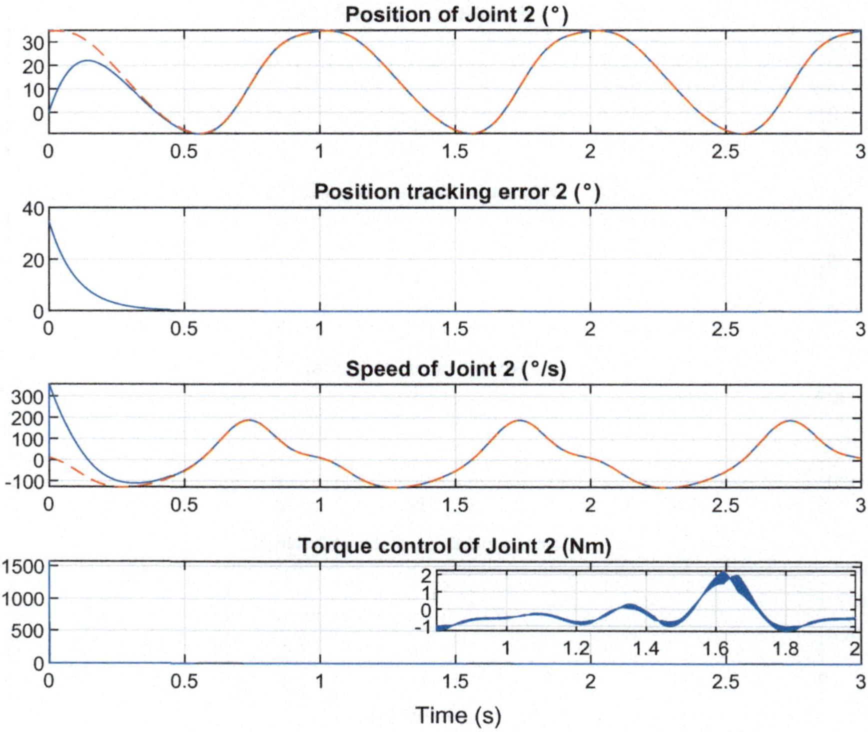


**Fig. 19** Influence of −50% of masses and −25% of length uncertainties on the knee joint tracking with augmented $L_1$ adaptive controller

Figures 14, 15, 16 and 17 present the evolution of positions, velocities, applied torques and position errors of the hip joint in the presence of mass variations of ±30 and ±50% and leg length variations of ±15 and ±25%.

Figures 18, 19, 20 and 21 present the evolution of positions, velocities, applied torques and position errors of the knee joint in the presence of mass variations of ±30 and ±50% and leg length variations of ±15 and ±25%.

In both cases, results show a good tracking of the desired trajectories in the presence of parametric variations with performant torques values. Thus, it is well obvious the good performances observed by the augmented $L_1$ adaptive control.

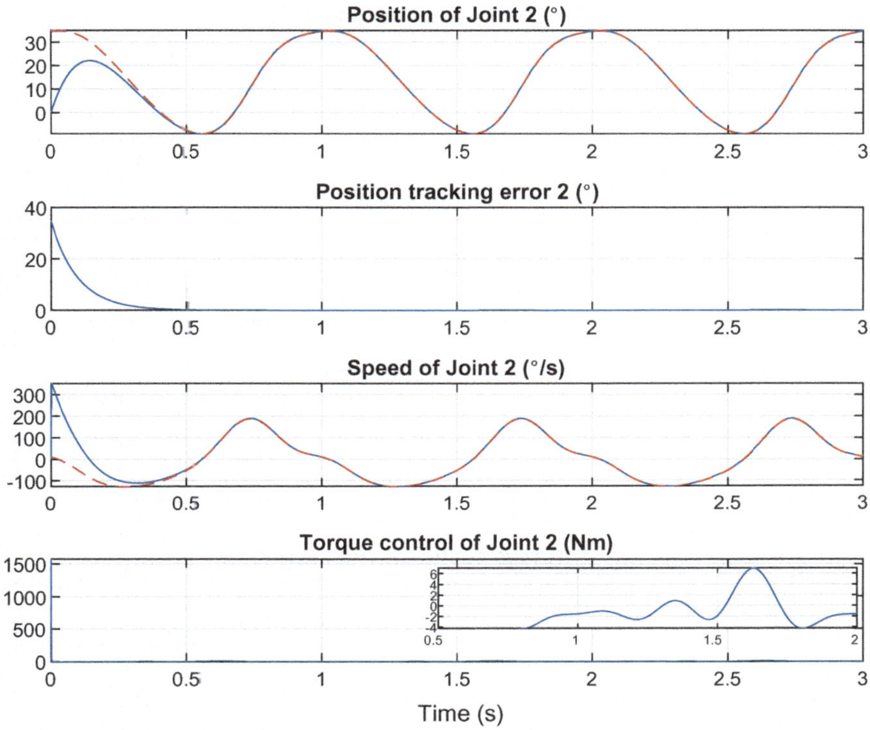


**Fig. 20** Influence of 30% of masses and 15% of length uncertainties on the knee joint tracking with augmented $L_1$ adaptive controller

## 5 Conclusion and Future Work

In this work, four adaptive controllers have been proposed and implemented to control an exoskeleton for kids rehabilitation. These control laws have been tested for the tracking of walking cycle desired trajectory. This study concerns kids who have between 2 and 13 years old (i.e. with different morphologies). Through numerical simulations, it has been shown that the augmented $L_1$ adaptive controller is the best one in terms of robustness against these parametric variations. Future works aim to validate the proposed approaches on a real exoskeleton, as well as the application to the targeted kids therapy.

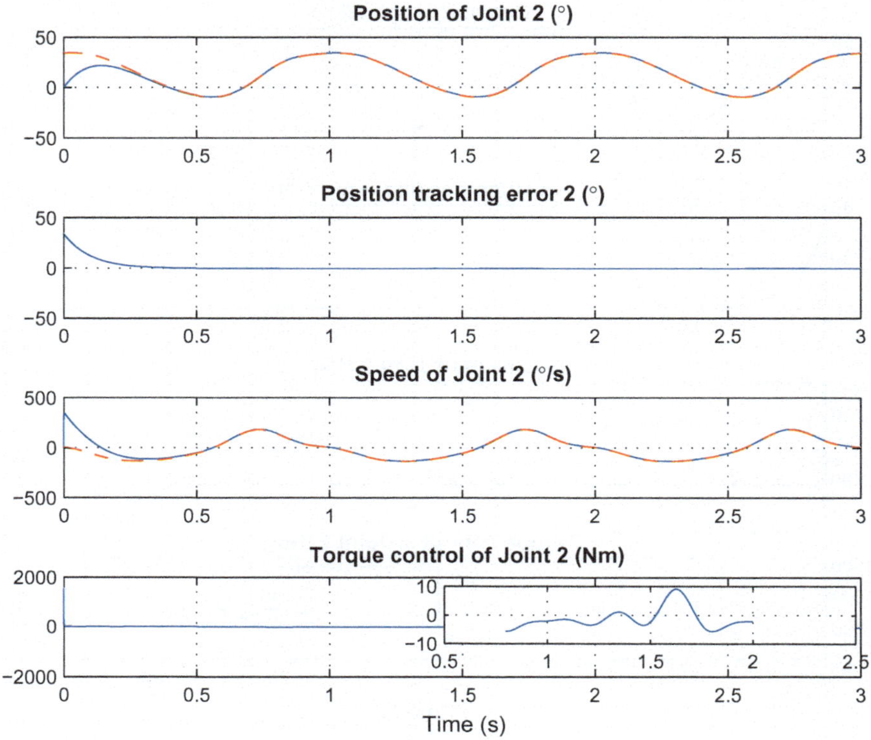


**Fig. 21** Influence of 50% of masses and 25% of length uncertainties on the knee joint tracking with augmented $L_1$ adaptive controller

**Acknowledgements** The present work is supported by (i) the "Association de Sauvegarde des Handicapés Moteurs - Sfax" (ASHMS), the (ii) Clinical Inverstigation Center (CIC) of the Hospitalo–University Center of Sfax (CHU) Tunisia, (iii) the Laboratory "Control & Energy Managements" (CEMLab) of the "National School of Engineering of Sfax", University of Sfax, Tunisia, and (iv) the Digital Research Center of Sfax, Tunisia (CRNS).

# References

1. Adams, E.: Power-multiplying exoskeletons are slimming down for use on the battlefield (2017). https://www.popsci.com/army-exoskeletons-lockheed-martin [online]
2. Duschau-Wicke, A., Brunsch, T., Lünenburger, L., Riener, R.: Adaptive support for patient-cooperative gait rehabilitation with the lokomat. In: IEEE International Conference on Intelligent Robots and Systems, pp. 2357–2361 (2007)
3. Zeilig, G., Weingarden, H., Obuchov, A., Bloch, A., Gaides, M., Reuveny, R., Ben-Dov, I.: Lokomat walking results in increased metabolic markers in individuals with high spinal cord injury. In: International Conference on Virtual Rehabilitation, ICVR, pp. 119–120 (2015)

4. Refai, H., Mohammed, S., Daachi, B., Amirat, Y.: Adaptive control of a human-driven knee joint orthosis. In: IEEE International Conference on Robotics and Automation, pp. 247.86–2491 (2012)
5. Slotine, J.-J.E., Li, W.: On the adaptive control of robot manipulators. Int. J. Robot. Res., 49–59 (1987)
6. Hovakimyan, N., Cao, C.: L1 adaptive control theory guaranted robustness with fast adaptation. Adv. Des. Control (2010)
7. Refai, H., Ben Abdessalem, M.S., Chemori, A., Mohammed, S., Amirat, Y.: Augmented L1 adaptive control of an actuated knee joint exoskeleton: from design to real-time experiments. In: IEEE International Conference on Robotics and Automation, ICRA, pp. 5707.8–5714 (2016)
8. Deep, A., Jaswal, R.:Role of management and virtual space for the rehabilitation of children affected with cerebral palsy: a review. In: IEEE International Conference on Signal Processing, Computing and Control, pp. 293–299 (2017)
9. Tucker, M., Olivier, J., Pagel, A., Bleuler, H., Bouri, M., Lambercy, O., Millán, J., Riener, R., Vallery, H., Gassert, R.: Control strategies for active lower extremity prosthetics and orthotics: a review. J. Neuroeng. Rehabil., 1–12 (2015)
10. Ghezal, M., Guiatni, M., Boussioud, I., Renane, C.S.: Design and robust control of a 2 DOFs lower limb exoskeleton. In: International Conference on Communications and Electrical Engineering (2018)
11. Bennehar, M., Chemori, A., Pierrot, F.: L1 adaptive control of parallel kinematic manipulators: design and real-time experiments. In: IEEE International Conference on Robotics and Automation, pp. 157.87–1592 (2015)
12. Hesse, S., Schmidt, H., Werner, C., Bardeleben, A.: Upper and lower extremity robotic devices for rehabilitation and for studying motor control. Curr. Opin. Neurol., 705–710 (2003)
13. Jamshidi, N., Rostami, M., Najarian, S., Menhaj, M.B., Saadatnia, M., Firooz, S.: Modelling of human walking to optimise the function of ankle-foot orthosis in Guillan-Barré patients with drop foot. Singap. Med. J. **50**(4), 412–737 (2009)
14. Rupal, B., Rafique, S., Singla, A., Singla, E., Isaksson, M., Virk, G.: Lower-limb exoskeletons: research trends and regulatory guidelines in medical and non-medical applications. Int. J. Adv. Robot. Syst., 1–27 (2017)

# Exoskeletons Control via Computed Torque for Lower Limb Rehabilitation

**Racem Jribi, Boutheina Maalej and Nabil Derbel**

**Abstract** Several are the diseases that cause difficulties in walking. This chapter concerns the assistance of deficient kids using a robotic system. In this context, the lokomat appears to be the most suitable system for kids suffering from cerebral palsy. Hence, PID controllers based on feedback linearization and adaptive feedback linearization approaches have been proposed to control 2 DOF exoskeletons which are considered as the fundamental part of the lokomat system. Besides, in order to test the effectiveness and the robustness of these controllers, several cases have been proposed. Simulations were carried out to compare and to show the well desired trajectory tracking in these cases.

**Keywords** Rehabilitation · Cerebral palsy · Robotic gait training · Exoskeletons · PID controllers · Perturbation · Adaptive feedback linearization

R. Jribi · B. Maalej (✉) · N. Derbel
Laboratory of Control & Energy Management, ENIS, University of Sfax, Sfax, Tunisia
e-mail: maalej.boutheina@gmail.com

R. Jribi
e-mail: jribi.racem@gmail.com

N. Derbel
e-mail: n.derbel@enis.rnu.tn

R. Jribi · B. Maalej
University of Gabes, Gabes, Tunisia

B. Maalej · N. Derbel
Digital Research Center of Sfax, Sfax, Tunisia

B. Maalej
Clinical Investigation Center, Sfax, Tunisia



J. Ghommam et al. (eds.), *New Trends in Robot Control*, Studies in Systems, Decision and Control 270, https://doi.org/10.1007/978-981-15-1819-5_7

# 1 Introduction

In the last few years, there has been a growing interest in robotic applications. Robots are nowadays present at home, providing different services in various fields such as supervision, leisure and even education [1]. The development of different categories of robots is on the rise, in particular medical robots used for rehabilitation. It is also proposed to maintain or even to enhance the quality of health services. Several studies show that the robotic rehabilitation is more efficient than traditional methods [2]. In fact, active assisted mode is better than passive mode when patient acts alone. It is to be mentioned that there are two types of robotic systems for rehabilitation. Lower limbs rehabilitation to improve walking and assist handicaps and upper limbs rehabilitation to allow patients to grasp and handle objects [3].

In this chapter, we are interested in lower limbs rehabilitation. Among the robotic devices used in this context, we cite:

- Andago is a mobile robot used for body-weight supported gait training [4].
- Ekso Bionics is an individual and autonomous portable assistance exoskeleton [5].
- Erigo pro includes four therapies: a motorized verticalization, a motorized mobilization, a programmable charging and a synchronized electro stimulation [6].
- Lokomat the most advanced system used for kids rehabilitation [7].

The objective of this chapter is to provide an overview of the robotic gait training for kids affected by cerebral palsy which is caused by the brain injury or the brain malformation is the common disease that affects kids before, during, or immediately after birth, while the infant's brain is under development [8]. In fact, many experiments are carried out on cats with a complete spinal cord section and in the absence of control of supraspinal centers and it has shown that it is possible to recover locomotor activity by motor training [9]. A short description of the lokomat system will be presented as well as different control modes existing to quantify and to modify the support amount that patients receive during walking, then simulation results of the proposed PID controllers have been presented [1].

# 2 Lokomat System and Different Control Modes

"Learning a specific task" is the best method for gait rehabilitation. In fact, this strategy provides an intensive functional locomotor [10]. Using Lokomat allows patients to combine treadmill rehabilitation which is synchronized with two exoskeletons with body weight suspension. So, to ensure a better autonomy for patients with muscular inability during their daily movements, exoskeletons seem to be an essential part which guide and support patients in a vertical position. This rehabilitation method is based principally on a cyclic motion automation process that allows for active and progressive sensor-motor (re)learning for patients [11].

## 2.1 System Description

### 2.1.1 System Components

Figure 1 shows the principal parts of the robotic system of rehabilitation:

- A motorized and programmable exoskeletons actuated at hip and knee joint level. They are composed of force sensors integrated into the motor at the joints, thus continuously measuring the degree of joint amplitude, the degree of participation or resistance of the subject during walking cycles. Locomotor movements are therefore guided according to a specified treadmill speed and time walking phases which correspond progressively to the normalized walking pattern [12].
- A suspension harness providing an adjustment of the patient's body weight. This body weight reduction, 40% of the weight of the maximum body overcomes the patient's muscle deficit [13].
- A motorized treadmill with a walking speed of 0.5–5 km/h (0.14–1.4 m/s). The treadmill is synchronized with the exoskeletons. Its speed can be adjusted during the session and exercises offering an adaptable walking frequency [14].
- A screen for visual feedback in order to make direct interventions on different exercises suggested by the virtual "Leader" robot.

Lokomat can be offered to kids aged from 2 to 13 years old (pediatric Lokomat). These new techniques make it possible to better understand and to better optimize potential therapeutic treatments to maintain the independence in daily activities for children until adulthood. Motor activity is therefore necessary for the general development of these children who, in addition to musculo-articular deformities, have impaired cardiorespiratory fitness for exercise combined with less muscular endurance compared to children of the same age.

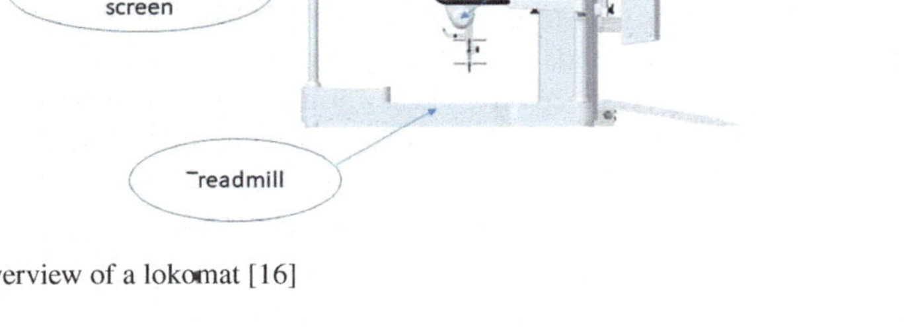


**Fig. 1** Overview of a lokomat [16]

### 2.1.2 Control Modes

The currently available control modes are the guidance force, the path control and FreeD module [15]. For the guidance force, knee and hip position joints are exactly specified, in fact, the lower limbs move in a predetermined cyclic trajectory. So, small deviations are allowed and when the deviation increases, the force that pushes the patient's leg back along its trajectory increases [16]. Contrary to the previous control mode, in path control mode, trajectories can be modified: it is about spatial and temporal freedom while a moving force guides patients to walk at the speed of the treadmill [17]. Finally, freeD module which makes the walking pattern more physiological and natural. The pelvis is movable with a lateral translation up to 4 cm (per side) and pelvic rotation up to 4° (per side) to ensure a natural lateral movement with a rhythmic movement of the weight to ensure more balance [20].

## *2.2 Advantages of the System*

Several are the advantages of this robotic system. It is used in a wide range of indications (Cerebrovascular accident, spinal cord injury, cognitive disorders, traumatic head injury or multiple sclerosis…) [21]. The case treated in this chapter is about kids suffered from cerebral palsy [22]. In fact, using this robotic system relief the therapeutic team especially in case of obese or spastic children. Second, processes plasticity activation with the high number of repetitions and the intensive work, hence, 40% approximately have the chance to improve their ability of walk. Moreover, through the coordination of the movements, it guarantees a natural and safe neuromuscular rehabilitation.

# 3 Exoskeleton Control

## *3.1 System Modeling*

The basic part of the robotic system is the exoskeletons. In the following, it will be considered as a system with two manipulator arms, where $l_1$ corresponds to the distance hip-knee and $l_2$ corresponds to the distance knee-foot, and $m_1$ and $m_2$ present thigh mass and the calf mass, respectively. These variables are calculated as the mean values of the masses and lengths regarding to the age range of children (between 2 and 13 years old).

The values of masses and lengths are found based on the following (Figs. 2 and 3):

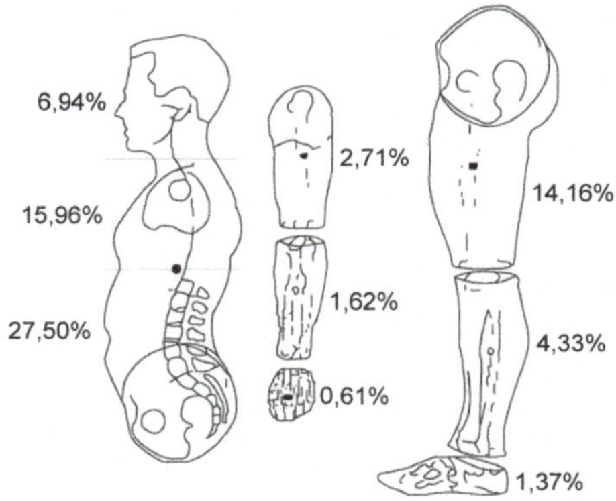


**Fig. 2** Mass coefficient (Cm) of DeLeva [18]

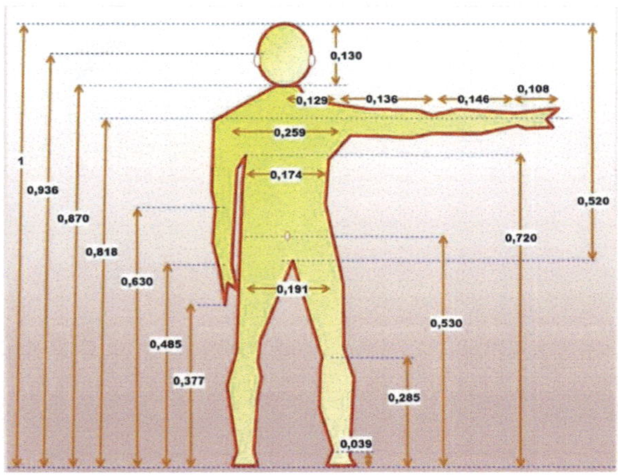


**Fig. 3** Human body proportions [19]

$$
\begin{aligned}
l_1 &= 0.245\, l_{total} \\
l_2 &= 0.246\, l_{total} \\
m_1 &= 0.1416\, m_{total} \\
m_2 &= 0.0433\, m_{total}
\end{aligned}
$$

where: $l_{total}$ and $m_{total}$ are 86 cm and 12 kg for a 2-year old child. They are equal to 145 cm and 38 kg for children who have 13 years old.

## 3.2 Dynamic Model

The dynamic model is written as (Fig. 4):

$$M(q)\ddot{q} + C(q, \dot{q})\dot{q} + G(q) = \tau \tag{1}$$

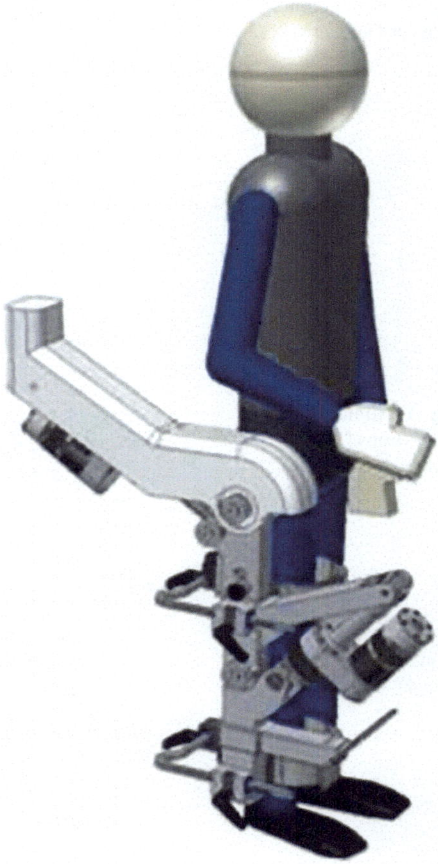

**Fig. 4** Exoskeleton with 2 joints [24]

where:

$q = [q_1\ q_2]^T \in \mathbb{R}^2$ is the position vector,
$\dot{q} = [\dot{q}_1\ \dot{q}_2]^T \in \mathbb{R}^2$ is the speed vector,
$\ddot{q} = [\dot{q}_1\ \dot{q}_2]^T \in \mathbb{R}^2$ is the acceleration vector,
$M(q) \in \mathbb{R}^2$ is the inertia matrix, which is symmetric, uniformly bounded and positive definite,
$C(q, \dot{q})\dot{q} \in \mathbb{R}^2$ represents the Coriolis, centrifugal forces and torques,
$G(q) \in \mathbb{R}^2$ denotes the gravity torque vector.
$\tau \in \mathbb{R}^2$ is the vector of actuator torques.

Components of matrix $M(q)$, and vectors $C(q,\dot{q})\dot{q}$ and $G(q)$ are:

$$M_{11} = J_1 + J_2 + m_2(l_1^2 + d_2^2 + 2l_1d_2\cos q_2) + m_1d_1^2$$
$$M_{12} = M_{21} = J_2 + m_2(d_2^2 + l_1d_2\cos q_2)$$
$$M_{22} = J_2 + m_2d_2^2$$

$$C_1 = -\left[2m_2l_1d_2\dot{q}_2\sin q_2\right]\dot{q}_1 - \left[m_2l_1d_2\dot{q}_2\sin q_2\right]\dot{q}_2$$
$$C_2 = \left[m_2l_1d_2\sin q_2\right]\dot{q}_1^2$$

$$G_1 = (m_1d_1 + m_2l_1)g\sin q_1 + m_2d_2g\sin(q_1 + q_2)$$
$$G_2 = m_2d_2g\sin(q_1 + q_2)$$

In the literature, there are several approaches to control these exoskeletons [23, 25–27].

In the following, PD and PID controllers are expressed by defining a desired dynamic of the trajectory error. The proposed control can be described as a feedback linearization with the pole placement procedure.

It is to be noted that a saturation block has been added before applying the control law given by the controller. In fact, the actuators present some limitations and cannot support high torques.

## 3.3 PD Controllers

The desired trajectory is defined by the position $q_d$, its speed $\dot{q}_d$ and its acceleration $\ddot{q}_d$.

The trajectory error is defined by the position error expressed by $e = q - q_d$, the speed error which is $\dot{e} = \dot{q} - \dot{q}_d$ and the acceleration error $\ddot{e} = \ddot{q} - \ddot{q}_d$.

The desired dynamic of the error is defined by:

$$\ddot{e} + k_d\dot{e} + k_pe = 0 \tag{2}$$

$k_d$ and $k_p$ are positive gains defining the dynamical behaviour and the stability of the closed loop system.

Using Eq. (1), one can easily write:

$$\ddot{q} - \ddot{q}_d + k_d\dot{e} + k_pe = 0 \tag{3}$$

Then:

$$[M(q)]^{-1}\left[\tau - C(q,\dot{q})\dot{q} - G(q)\right] - \ddot{q}_d + k_d\dot{e} + k_p e = 0 \tag{4}$$

This gives the required control expressed as:

$$\tau = M(q)\left[\ddot{q}_d - k_d\dot{e} - k_p e\right] + C(q,\dot{q})\dot{q} + G(q) \tag{5}$$

## 3.4 PID Controllers

The desired dynamic of the trajectory error is defined by:

$$\dddot{e} + k_d\ddot{e} + k_p\dot{e} + k_i e = 0 \tag{6}$$

$k_d$, $k_p$ and $k_i$ are positive gains defining the dynamical behaviour and the stability of the closed loop system. This equation is yielded from:

$$\ddot{e} + k_d\dot{e} + k_p e + k_i\int e\,dt = 0 \tag{7}$$

Substituting $\ddot{q}$ by its expression given by Eq. (1) yields:

$$[M(q)]^{-1}\left[\tau - C(q,\dot{q})\dot{q} - G(q)\right] - \ddot{q}_d + k_d\dot{e} + k_p e + k_i\int e\,dt = 0 \tag{8}$$

This gives the required control expressed as:

$$\tau = M(q)\left[\ddot{q}_d - k_d\dot{e} - k_p e - k_i\int e\,dt\right] + C(q,\dot{q})\dot{q} + G(q) \tag{9}$$

## 3.5 Adaptive Feedback Linearization Controllers

In this section an adaptive feedback linearization control is proposed. In fact, since kids parameters vary, a robust controller is proposed. The objective of this approach, first, is to transform the nonlinear dynamic system into a linear dynamic one, then, an online estimation of the parameters is done. The desired dynamic is defined by Eq. (6).

Considering now that the system parameters are unknown, the system Eq. (1) becomes:

$$M(q,\theta)\ddot{q} + C(q,\dot{q},\theta)\dot{q} + G(q,\theta) = \tau \tag{10}$$

where $\theta \in \mathbb{R}$ is an unknown (or ill known) parameter. The case where $\theta$ becomes a vector can be easily deduced from the following reasoning. Hence, the idea consists to implement an adaptive approach using the estimation parameter $\widehat{\theta}$. An adaptation law of $\widehat{\theta}$ will determined later. The applied control law becomes:

$$\tau = M(q,\widehat{\theta})u + C(q,\dot{q},\widehat{\theta})\dot{q} + G(q,\widehat{\theta}) \tag{11}$$

with:

$$u = \ddot{q}_d - k_d\dot{e} - k_p e - k_i \int edt \tag{12}$$

Substituting Eq. (11) into Eq. (10) gives:

$$M(q,\theta)\ddot{q} + C(q,\dot{q},\theta)\dot{q} + G(q,\theta) = M(q,\widehat{\theta})u + C(q,\dot{q},\widehat{\theta})\dot{q} + G(q,\widehat{\theta}) \tag{13}$$

Then,

$$\begin{aligned} M(q,\theta)\ddot{q} + C(q,\dot{q},\theta)\dot{q} + G(q,\theta) = M(q,\widehat{\theta})\left(\ddot{q}_d - k_d\dot{e} - k_p e - k_i \int edt\right) \\ +C(q,\dot{q},\widehat{\theta})\dot{q} + G(q,\widehat{\theta}) \end{aligned} \tag{14}$$

This gives,

$$\begin{aligned} &M(q,\theta)\ddot{q} - M(q,\widehat{\theta})\ddot{q}_d + \left[C(q,\dot{q},\theta) - C(q,\dot{q},\widehat{\theta})\right]\dot{q} + \left[G(q,\theta) - G(q,\widehat{\theta})\right] = \\ &M(q,\widehat{\theta})\left(-k_d\dot{e} - k_p e - k_i \int edt\right) \end{aligned} \tag{15}$$

Hence:

$$\begin{aligned} &\ddot{q} - M^{-1}(q,\theta)M(q,\widehat{\theta})\ddot{q}_d \\ &+M^{-1}(q,\theta)\left[C(q,\dot{q},\theta) - C(q,\dot{q},\widehat{\theta})\right]\dot{q} + M^{-1}(q,\theta)\left[G(q,\theta) - G(q,\widehat{\theta})\right] \\ &+M^{-1}(q,\theta)M(q,\widehat{\theta})\left[k_d\dot{e} + k_p e + k_i \int edt\right] = 0 \end{aligned} \tag{16}$$

Let's use the Taylor expansion of $M(q,\theta)$, $C(q,\dot{q},\theta)$ and $G(q,\theta)$:

$$M(q,\theta) - M(q,\widehat{\theta}) = \frac{\partial M}{\partial \theta}\Delta\theta \tag{17}$$

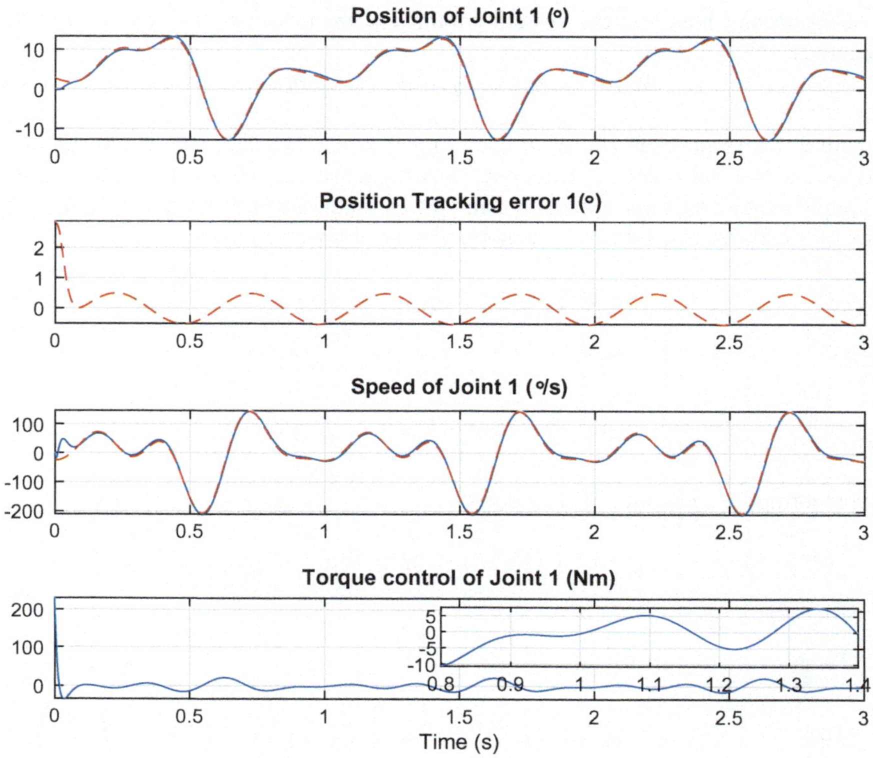


**Fig. 5** Position, speed and torque control of joint 1, yielded with PD controllers without external perturbations. Continued line actual trajectory, dashed line: desired trajectory

$$C(q,\dot{q},\theta) - C(q,\dot{q},\widehat{\theta}) = \frac{\partial C}{\partial \theta}\Delta\theta \tag{18}$$

$$G(q,\theta) - G(q,\widehat{\theta}) = \frac{\partial G}{\partial \theta}\Delta\theta \tag{19}$$

Moreover, using $MM^{-1} = I$ ($I$ is the identity matrix with an appropriate dimension), we can write successively:

$$\frac{\partial M}{\partial \theta}M^{-1}(q,\widehat{\theta}) + M(q,\widehat{\theta})\frac{\partial M^{-1}}{\partial \theta} = 0 \tag{20}$$

and:

$$\frac{\partial M^{-1}}{\partial \theta} = -M^{-1}(q,\widehat{\theta})\frac{\partial M}{\partial \theta}M^{-1}(q,\widehat{\theta}) \tag{21}$$

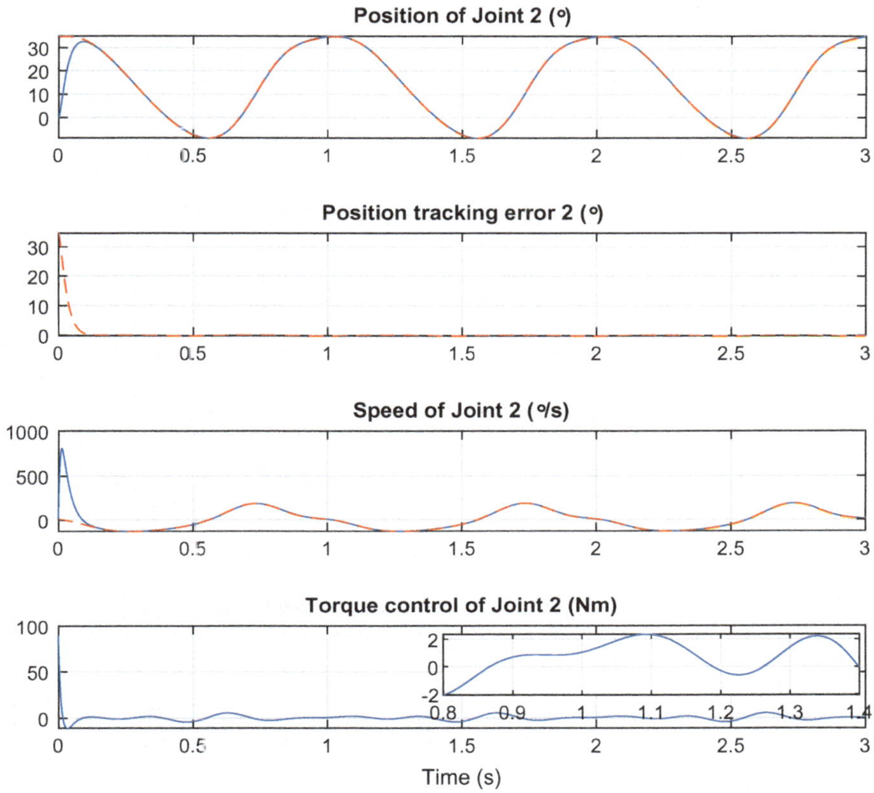


**Fig. 6** Position, speed and torque control of joint 2, yielded with PD controllers without external perturbations. Continued line actual trajectory, dashed line: desired trajectory

Thus, we obtain:

$$\ddot{e}+k_d\dot{e}+k_pe+k_i\int edt+M^{-1}(q,\widehat{\theta})\left(\frac{\partial C}{\partial\theta}\dot{q}+\frac{\partial G}{\partial\theta}\right)\Delta\theta$$
$$-M^{-1}(q,\widehat{\theta})\frac{\partial M}{\partial\theta}\left[\left(-k_d\dot{e}-k_pe-k_i\int edt\right)+\ddot{q}_d\right]\Delta\theta=0 \tag{22}$$

In the following, we consider the stability analysis using the Lyapunov theory. For that, let's consider the state vector $E$.

$$E=\begin{pmatrix}\int edt\\ e\\ \dot{e}\end{pmatrix} \tag{23}$$

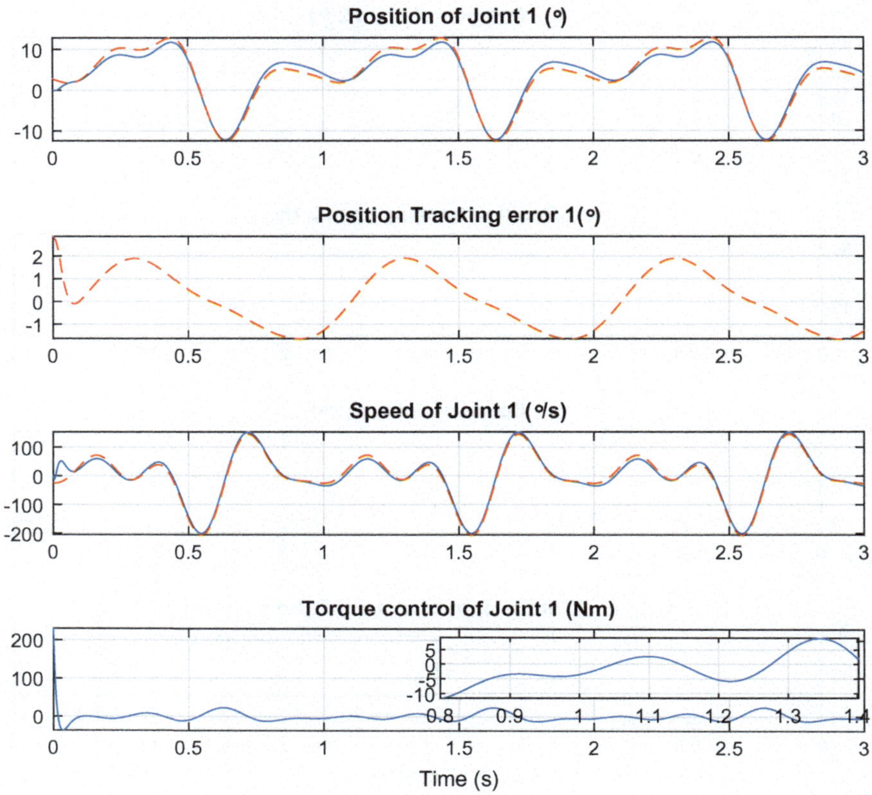


**Fig. 7** Position, speed and torque control of joint 1, yielded with PD controllers in presence of external perturbations. Continued line actual trajectory, dashed line: desired trajectory

The state equation of the system becomes:

$$\dot{E} = AE + B\Delta\theta \tag{24}$$

with:

$$A = \begin{pmatrix} 0 & I & 0 \\ 0 & 0 & I \\ -k_i I & -k_p I & -k_d I \end{pmatrix}, \quad B = \begin{pmatrix} 0 \\ 0 \\ L \end{pmatrix} \tag{25}$$

where:

$$\begin{aligned} L = & -M^{-1}(q, \widehat{\theta}) \left( \frac{dC}{d\theta}\dot{q} + \frac{G}{d\theta} \right) \\ & + M^{-1}(q, \widehat{\theta}) \frac{\partial M}{\partial \theta} \left[ \left( -k_d \dot{e} - k_p e - k_i \int edt \right) + \ddot{q}_d \right] \end{aligned} \tag{26}$$

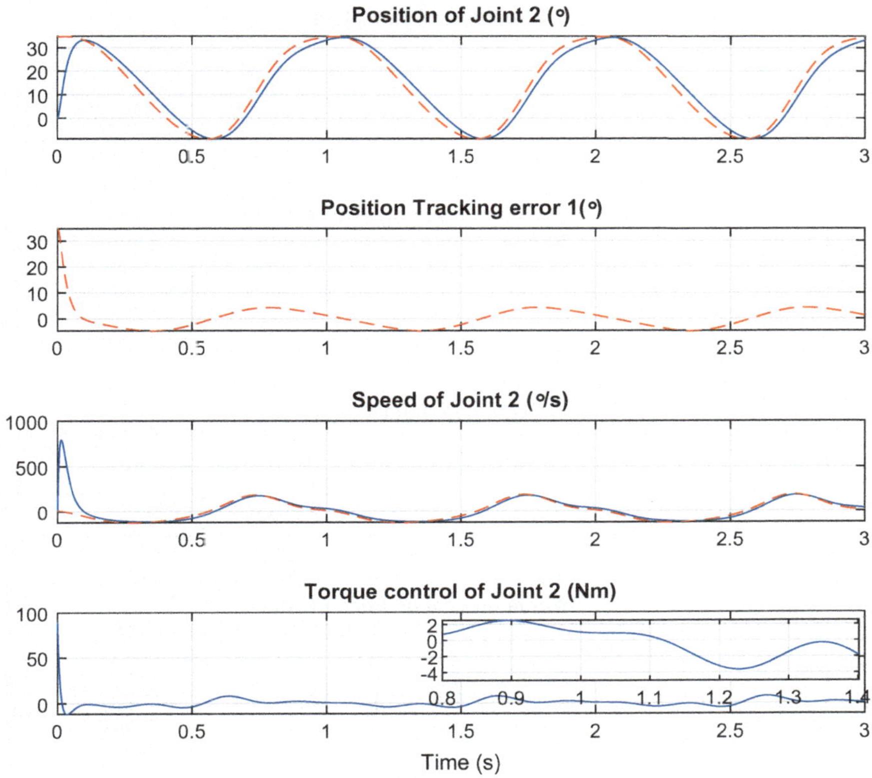


**Fig. 8** Position, speed and torque control of joint 2, yielded with PD controllers in presence of external perturbations. Continued line actual trajectory, dashed line: desired trajectory

The Lyapunov function associated to the system is

$$V = E^T P E + \frac{1}{\rho} \Delta\theta^2 \tag{27}$$

Its differential with respect to time becomes:

$$\begin{aligned} \dot{V} &= \dot{E}^T (A^T P + P A) E + 2 E^T P B \Delta\theta + \frac{2}{\rho} \Delta\theta \Delta\dot{\theta} \\ &= -E^T Q E + 2\Delta\theta \left[ \frac{\Delta\dot{\theta}}{\rho} + E^T P B \right] \end{aligned} \tag{28}$$

with the adaptation law:

$$\Delta\dot{\theta} = -\rho B^T P E \tag{29}$$

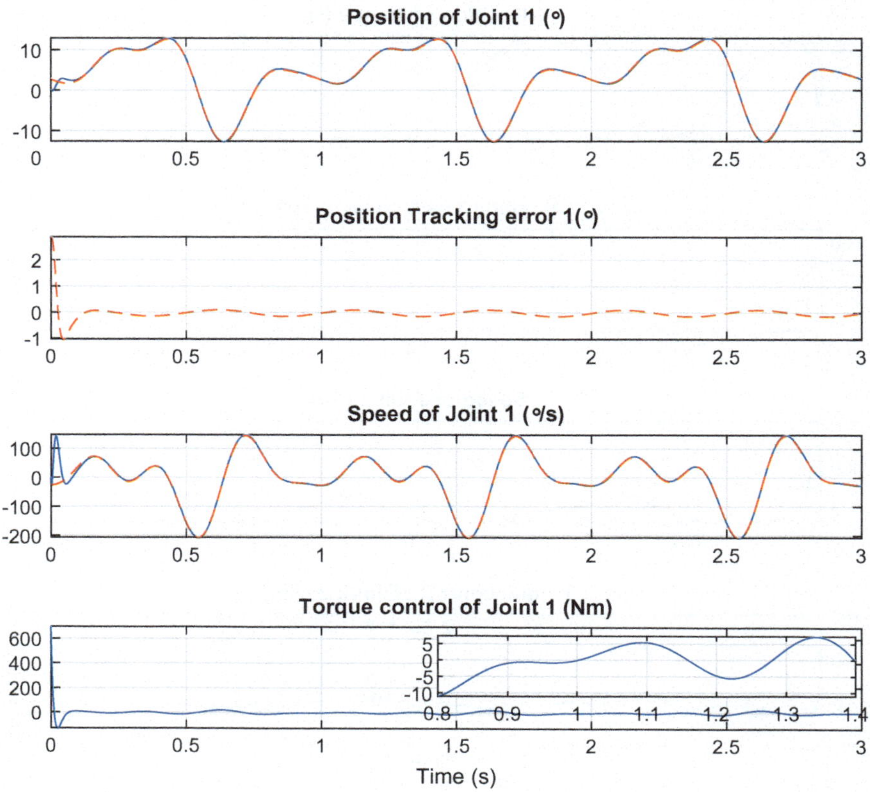


**Fig. 9** Position, speed and torque control of joint 1, yielded with PID controllers without external perturbations. Continued line actual trajectory, dashed line: desired trajectory

Assuming that the unknown parameter $\theta$ is constant or it presents slow variations, we can write $\Delta\dot{\theta} = \dot{\theta} - \dot{\hat{\theta}} \simeq -\dot{\hat{\theta}}$, the adaptation law becomes:

$$\dot{\hat{\theta}} = \rho B^T P E \tag{30}$$

Using Theorem 1 elaborated by Orlov [28] based on Krasovskii–LaSalle's invariance principle [29, 30], it has been shown that for nonautonomous systems with nonsmooth Lyapunov functions for which their time derivatives along the system trajectories are negative-semidefinite, the extended invariance principle involves a coupled indefinite function to ensure the asymptotic stability of the closed loop system.

This confirms that the adaptive closed loop system is asymptotically stable. Moreover, the proposed adaptive feedback linearization controller becomes robust, instead of the classical feedback linearization, which is well known non robust.

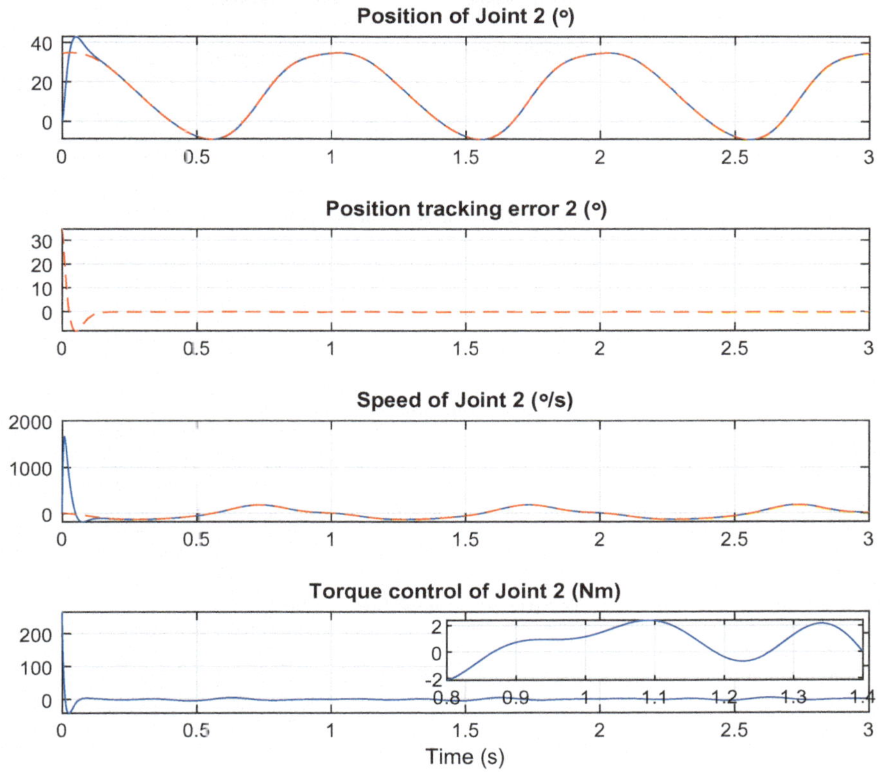


**Fig. 10** Position, speed and torque control of joint 2, yielded with PID controllers without external perturbations. Continued line actual trajectory, dashed line: desired trajectory

# 4 Simulation Results

## 4.1 PD Controllers

Figures 5 and 6 represent the evolution of positions, speeds and torques of joint 1 and joint 2, respectively. The desired trajectory describe legs'movement. It is clear that the system reaches the desired trajectory after a short time.

In order to check the robustness of the computed torque control, we consider the case of excited kids when they are afraid so they apply a perturbation torque vector $\tau_p$ on the system. The equation defining the dynamical behaviour of the system becomes:

$$M(q)\ddot{q} + C(q, \dot{q})\dot{q} + G(q) = \tau + \tau_p \tag{31}$$

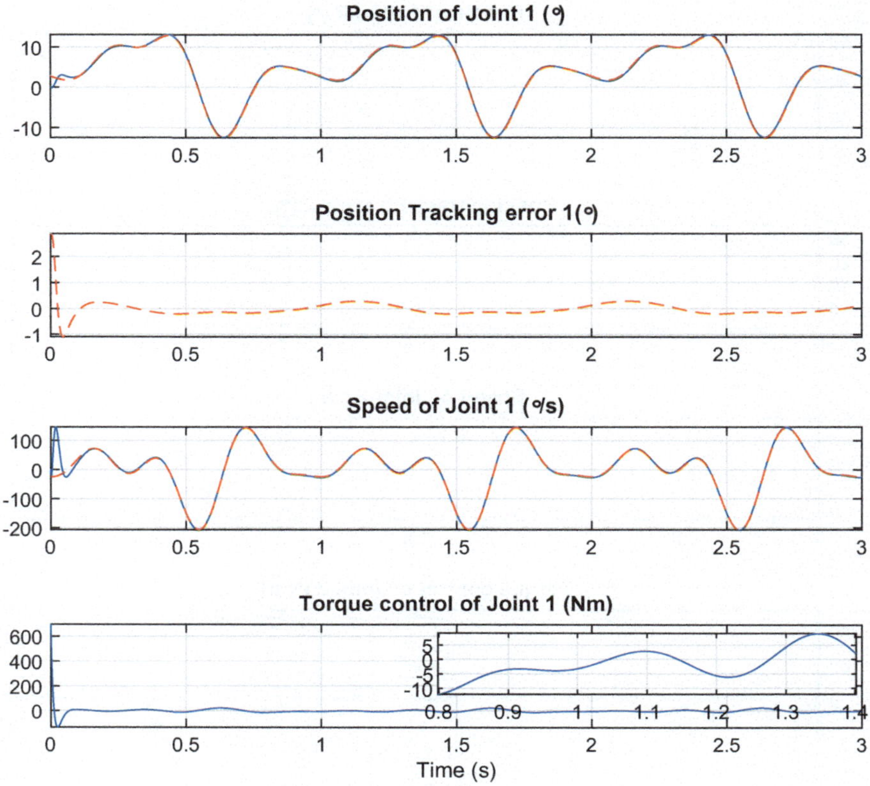


**Fig. 11** Position, speed and torque control of joint 1, yielded with PID controllers in presence of external perturbations. Continued line actual trajectory, dashed line: desired trajectory

The perturbed torque vector $\tau_p$ has been chosen having a sinusoidal evolution, explaining the rapid variation torques. Results are presented in Figs. 7 and 8. These figures illustrate the evolution of the state variables (positions and speeds of the two joints). Regarding these figures, it is clear that the system performances have been affected by the external perturbation.

## 4.2 PID Controllers

In order to improve the obtained results, we propose to use PID controllers. In fact, the integral action forces the convergence of the error to zero. Figures 9 and 10 represent the evolution of positions, speeds and torques of hip and knee joint, without applying any perturbation. Results are comparable to those obtained by PD controllers.

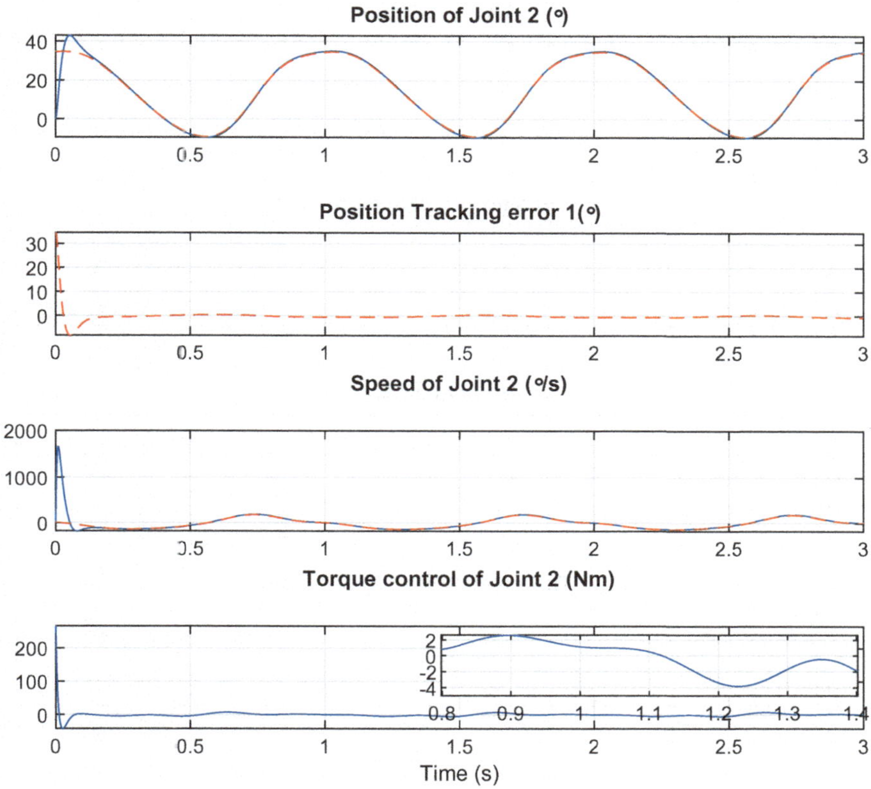


**Fig. 12** Position, speed and torque control of joint 2, yielded with PID controllers in presence of external perturbations. Continued line actual trajectory, dashed line: desired trajectory

In order to check the robustness of the used controller, we consider the perturbed case by applying a perturbation torque $\tau_p$ on the system. Results are presented in Figs. 11 and 12. These figures illustrate the evolution of positions, speeds and torques' control. Based on these figures, it is clear that the system performances become unaffected by torque perturbations comparing to those given by PD controllers.

### *4.3 Adaptive Feedback Linearization Based on PID Controller*

Figures 13 and 14 represent the evolution of positions, speeds and torques of hip and knee joint, in the presence of an additional perturbation as a bolt unscrewing [31]. The variation of the parameter $a_1$ has no effect clearly and the system converges to the desired trajectories in a short time.

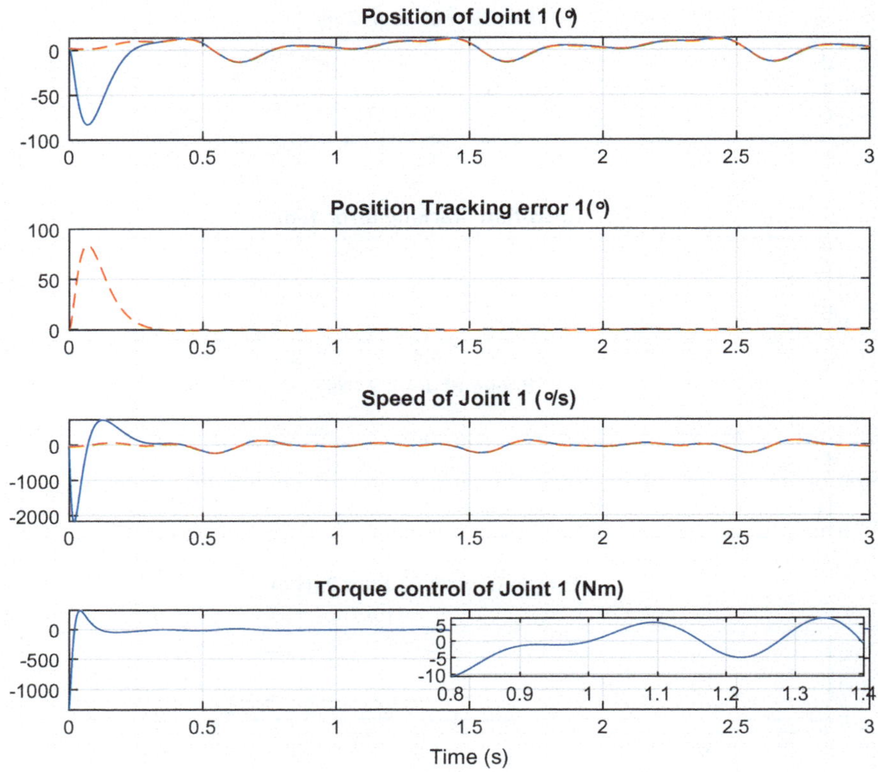


**Fig. 13** Position, speed and torque control of hip joint, in case of parameter variation. Continued line actual trajectory, dashed line: desired trajectory

# 5 Conclusion

This chapter concerns the assistance of children suffering from cerebral palsy. The object has been devoted to design and to realise a robotic system for rehabilitation. PID controllers based on the feedback linearization approach are developed to control the exoskeletons. Despite that the feedback linearization approach is not robust against perturbations, we have considered the case where kids are not totally cooperative, and we have considered that they apply external torques. Then, we have implemented an adaptive feedback linearization controller in order to deal with the perturbation case. Simulation results show the effectiveness of these controllers even with external perturbations provided by kids or perturbations such as a bolt unscrewing.

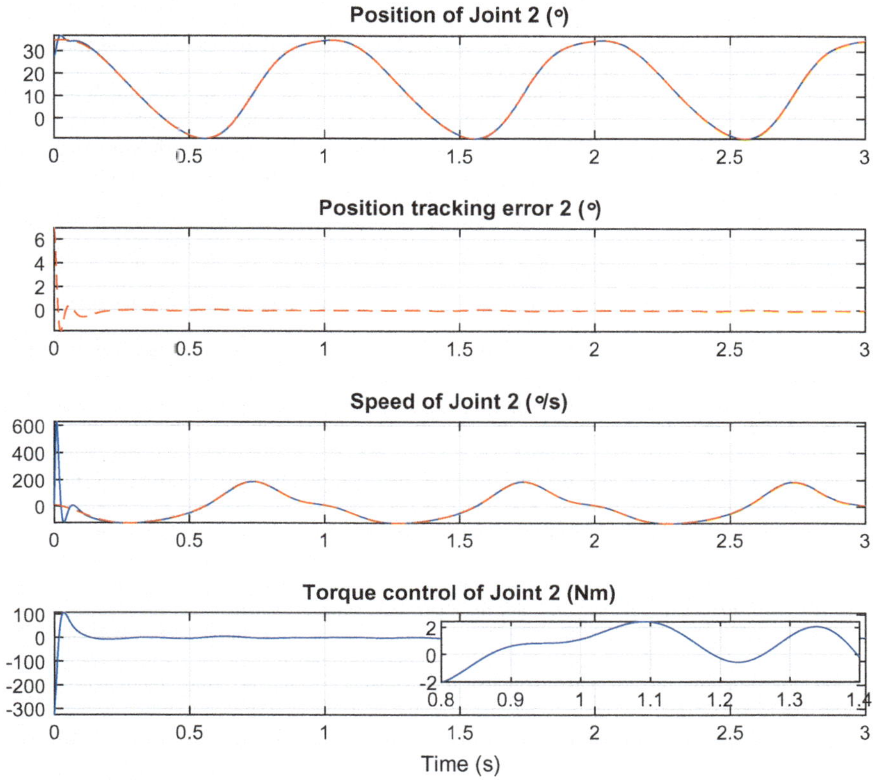


**Fig. 14** Position, speed and torque control of knee joint, in case of parameter variation. Continued line actual trajectory, dashed line: desired trajectory

**Acknowledgements** The present work is supported by (i) the "Association de Sauvegarde des Handicapés Moteurs - Sfax" (ASHMS), the (ii) Clinical Investigation Center (CIC) of the Hospitalo—University Center of Sfax (CHU) Tunisia, (iii) the Laboratory "Control & Energy Managements" (CEMLab) of the "National School of Engineering of Sfax", University of Sfax, Tunisia, and (iv) the Digital Research Center of Sfax, Tunisia (CRNS).

# References

1. Maalej, B., Jribi, R., Ayadi, N., Abdelhedi F., Derbel, N.: On a robotic application for rehabilitation systems dedicated to kids affected by cerebral palsy. In: International Multi-Conference on Systems, Signals & Devices, pp. 414–419 (2018)
2. Bouteraa, Y., Ben Abdallah, I.: Exoskeleton robots for upper-limb rehabilitation. In: International Multi-Conference on Systems, Signals & Devices (2016)
3. Hesse, S., Schmidt, H., Werner, C., Bardeleben, A.: Upper and lower extremity robotic devices for rehabilitation and for studying motor control. Curr. Opin. Neurol. **16**, 705–710 (2003)

4. Fuhrmann, M.: Andago - the first of its kind mobile robot for rehabilitation is now available in USA. https://www.businesswire.com/news/home/20160322006027/en/Andago---Kind-Mobile-Robot-Rehabilitation-USA' (2016)
5. Strickland, E.: Good-bye, Wheelchair, Hello Exoskeleton. https://spectrum.ieee.org/biomedical/bionics/goodbye-wheelchair-hello-exoskeleton (2011)
6. Colombo, G., Schreier, R., Mayr, A., Plewa, H., Rupp, R.: Novel tilt table with integrated robotic stepping mechanism: design principles and clinical application. In: 9th International Conference on Rehabilitation Robotics, pp. 227–230 (2005)
7. Bernhardt, M., Frey, M., Colombo, G., Riener, R.: Hybrid force-position control yields cooperative behaviour of the rehabilitation robot lokomat. In: International Conference on Rehabilitation Robotics, pp. 536–539 (2005)
8. Novak, I.: Evidence-based diagnosis, health care, and rehabilitation for children with cerebral palsy. J. Child Neurol. **29**, 1141–1156 (2014)
9. Whelan, P.J.: Control of locomotion in the decerebrate cat. Prog. Neurobiol. **49**, 481–515, 199 (1996)
10. Marder, E., Bucher, D.: Central pattern generators and the control of rhythmic movements. Elsevier Sci. **11**, 986–996 (2001)
11. Riener, R., Duschau-Wicke, A., König, A., Bolliger, M., Wieser, M., Vallery, H.: Automation in rehabilitation: how to include the human into the loop. In: World Congress on Medical Physics and Biomedical Engineering, pp. 180–183 (2009)
12. Zhang, X., Yue, Z., Wang, J.: Robotics in lower-limb rehabilitation after stroke. Behav. Neurol. (2017)
13. Arakelian, V., Briot, S.: Balancing of Linkages and Robot Manipulators: Advanced Methods with Illustrative Examples, pp. 44–48. Springer, Berlin (2015)
14. Colombo, G., Joerg, M., Schreier, R., Dietz, V.: Treadmill training of paraplegic patients using a robotic orthosis. J. Rehabil. Res. Dev. **37**(6), 693–700 (2000)
15. Système de rééducation démarche robotisé. http://www.medicalexpo.fr/prod/hocoma/product-68750-575453.html
16. Beyl, P., Van Damme, M., Van Ham, R., Vanderborght, B., Lefeber, D.: Design and control of a lower limb exoskeleton for robot-assisted gait training. Appl. Bionics Biomech. **6**(2), 229–243 (2009)
17. Long, Y., Du, Z.-j., Wang, W., Dong, W.: Development of a wearable exoskeleton rehabilitation system based on hybrid control mode. Int. J. Adv. Robot. Syst. 1–10 (2016)
18. Biomécanique 3D appliquée aux STAPS. http://biomecanique3d.univ-lyon1.fr/webapp/website/website.html?id=1599660&pageId=1898
19. Nombres - curiosités, théorie et usages. http://villemin.gerard.free.fr/Biologie/CorpsPro.htm
20. Duschau-Wicke, A., von Zitzewitz, J., Caprez, A., Lünenburger, L., Riener, R.: Path control: a method for patient-cooperative robot-aided gait rehabilitation. IEEE Trans. Neural Syst. Rehabil. Eng. **18**(1) (2010)
21. Aurich-Schuler, T., Grob, F., van Hedel, H.J.A. Labruyère, R.: Can Lokomat therapy with children and adolescents be improved? An adaptive clinical pilot trial comparing Guidance force, Path control, and FreeD. J. Neuro Eng. Rehabil. (2017)
22. Barbeau, H., Visitin, M.: Optimal outcomes obtained with body-weight support combined with treadmill training in stroke subjects. Arch. Phys. Med. Rehabil. **84**(10), 1458–1465 (2003)
23. Wright, F.V.: Adopting new technologies, "Techno-partnering" with a new robotic-assisted treadmill gait trainer for children with cerebral palsy: what paediatric physiotherapists need to consider. Physiotherapy Practice, pp. 22–24 (2016)
24. http://www.medicalexpo.fr/prod/hocoma/product-68750-773915.html
25. Abdelhedi, F., Bouteraa, Y., Chemori, A., Derbel, N.: Nonlinear PID and feedforward control of robotic manipulators. In: 15th International Conference on Sciences and Techniques of Automatic Control & Computer Engineering, pp. 349–354 (2015)
26. Lama, M.A., Kelly, R., Santibañez, V.: Stable computed-torque control of robot manipulators via fuzzy self-tuning. IEEE Trans. Syst. Man Cybern. Part B: Cybern. **30**(1) (2000)

27. Zhao, D., Liang, H., Zhu, Q.: Robust control for robotic manipulators with non-smooth strategy. Int. J. Model. Identif. Control **23**(2), 112–120 (2015)
28. Orlov, Y.: Extended invariance principle for nonautonomous switched systems. IEEE Trans. Autom. Control **48**, 1448–1452 (2003)
29. Krassovskii, N.N.: Problems of the theory of stability of motion. "Moscow, Russia: Fizmatigiz, 1959. In Russian, English translation, Stanford, CA: Stanford University Press, 1963"
30. LaSalle, J.P.: Some extensions of Lyapunov's second method. IRE Trans. Circuit Theory **7**, 520–527 (1960)
31. Jribi, R., Maalej, B., Derbel, N.: Robust adaptive feedback linearization control of human exoskeletons. In: International Conference on Systems, Signals and Devices (2019)

# Part II

# Intelligent Control for an Uncertain Mobile Robot with External Disturbances Estimator

**Yasmine Koubaa, Mohamed Boukattaya and Tarak Damak**

**Abstract** In this chapter, a new control approach for trajectory tracking problem of nonholonomic wheeled mobile robot (WMR) is proposed to cope with both uncertainties and external torque disturbances. The main contribution is the simultaneous exact estimation and cancelation of uncertainties and external torque disturbances without the requirement of torque measurement. First, a kinematic backstepping controller is proposed to achieve perfect velocity tracking. Then, a robust dynamic adaptive control algorithm with two update laws is developed to estimate and compensate the dynamic uncertainties and the unmeasured external torque disturbances. The design of the update laws use only position and velocity measurements and are derived from the Lyapunov stability theorem. Consequently, the proposed controllers prove that they not only can guarantee the stability and the trajectory tracking error is as small as possible but also the boundedness of all the states and signals of the closed-loop system and the convergence of the estimated disturbance to the real values. Finally, the simulation results demonstrate good tracking performance and robustness of the proposed controller.

**Keywords** Nonholonomic wheeled mobile robot · Trajectory tracking · Kinematic control · Adaptive dynamic control · External disturbances · Uncertain parameters

---

Y. Koubaa (✉) · M. Boukattaya · T. Damak
Laboratory of Sciences and Techniques of Automatic Control and Computer Engineering (Lab-STA), National School of Engineering of Sfax, 1173, 3038 Sfax, Tunisia
e-mail: yasminekoubaa@yahoo.fr

M. Boukattaya
e-mail: mohamed.boukattaya@gmail.com

T. Damak
e-mail: tarak.damak@enis.rnu.tn

 
J. Ghommam et al. (eds.), *New Trends in Robot Control*, Studies in Systems, Decision and Control 270, https://doi.org/10.1007/978-981-15-1819-5_8

# 1 Introduction

Wheeled mobile robot (WMR) has been widely used in various industrial and service applications, such as: transportation, inspection, security and planetary exploration etc. These applications require mobile robot to be able to track a specified path stably. Control architecture for WMR depends on whether the mobile robot is represented using kinematic model or dynamic model. Several studies that rely on kinematic model of mobile robot, such as Lyapunov control [3], neural networks [23], Backstepping [27], linearization [10], have been investigated to solve the path tracking problem. The objective of these controllers is to produce velocity outputs for the mobile robot to make the tracking error between the actual and reference trajectory converges to zeros. The above works ignore models that include dynamic effects. However, in real applications, good tracking performance cannot be achieved through considering only kinematic model, because mobile robot suffers from plant uncertainties and external disturbances. Thus, it is more realistic to introduce a torque controller based on system dynamics, such that the actual velocity of the robot system achieves the produced desired velocity. Various approaches have been interested in trajectory tracking at dynamic level using Sliding mode control [7, 9, 25, 26], Artificial intelligence control [1, 8, 11, 12, 21], Backstepping control [14], or Adaptive control [2, 4, 5, 13, 17, 22].

Artificial intelligence control using neural networks [8, 12, 15, 21] and fuzzy logic [1, 11] are regarded as an efficient strategy to design robot controller. Fierro and Lewis [8] proposed an artificial neural networks controller that integrates a Backstepping kinematic controller and a multilayer feedforward neural network dynamic controller to cope with unknown dynamics and disturbances. Mohareri et al. [12] designed an adaptive neural network controller, in which the control gains were tuned on line, to improve the tracking performance in the presence of unknown dynamics and uncertain parameters. However, neural networks have many drawbacks such as slow convergence, involved structure and difficulties in choosing the appropriate network. Abadi and Khooban [1] presented a mamdani type fuzzy logic controller to deal with both modeled and unmodeled dynamics in the mobile robot model. The significant drawbacks of fuzzy logic control method is the difficulty in determining fuzzy membership functions and fuzzy rules, especially when the controlled system is too complicated. The sliding-mode control (SMC), which provides many proprieties such as insensibility to parameters variation, fast response and external disturbances rejection, is considered to be an effective technique for the control of robotics systems. However, priori knowledge of the upper bounds of the norm of perturbations is required to obtain convergence. Moreover, SMC method often needs large control gains and can easily cause chattering phenomenon. In order to overcome the drawbacks of SMC, many researchers are interested in integrating sliding mode control with fuzzy logic control [18, 24] or with neural networks control [19, 20]. Yue et al. [24] developed an adaptive fuzzy sliding-mode dynamic control to solve the trajectory tracking problem of mobile robot. The proposed controller not only removed the chattering phenomenon in the sliding mode control, but also compensated for uncertain dynamics. Su [18] presented an adaptive controller using fuzzy basis function networks to approximate model uncertainties and unknown friction between the wheel and terrain surface. Sun et al. [19] proposed hybrid sliding-mode neural network method to solve the path tracking problem of wheeled mobile manipulator. A sliding-mode equivalent controller, composed of neural network control, robust scheme and proportional control, was constructed in the dynamic level to remove the effect of uncertainties and disturbances. Despite the interesting features of the aforementioned controllers, they are difficult to tune and computationally expensive, in contrast to the simplicity that pure adaptive control provides. Adaptive control is an interesting strategy to be applied to the problem of path tracking since it is easy to implement. In this chapter, an adaptive control law will be proposed to deal with unknown dynamics, such as mass and inertia, and external torque disturbances. The uncertain dynamics and disturbances are taking into account in the dynamic model. Compared with the previous control approaches in the presence of uncertainties and external torque disturbances, the contributions and novelty of this chapter are as follows:

- Formulation of a new adaptive algorithm with two updates laws to estimate and compensate both the uncertainties and external torque disturbances. In contrast to the other works which are focused only on the good tracking performance and do not give any importance to the estimation process, the one proposed in this paper is able to generate a precise estimation of external torque disturbances. Consequently, the proposed controller not only guarantees the good tracking performance but also the convergence of the estimated disturbances to the real values and the boundedness of all the states and signals of the closed-loop system.
- Adaptive laws for compensating uncertainties and external torque disturbances are derived using only position and velocity measurements, which suppress the use of the torque sensor.
- The stability of the closed loop system and the convergence of tracking errors to zeros are rigorously proven using Lyapunov stability theory. In addition, the convergence of the estimation errors to zero is also shown.

The outline of the rest of the paper is as follows: Sect. 2 is devoted to kinematic and dynamic modeling of the mobile robot with nonholonomic constraints. Section 3 presents the kinematic control design. Section 4 illustrates the main results of dynamic control design. Section 5 presents computer simulation results to illustrate the effectiveness of the proposed theory. Conclusions are formulated in Sect. 6.

## 2 Kinematics and Dynamics of Wheeled Mobile Robot

A typical model of nonholonomic WMR is shown in Fig. 1. The robot has two driving wheels mounted on the same axis and a front free wheel. The two driving wheels are independently driven by two actuators (e.g., DC motors) to achieve the motion and orientation. $\{W\} = \left\{ O \ x \ y \right\}$ is the word coordinate system and $\{R\} = \left\{ P_0 \ X \ Y \right\}$

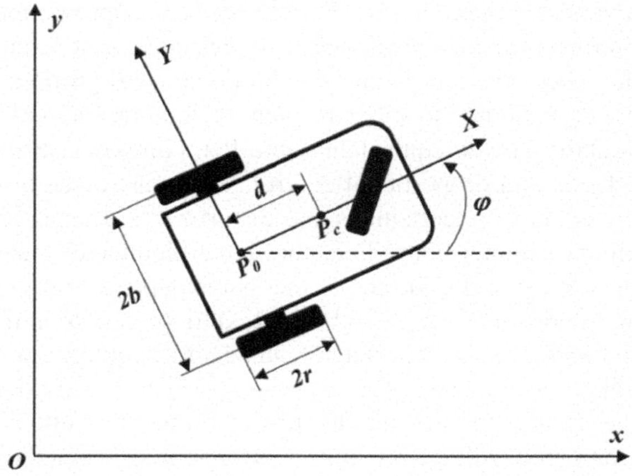


**Fig. 1** A differential mobile robot with two actuated wheels

is the coordinate system fixed to the robot. $P_0$ is the origin of the robot coordinate system and the middle between the right and left driving wheels. $P_c$ is the center of mass of the robot which is located at a distance $d$ from $P_0$ in the $X$ axis direction. $r$ is the radius of the driving wheel. The two driving wheels are separated by $2b$. $\varphi$ is the angle between $x$ axis of $\{W\}$ and $X$ axis of $\{R\}$.

## 2.1 Kinematic Model

The configuration of the mobile robot can be described by three generalized coordinates ($n = 3$)

$$q = \begin{bmatrix} x_c & y_c & \varphi \end{bmatrix}^T \quad (1)$$

$(x_c, y_c)$ are the coordinates of $P_c$ and $\varphi$ is the heading angle of the WMR.

In the kinematic model, it is supposed that in each contact, there exists a pure rolling motion. Assuming that the velocity of $P_c$ is in the direction of the axis of symmetry ($X$ axis) and the wheels do not slip, the following constraint set ($m = 1$), with respect to $P_c$ are obtained

$$\dot{y}_c \cos\varphi - \dot{x}_c \sin\varphi - \dot{\varphi} d = 0 \quad (2)$$

This constraint can be rewritten in the form of:

$$A(q)\dot{q} = 0 \quad (3)$$

where

$$A(q) = \begin{bmatrix} -\sin\varphi & \cos\varphi & -d \end{bmatrix} \quad (4)$$

Let $N(A)$ be the null space of $A(q)$. Then, by spanning the null space of $A(q)$ a set of smooth and linearly independent vector fields $s_1(q), \ldots, s_{n-m}(q)$ can be obtained. If we let $S(q) = [s_1(q), \ldots, s_{n-m}(q)]$ be the full rank matrix consists of these vectors, then the following equation hold

$$A(q)S(q) = 0 \tag{5}$$

From (3) and (5), we can deduce the auxiliary velocity vector $\upsilon \in R^{n-m}$ such that

$$\dot{q} = S(q)\upsilon \tag{6}$$

where

$$S(q) = \begin{bmatrix} \cos\varphi & -d\sin\varphi \\ \sin\varphi & d\cos\varphi \\ 0 & 1 \end{bmatrix} \tag{7}$$

$S(q)$ is the kinematic matrix and $\upsilon = [u \;\; w]^T$ is the velocity vector. $u$ is the linear velocity of WMR and $w$ is the angular velocity of WMR.

## 2.2 Dynamic Model

The WMR dynamic model can be derived using Euler-Lagrange formulation [16]:

$$\frac{d}{dt}\left(\frac{\partial L}{\partial \dot{q}}\right) - \frac{\partial L}{\partial q} = B(q).(\tau + \tau_d) - A(q)^T\lambda \tag{8}$$

where $L = T - U$ is the Lagrangian which is the difference between kinetic and potential energy of the system, $B(q)$ is the input transformation matrix, $\tau$ is the driving torque vector, $\tau_d$ is the vector of external disturbance, $A(q)$ is the matrix associated with nonholonomic constraints and $\lambda$ is the Lagrangian multiplier vector.

Under assumption that mobile robot is composed by $n_i$ elements and moves only on a plane surface, potential energy of the robot is zero ($U = 0$) and the kinetic energy of the whole structure is given by the following equation:

$$T = \frac{1}{2}\sum_{i=1}^{n_i}\left(u_i^T m_i u_i + w_i^T I_i w_i\right) \tag{9}$$

where $m_i$ and $I_i$ are the mass and the inertial moment for the $i$th element with respect to frame $i$, $u_i$ is the linear velocity of the center of mass for the $i$th element, $w_i$ is the angular velocity of the $i$th element respect to frame $i$.

After calculating (8) and considering (9), it is possible to deduce the dynamic model of the robot system:

$$M(q)\ddot{q} + C(q,\dot{q})\dot{q} + A^T(q)\lambda = B(q)(\tau + \tau_d) \tag{10}$$

where $M(q) \in R^{n\times n}$ is a symmetric positive definite inertia matrix, $C(q,\dot{q}) \in R^{n\times n}$ is the centripetal-Coriolis matrix, $A(q) \in R^{n\times m}$ is the matrix associated with nonholonomic constraints, $\lambda \in R^{m\times 1}$ is the vector of Lagrange multipliers, $B(q) \in R^{n\times(n-m)}$ defines the input transformation matrix, $\tau \in R^{(n-m)\times 1}$ presents the vector of input torque, $\tau_d \in \mathbb{R}^{(n-m)\times 1}$ defines the vector of external disturbances.

The variables in (10) are defined as follows:

$$\begin{aligned}
M(q) &= \begin{bmatrix} m & 0 & md\sin\varphi \\ 0 & m & -md\cos\varphi \\ md\sin\varphi & -md\cos\varphi & I \end{bmatrix}, \\
C(q,\dot{q}) &= \begin{bmatrix} 0 & 0 & md\dot{\varphi}\cos\varphi \\ 0 & 0 & md\dot{\varphi}\sin\varphi \\ 0 & 0 & 0 \end{bmatrix}, \\
B(q) &= \frac{1}{r}\begin{bmatrix} \cos\varphi & \cos\varphi \\ -\sin\varphi & \sin\varphi \\ b & -b \end{bmatrix}, \\
\lambda &= -m\left(\dot{x}_c\cos\varphi + \dot{y}_c\sin\varphi\right)\dot{\varphi}, \\
\tau &= \begin{bmatrix} \tau_r \\ \tau_l \end{bmatrix}.
\end{aligned} \tag{11}$$

where $\tau_r$ and $\tau_l$ defines right and left wheel torques, respectively. In the above robot dynamics equation, $m = m_c + 2m_w$ and $I = m_c d^2 + I_c + 2m_w b^2 + 2I_m$, where $m_c$ is the mass of the robot platform, $m_w$ is the mass of each driving wheel with its motor, $I_c$ is the inertia of the robot platform about a vertical axis through the center of mass, $I_w$ is the inertia of each wheel with its motor about the wheel axis and $I_m$ is the inertia of each wheel with its motor about the wheel diameter.

The dynamic equation (10) has many proprieties [4]:

**Property 1** *$M(q)$ is symmetric and positive definite matrix, such that:*

$$\lambda_{\min}(M)I \le M(q) \le \lambda_{\max}(M)I \tag{12}$$

*where $\lambda_{\min}(M)$ and $\lambda_{\max}(M)$ define the minimum and the maximum eigenvalues of the inertia matrix $M$.*

**Property 2** *The inertia matrix $M(q)$ and the centripetal-Coriolis matrix $C(q)$ present a skew-symmetric relationship:*

$$x^T\left(\dot{M} - 2C\right)x = 0, \forall x \in \mathbb{R}^{n-m} \tag{13}$$

*where $\dot{M}$ represents the time derivative of the inertia matrix.*

## 2.3 Reduced Dynamics

The system (10) can be transformed into a more appropriate representation for controls purposes. Differentiating (6), substituting this result into (10) and then multiplying by $S^T$, we can eliminate $A^T\lambda$. The reduced dynamic model of the nonholonomic mobile robot is described by:

$$M_\upsilon(q)\dot{\upsilon} + C_\upsilon(q,\dot{q})\upsilon = B_\upsilon(q)(\tau + \tau_d) \tag{14}$$

where $M_\upsilon = S^T M S$, $C_\upsilon = S^T(M\dot{S} + CS)$ and $B_\upsilon = S^T B$.

The variables in (14) are defined as

$$\begin{aligned} M_\upsilon(q) &= \begin{bmatrix} m & 0 \\ 0 & I - md^2 \end{bmatrix}, \\ C_\upsilon(q) &= \begin{bmatrix} 0\,0 \\ 0\,0 \end{bmatrix}, \\ B_\upsilon(q) &= \tfrac{1}{r}\begin{bmatrix} 1 & 1 \\ b & -b \end{bmatrix}. \end{aligned} \tag{15}$$

*Remark 1* The dynamic equation (14), which present the behavior of the mobile robot system in a new set of local coordinates, is formulated using jacobian matrix $S(q)$. This matrix transforms velocities $\upsilon$ in mobile base coordinates to velocities in Cartesian coordinates $\dot{q}$. Hence, the properties of the original dynamics (10) are satisfied for a new set of coordinates [8].

**Property 3** *$M_\upsilon(q)$ is symmetric and positive definite matrix, such that:*

$$\lambda_{\min}(M_\upsilon)I \le M_\upsilon(q) \le \lambda_{\max}(M_\upsilon)I \tag{16}$$

*where $\lambda_{\min}(M_\upsilon)$ and $\lambda_{\max}(M_\upsilon)$ define the minimum and the maximum eigenvalues of $M_\upsilon$.*

**Property 4** *The inertia matrix $M_\upsilon$ and the centripetal-Coriolis matrix $C_\upsilon$ present a skew-symmetric relationship:*

$$x^T\left(\dot{M}_\upsilon - 2C_\upsilon\right)x = 0, \forall x \in \mathbb{R}^{n-m} \tag{17}$$

*where $\dot{M}_\upsilon$ represents the time derivative of $M_\upsilon$.*

**Property 5** *The left hand side (14) can be transformed in the following form as shown below:*

$$M_\upsilon(q)\dot{\upsilon} + C_\upsilon(q,\dot{q})\upsilon = Y(q,\dot{q},\upsilon,\dot{\upsilon})\,p, \forall \upsilon, \dot{\upsilon} \in \mathbb{R}^{n-m} \tag{18}$$

*where $p \in \mathbb{R}^1$ denotes the vector of unknown (or uncertain) variables and $Y \in \mathbb{R}^{(n-m)\times 1}$ defines the "regressor" matrix, which incorporates known functions.*

In the case of the WMR considered in our study, the robot regressor $Y$ and the vector of uncertain parameters $p$ are introduced as follows:

$$Y = \begin{bmatrix} \dot{u} & 0 & 0 \\ 0 & \dot{w} & \dot{w} \end{bmatrix}, \quad p = \begin{bmatrix} m & I & -md^2 \end{bmatrix}^T \tag{19}$$

## 3 Kinematic Controller Design

To resolve the trajectory tracking problem, a reference mobile robot that produces a trajectory for the actual one to follow should be defined

$$\begin{aligned} \dot{x}_r &= u_r \cos\varphi_r, \\ \dot{y}_r &= u_r \sin\varphi_r, \\ \dot{\varphi}_r &= w_r, \\ q_r &= [x_r \;\; y_r \;\; \varphi_r]^T, \upsilon_r(t) = [u_r \;\; w_r]^T. \end{aligned} \tag{20}$$

where $q_r$ denotes the reference time varying position and orientation trajectory, and $\upsilon_r(t)$ denote the reference time varying linear and the angular velocity.

The posture tracking error between the reference robot and the actual robot can be expressed in the mobile basis frame as:

$$q_e = [e_1 \;\; e_2 \;\; e_3]^T = T\tilde{q} \tag{21}$$

where: $T = \begin{bmatrix} \cos\varphi & \sin\varphi & 0 \\ -\sin\varphi & \cos\varphi & 0 \\ 0 & 0 & 1 \end{bmatrix}$ and $\tilde{q} = \begin{bmatrix} x_r - x_0 \\ y_r - y_0 \\ \varphi_r - \varphi \end{bmatrix}$.

The derivative of the posture tracking error given in (21) can be written as follows:

$$\dot{q}_e = \begin{bmatrix} \dot{e}_1 \\ \dot{e}_2 \\ \dot{e}_3 \end{bmatrix} = \begin{bmatrix} -u + we_2 + u_r \cos e_3 \\ -we_1 + u_r \sin e_3 \\ -w + w_r \end{bmatrix} \tag{22}$$

Based on Backstepping method, the kinematic controller is developed to select the desired velocity input [8]:

$$\upsilon_c = \begin{bmatrix} u_c \\ w_c \end{bmatrix} = \begin{bmatrix} u_r \cos e_3 + k_1 e_1 \\ w_r + k_2 u_r e_2 + k_3 u_r \sin e_3 \end{bmatrix} \tag{23}$$

where $k_1$, $k_2$ and $k_3$ are positive constants.

**Assumption 1** It is assumed that the reference posture $q_r$ and the reference velocity $\upsilon_r(t)$ are bounded and uniformly continuous. Moreover, they have uniformly continuous and bounded derivative reaching the second order.

**Assumption 2** It is assumed that the velocity $v(t)$ as well as its derivative reaching the second order are uniformly continuous and bounded.

To prove the stability of the kinematic controller design, we choose the following Lyapunov candidate function:

$$V_1 = \frac{1}{2}({e_1}^2 + {e_2}^2) + \frac{1 - \cos e_3}{k_2} \tag{24}$$

Clearly $V_1 \geq 0$, differentiating (24), we obtain that:

$$\dot{V}_1 = e_1\dot{e}_1 + e_2\dot{e}_2 + \frac{\sin e_3}{k_2}\dot{e}_3 \tag{25}$$

Substituting (22) and (23) in the time derivative of $V_1$, we obtain

$$\begin{aligned}\dot{V}_1 &= e_1(-u + we_2 + u_r \cos e_3) + e_2(-we_1 + u_r \sin e_3) + \frac{\sin e_3}{k_2}(w_r - w) \\ &= -k_1{e_1}^2 - \frac{k_3 u_r \sin^2 e_3}{k_2} \leq 0 \end{aligned} \tag{26}$$

$V_1$ is positive. Equation (26) shows that $V_1$ is decreasing. Thus, $V_1$ is bounded, i.e. $V_1 \in L_\infty$. This implies that all the signals of $V_1$ are bounded, i.e. $e_1, e_2, e_3 \in L_\infty$. Moreover, according to Assumption 1, the control law (23) is bounded. Hence, all the system variables are bounded. Furthermore, Eq. (22) and Assumption 2 can be used to illustrate that $\dot{e_1}, \dot{e_2}, \dot{e_3} \in L_\infty$. Finally, using the Barbalat's Lemma [4], we can deduce that $e_1, e_2, e_3 \to 0$ as $t \to \infty$. It follows from (21) that $\tilde{q} \to 0$ as $t \to \infty$ i.e. $x_0 \to x_r$, $y_0 \to y_r$ and $\varphi \to \varphi_r$ as $t \to \infty$.

## 4 Dynamic Controller Design

In this subsection, dynamic tracking controllers are designed, such that the actual velocities of robot system converge to the desired velocities produced by the kinematic controller.

We introduce these assumptions which are needed to design of the dynamic control:

**Assumption 3** The matrices $M_v$ and $C_v$ of the dynamics (14) are unknown.

**Assumption 4** The matrices $M_v$ and $C_v$ are bounded and the Jacobian matrix $S(q)$ is full rank.

**Assumption 5** $\tau_d$ is assumed to be bounded. Also, it is assumed to be arbitrarily large and slowly varying with time, which implies that $\dot{\tau}_d$ is zero or negligible.

## 4.1 Non Adaptive Controller

Assuming that the external disturbances are not taken into account and the dynamic parameters of the robot system are exactly known, the non-adaptive method is presented. To design the input torques, we select the auxiliary velocity tracking error as well as its derivative:

$$e_{\upsilon}(t) = \upsilon_c(t) - \upsilon(t) = \begin{bmatrix} u_c - u \\ w_c - w \end{bmatrix} \tag{27}$$

$$\dot{e}_{\upsilon}(t) = \dot{\upsilon}_c(t) - \dot{\upsilon}(t) \tag{28}$$

The following control law is considered to provide the input torque for the mobile robot:

$$\tau_{\upsilon} = B_{\upsilon}\tau = M_{\upsilon}\dot{\upsilon}_c + C_{\upsilon}\upsilon_c + K_p e_{\upsilon} \tag{29}$$

where $K_p$ is a positive definite matrix.

Substituting (29) into (14), the closed-loop dynamic is given by

$$M_{\upsilon}\dot{e}_{\upsilon} + C_{\upsilon}e_{\upsilon} + K_p e_{\upsilon} = 0 \tag{30}$$

Theorem 1 is presented to demonstrate the stability of the closed loop system (30).

**Theorem 1** *We assume that the nonholonomic wheeled mobile robot (14) is not affected by external disturbances ($\tau_d = 0$) and dynamic parameters are exactly known. The closed-loop system (30) is asymptotically stable and the velocity tracking error $e_{\upsilon}$ asymptotically converge to zero as $t \longrightarrow \infty$, if the kinematic controller (23) and the dynamic controller (29) are applied.*

***Proof*** We define the following Lyapunov function candidate as:

$$V_2 = \frac{1}{2} e_{\upsilon}^T M_{\upsilon} e_{\upsilon} \tag{31}$$

We calculate the time derivative of (31) and after that we substitute the closed-loop dynamic (30) in (31), we have

$$\begin{aligned} \dot{V}_2 &= e_{\upsilon}^T \left( M_{\upsilon}\dot{e}_{\upsilon} + \tfrac{1}{2}\dot{M}_{\upsilon}e_{\upsilon} \right) \\ &= e_{\upsilon}^T \left( -K_p e_{\upsilon} + \tfrac{1}{2} e_{\upsilon}(\dot{M}_{\upsilon} - 2C_{\upsilon}) \right) \end{aligned} \tag{32}$$

Applying the skew-symmetric Property 4, we obtain

$$\dot{V}_2 = -e_{\upsilon}^T K_p e_{\upsilon} \leq 0 \tag{33}$$

$V_2$ is a positive function. It is clear from (33) that $V_2$ is decreasing. Thus, $V_2$ is bounded, i.e. $V_2 \in L_{\infty}$. This implies that all the signals of $V_2$ are bounded, i.e.

$e_v \in L_\infty$, and hence $v, v_c \in L_\infty$. From the stability of the kinematic controller, we can assume that $\dot{v}_c \in L_\infty$. In addition, according to Assumption 4, the control law in (29) is bounded. As consequence, all the system signals are bounded. Moreover, the closed-loop dynamics (30) and the Assumption 4 are used to illustrate that $\dot{e}_v \in L_\infty$. Finally, using Barbalat's Lemma [4], we can deduce that $e_v \to 0$ as $t \to \infty$.

## 4.2 Adaptive Controller with External Torque Disturbance Estimation

In real applications, external disturbance must exist $\tau_d \neq 0$. It may be caused by vibration of the driving motors, ground irregularities when the robot navigates on uneven surfaces or instantaneous collision with an obstacle. To avoid these issues, we propose an adaptive control law to deal with the effect of external disturbance. $\hat{\tau}_d$ is supposed to be the estimated value of actual external disturbance $\tau_d$. The developed control law is presented as follows:

$$\tau_v = B_v \tau = M_v \dot{v}_c + C_v v_c + K_p e_v - B_v \hat{\tau}_d \tag{34}$$

Substituting (34) into (14) to obtain the closed-loop dynamic:

$$M_v \dot{e}_v + C_v e_v + K_p e_v + B_v \tilde{\tau}_d = 0 \tag{35}$$

where $\tilde{\tau}_d = \tau_d - \hat{\tau}_d$ defines the vector of external disturbance error, which is the difference between the real external disturbance and the estimated one.

Theorem 2 is presented to demonstrate the stability of the closed loop system (35).

**Theorem 2** *We assume that the nonholonomic wheeled mobile robot (14) is affected by external disturbances ($\tau_d \neq 0$) and dynamic parameters are exactly known. The closed-loop system in (35) is asymptotically stable and the velocity tracking error $e_v$ asymptotically converge to zero as $t \longrightarrow \infty$, if the kinematic controller (23) is designed, and the adaptive dynamic controller (34) is developed, such that the adaptive law for the external disturbances is:*

$$\dot{\hat{\tau}}_d = K_d^{-T} B_v^T e_v \tag{36}$$

*where $K_d$ is a positive definite matrix.*

***Proof*** We consider the following Lyapunov function candidate:

$$V_2 = \frac{1}{2} e_v^T . M_v . e_v + \frac{1}{2} \tilde{\tau}_d^T . K_d . \tilde{\tau}_d \tag{37}$$

The time derivative of $V_2$ is

$$\dot{V}_2 = e_v^T.M_v.\dot{e}_v + \frac{1}{2}e_v^T.\dot{M}_v.e_v + \dot{\tilde{\tau}}_d^{\,T}.K_d.\tilde{\tau}_d \tag{38}$$

Substituting the closed-loop dynamic (35) in (38), we have:

$$\begin{aligned}\dot{V}_2 = &-e_v^T K_p e_v + \frac{1}{2}e_v^T.\left(\dot{M}_v - 2C_v\right).e_v \\ &+ \left(-e_v^T.B_v + \dot{\tilde{\tau}}_d^{\,T}.K_d\right)\tilde{\tau}_d\end{aligned} \tag{39}$$

Applying the skew-symmetric Property 4 and the adaption law (36), we obtain:

$$\dot{V}_2 = -e_v^T K_p e_v \leq 0 \tag{40}$$

$V_2$ is a positive function. It is clear from (40) that $V_2$ is decreasing. Thus, $V_2$ is bounded, i.e. $V_2 \in L_\infty$. This implies that all the signals of $V_2$ are bounded, i.e. $e_v$ and $\tilde{\tau}_d \in L_\infty$, and hence $v, v_c \in L_\infty$. From the stability of the kinematic controller, we can assume that $\dot{v}_c \in L_\infty$. Furthermore, since the vector of external disturbance $\tau_d$ is bounded, we can deduce that $\hat{\tau}_d$ is also bounded, i.e., $\hat{\tau}_d \in L_\infty$. In addition, according to Assumption 4 the control law (34) that contains the update law (36) is also bounded. As consequence, all the system signals are bounded. Moreover, the closed-loop dynamics (35) and the Assumption 4 are used to show that $\dot{e}_v \in L_\infty$. Finally, using the Barbalat's Lemma [4], we can deduce that $e_v \to 0$ as $t \to \infty$.

### *4.3 Adaptive Controller with External Torque Disturbance and Uncertainty Estimation*

We have supposed that the dynamics of the mobile robot are exactly known in designing the above dynamic controller, which is able to estimate external disturbances. However, dynamic parameters of the robot system can be uncertain or unknown in practise. This is due to inexact measurement of the mobile robot parameters, particularly the mass and the moment of inertia. In this case, considering that the dynamic parameters are exactly known may lead to mediocre tracking performance and may incur instability. Hence, a new adaptive control law is developed such that it estimates unknown parameters and compensates for uncertainties.

Consider $\hat{M}_v$ and $\hat{C}_v$ the estimated values of $M_v$ and $C_v$ respectively.

Hence the control law can be expressed as:

$$\tau_v = B_v \tau = \hat{M}_v \dot{v}_c + \hat{C}_v v_c + K_p e_v - B_v \hat{\tau}_d \tag{41}$$

Using the regression Property 5 and assuming $\hat{p}$ to be the estimated value of real dynamic parameters vector $p$, the control law is expressed as follow:

$$\tau_v = B_v \tau = Y(q, \dot{q}, v_c, \dot{v}_c)\hat{p} + K_p e_v - B_v \hat{\tau}_d \tag{42}$$

Substituting (42) into (14) to obtain the closed-loop dynamic

$$M_v \dot{e}_v + C_v e_v + K_p e_v = Y(q, \dot{q}, v_c, \dot{v}_c)\tilde{p} - B_v \tilde{\tau}_d \tag{43}$$

where $\tilde{p} = p - \hat{p}$ is the vector of parameter uncertainties error, which is the difference between the real parameters and the estimated ones.

Theorem 3 is presented to demonstrate the stability of the closed loop system (43).

**Theorem 3** *We assume that the nonholonomic wheeled mobile robot (14) is affected by external disturbances ($\tau_d \neq 0$) and unknown parameters. If the kinematic controller (23) is designed, and the dynamic controller (42) is developed, such that adaptive laws for unknown parameters and external disturbances are defined as:*

$$\dot{\hat{p}} = -\Gamma^{-T} Y^T e_v \tag{44}$$

$$\dot{\hat{\tau}}_d = K_d^{-T} B_v^T e_v \tag{45}$$

*where $\Gamma$ and $K_d$ are positive definite matrices, then,*

- *(i) the closed-loop system in (43) is asymptotically stable and the velocity tracking error $e_v$ asymptotically converge to zero as $t \longrightarrow \infty$.*
- *(ii) In addition, assuming that $\eta_1$, $\eta_2$ and $\lambda$ are positive constants, therefore, for all $t_0 \geq 0$ we have*
  $\eta_1 I \leq \int_{t_0}^{t_0+\lambda} Z_a^T Z_a dt \leq \eta_2 I$
  *where $Z_a = Z(v, \dot{v})$ and $Z(v_c, \dot{v}_c) = [-B_v(q) \;\; Y(v_c, \dot{v}_c)]$, then the external disturbances and parameter estimation errors converge to zero.*

***Proof*** (i) We define the Lyapunov function candidate as:

$$V_2 = \frac{1}{2} e_v^T . M_v . e_v + \frac{1}{2} \tilde{p}^T . \Gamma . \tilde{p} + \frac{1}{2} \tilde{\tau}_d^T . K_d . \tilde{\tau}_d \tag{46}$$

We calculate the time derivative of $V_2$:

$$\dot{V}_2 = e_v^T . M_v . \dot{e}_v + \frac{1}{2} e_v^T . \dot{M}_v . e_v + \dot{\tilde{p}}^T . \Gamma . \tilde{p} + \dot{\tilde{\tau}}_d^T . K_d . \tilde{\tau}_d \tag{47}$$

Substituting the closed-loop dynamic (43) in (47), we have:

$$\begin{aligned} \dot{V}_2 = & -e_v^T K_p e_v + \frac{1}{2} e_v^T . \left(\dot{M}_v - 2C_v\right) . e_v \\ & + \left(e_v^T . Y + \dot{\tilde{p}}^T . \Gamma\right) \tilde{p} + \left(-e_v^T . B_v + K_d . \dot{\tilde{\tau}}_d^T\right) \tilde{\tau}_d \end{aligned} \tag{48}$$

Applying the skew-symmetric Property 4 and the adaption law for unknown parameters (44) and adaptation law for external disturbance (45), we obtain:

$$\dot{V}_2 = -{e_v}^T K_p e_v \leq 0 \tag{49}$$

$V_2$ is a positive function. It is clear from (49) that $V_2$ is decreasing. Thus $V_2$ is bounded, i.e. $V_2 \in L_\infty$. This implies that all the signals of $V_2$ are bounded, i.e. $e_v, \tilde{p}, \tilde{\tau}_d \in L_\infty$, and hence $v, v_c \in L_\infty$. From the stability of the kinematic controller, we can assume that $\dot{v}_c \in L_\infty$. Furthermore, since the vector of external disturbance $\tau_d$ is bounded, we can deduce that $\hat{\tau}_d$ is also bounded, i.e., $\hat{\tau}_d \in L_\infty$. Also according to Assumption 4, the regression dynamic expression (18) is utilized to illustrate that $p \in L_\infty$, hence $\hat{p} \in L_\infty$ In addition, from Assumption 4, we can deduce that the control law (42) that contains the update laws (44) and (45) is also bounded. As consequence, all the system signals are bounded. Moreover, we use the closed-loop dynamics (43) and the Assumption 4 to illustrate that $\dot{e}_v \in L_\infty$. Finally, using the Barbalat's Lemma [4], we can deduce that $e_v \to 0$ as $t \to \infty$.

(ii) If $V_2 \equiv 0$ then $e_v \equiv 0$ and $\dot{e_v} \equiv 0$, as consequence, the left side of (43) is assumed to be zero. Therefore, we have $Y\tilde{p} - B_v\tilde{\tau}_d \equiv 0$ or equivalently

$$\begin{bmatrix} -B_v(q) & Y(v_c, \dot{v}_c) \end{bmatrix} \begin{bmatrix} \tilde{\tau}_d \\ \tilde{p} \end{bmatrix} \equiv 0 \tag{50}$$

Moreover, according to adaptation laws, if $e_v \equiv 0$, it is easy to show that $\dot{\hat{\tau}}_d \equiv 0$ and $\dot{\hat{p}} \equiv 0$. This implies that they are constant vectors. Thus, using the fact that all columns of $Z(v_c, \dot{v}_c) = \begin{bmatrix} -B_v(q)Y(v_c, \dot{v}_c) \end{bmatrix}$ are independent (i.e. Z defines a full column rank matrix or $Z^T$ defines a full column rank matrix), we can deduce from (50) that $\begin{bmatrix} \tilde{\tau}_d \ \tilde{p} \end{bmatrix} \equiv 0$. Since $v_c$ is a function of time, Z is also considered to be a function of time, $Z(v_c, \dot{v}_c) = Z(t)$. As stated in [6], $Z^T$ is assumed to be a full row matrix rank during $[t_1, t_2]$ if:

$$D(t_1, t_2) = \int_{t_1}^{t_2} Z^T(t)Z(t)dt \tag{51}$$

is a nonsingular matrix.

Where $D$ denotes a Hermitian matrix. This is used to show that $D$ is a symmetric positive semi-definite matrix. Therefore, $D$ will be positive definite using the fact that it is nonsingular. Using the theorem of Rayleigh–Ritz [6], we obtain

$$\beta_1 I \leq D \leq \beta_2 I \tag{52}$$

$\beta_1$ and $\beta_2$ which denote the minimum as well as the maximum eigenvalues of $D$, respectively, are positive scalars. According to (51) and (52), we can write

$$\beta_1 I \le \int_{t_1}^{t_2} Z^T(t)Z(t)dt \le \beta_2 I \tag{53}$$

Since $D$ depends on $t_1$ and $t_2$, Hence $\beta_1$ and $\beta_2$ changes during $[t_1, t_2]$. By selecting $t_1 = t_0$ and $t_2 = t_0 + \lambda$, and choosing $\eta_1 = \sup_{t_0 \in (-\infty,\infty)} \beta_1(t_0, t_0 + \lambda)$ and $\eta_2 = \inf_{t_0 \in (-\infty,\infty)} \beta_2(t_0, t_0 + \lambda)$, $\eta_1$ and $\eta_2$ are considered to be constant scalars. Therefore the following expression is obtained:

$$\eta_1 I \le \int_{t_0}^{t_0+\lambda} Z^T(t)Z(t)dt \le \eta_2 I \tag{54}$$

or in terms of $q$

$$\eta_1 I \le \int_{t_0}^{t_0+\lambda} Z^T(\upsilon_c, \dot{\upsilon}_c)Z(\upsilon_c, \dot{\upsilon}_c)dt \le \eta_2 I \tag{55}$$

Since velocity tracking error converge to zero, the following condition is satisfied:

$$\eta_1 I \le \int_{t_0}^{t_0+\lambda} {Z_a}^T Z_a dt \le \eta_2 I \tag{56}$$

## 5 Simulation Results

In this section, some computer simulations were performed to show the tracking performance and effectiveness of the proposed controller under unknown dynamics and external torque disturbances.

In the simulation, we assume that the parameters of the mobile robot are

$b = 0.75\,\text{m}, r = 0.15\,\text{m}, d = 0.3\,\text{m}, m_c = 30\,\text{kg},$
$m_w = 1\,\text{kg}, I_c = 15.625\,\text{kg}\cdot\text{m}^2, I_w = 0.005\,\text{kg}\cdot\text{m}^2,$
$I_m = 0.0025\,\text{kg}\cdot\text{m}^2.$

We select the following reference trajectory as a function of time:

$x_r(t) = 0.2t + 0.3,\ y_r(t) = 0.5 + 0.25\sin(0.2\pi t).$
The initial posture of the WMR is $q(0) = [x_c(0)\ \ y_c(0)\ \ \varphi(0)]^T = [0\ \ 0\ \ \pi/3]^T$.

The control gains used in simulation were selected as

$k_1 = 2, k_2 = 50, k_3 = 0.1,$
$K_p = [250\ \ 0;\ 0\ \ 250],\ \Gamma = 0.001 I_3,\ K_d = [0.01\ \ 0;\ 0\ \ 0.01].$

It is assumed that two cases of external disturbances imposed on the wheels of mobile robot are set as
case i: $\tau_d = \left[5 rectpuls(t-10, 1)\ \ 5\text{rectpuls}(t-15, 1)\right]^T$
case ii: $\tau_d = [5\sin(2t)\ \ 5\cos(2t)]^T$.

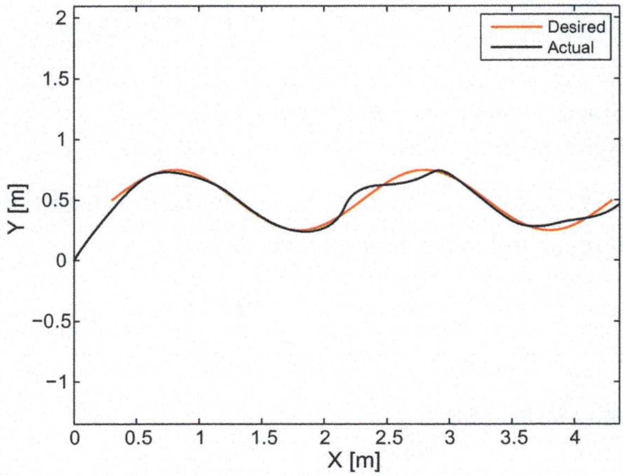


**Fig. 2** Trajectory tracking of the non adaptive controller for case i

In the first case the square external disturbance illustrates the conditions of WMR in which the wheel is confronting an instantaneous collision with an obstacle, or the robot system is hitting a 'hole' or a 'bump' while navigating. In the second case, the sinusoidal external disturbance reveals vibration of the driving motor of the WMR.

The tracking results of the non adaptive controller, the adaptive controller with external torque disturbance estimation and the adaptive controller with external torque disturbance and uncertainty estimation are depicted in Figs. 2, 3, 4, 5, 6 and 7 for case i, and in Figs. 8, 9, 10 and 11 for case ii.

It can be seen from Figs. 2 and 8 that the external disturbances have undesirable effects on the tracking performance of the non adaptive controller. Accordingly, the non adaptive approach cannot overcome the effect of external disturbances. However, the proposed adaptive controller which is able to estimate parameters and external torque disturbances, ensure an acceptable path tracking.

Figure 3 shows that using the adaptive controller with external torque disturbance estimation, the mobile robot can successfully track the desired trajectory. The tracking errors converge to zero (Fig. 3b). Actual velocity of the mobile robot converge to the desired velocity generated by the kinematic controller (Fig. 4). In addition, Fig. 5 illustrates the estimated external disturbances are very close to the actual ones. Indeed, the estimated disturbance vector is used to deal with external disturbances in order to enhance the tracking performance.

In order to show the efficiency of the adaptive controller with external torque disturbance and uncertainty estimation, actual mass ($m_c$) and inertia ($I_c$) of the mobile robot increase as following:

$$\begin{aligned} m_c &= 30\,\text{kg} \mapsto 45\,\text{kg} \\ I_c &= 15.625\,\text{kg/m}^2 \mapsto 30\,\text{kg/m}^2. \end{aligned}$$

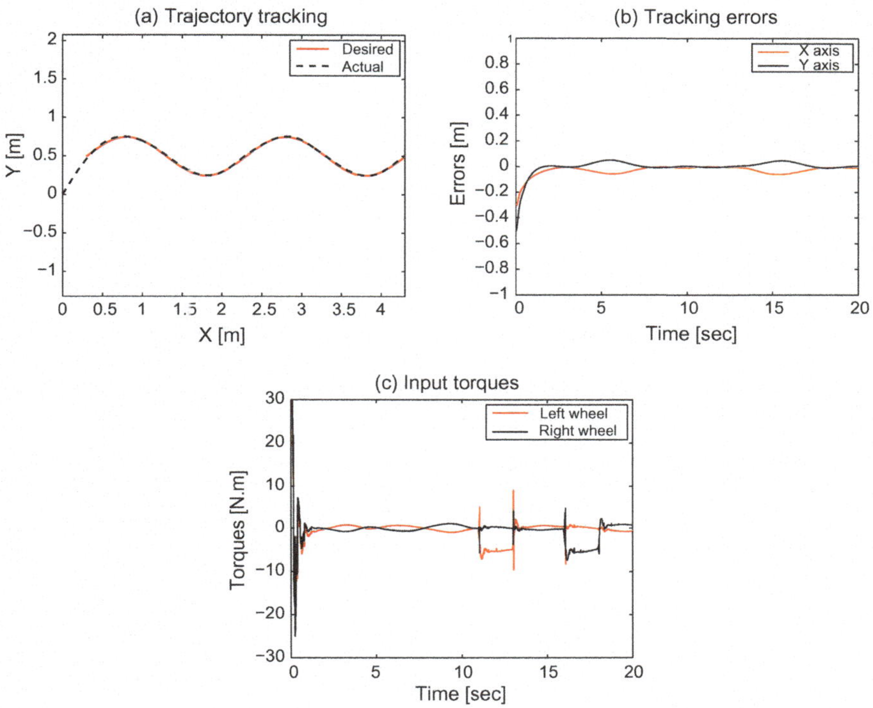


**Fig. 3** Response of the proposed adaptive controller with external torque disturbance estimation for case i

The proposed adaptive controller which contains two update laws, ensures good estimation of the square disturbances (Fig. 7), and provides a good tracking performance (Fig. 6a, b). Moreover, the input torques in Fig. 3c obtained when robots' dynamic parameters are nominal is very similar to the one obtained when robots' dynamic parameters are changed in Fig. 6c. This verifies the effectiveness of the proposed controller and its indifference toward dynamic uncertainties.

Now, we show that when the mobile robot is affected by sinusoidal disturbances, good tracking performance is maintained using the adaptive controller with disturbances and uncertainty estimation (Fig. 9a). The tracking errors converge to zero (Fig. 9b). Actual velocity of the mobile robot converge to the desired velocity generated by the kinematic controller (Fig. 10). Moreover, Fig. 11 shows an acceptable estimation of the sinusoidal disturbances.

From Figs. 2, 3, 4, 5, 6, 7, 8, 9, 10 and 11, we draw a conclusion that the designed control method can compensate not only for external disturbances, but also for change of robot dynamics.

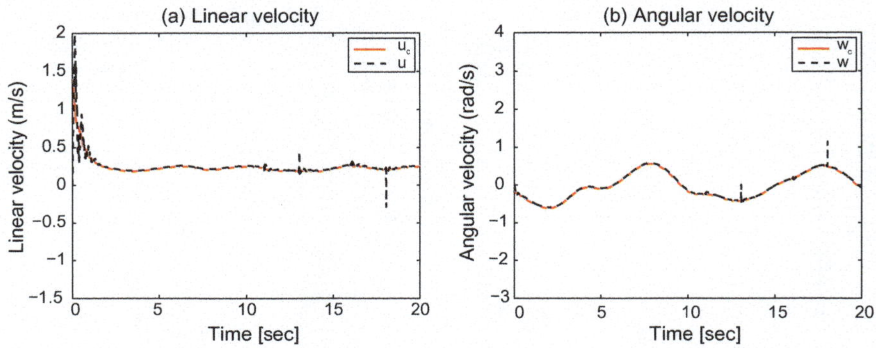


**Fig. 4** Actual velocity and desired velocity using adaptive controller with external torque disturbance estimation for case i

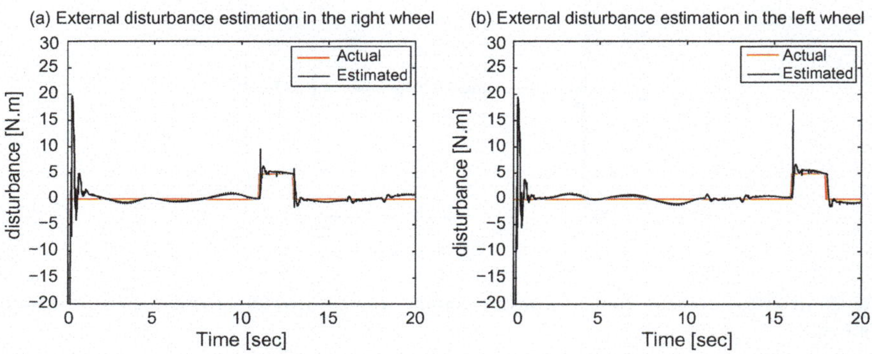


**Fig. 5** Estimated external disturbances using adaptive controller with external torque disturbance estimation for case i

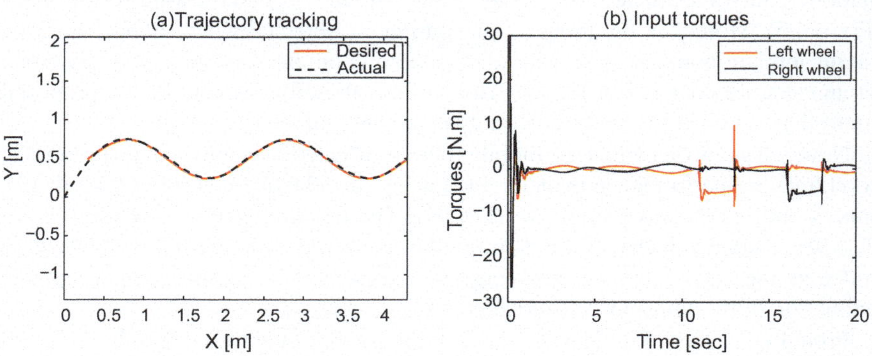


**Fig. 6** Response of the proposed adaptive controller with external torque disturbance and uncertainty estimation for case i

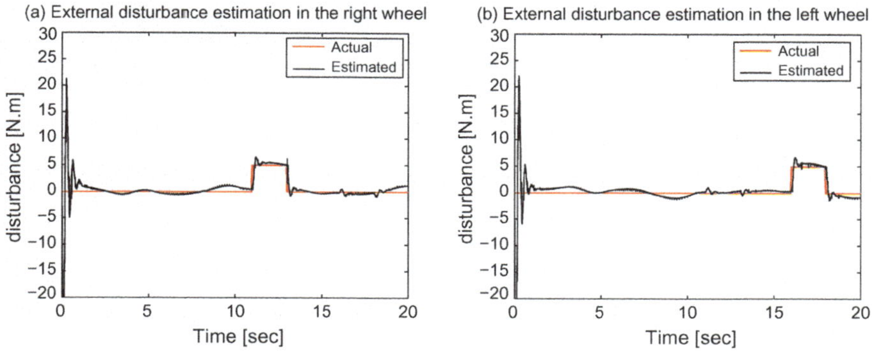


**Fig. 7** Estimated external disturbances using adaptive controller with external torque disturbance and uncertainty estimation for case i

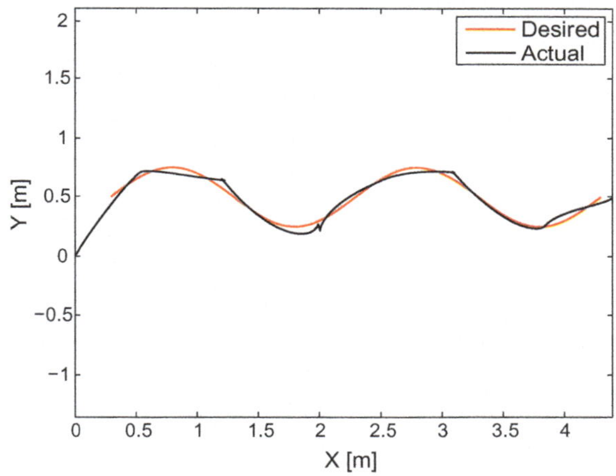


**Fig. 8** Trajectory tracking of the non adaptive controller for case ii

# 6 Conclusions

In this paper, we have designed an adaptive mobile robot controller affected by uncertain dynamics and external disturbances. The proposed controller incorporates two estimators in order to estimate and compensate for unknown dynamics and external disturbances applied to mobile robot. Our proposed method guaranties not only cancelation of the effect of uncertainties, but also the estimation of uncertain parameters as well as external disturbances. Thus, good tracking performance is achieved. The stability of the controlled system including estimators has been proven using Lyapunov stability theory. The presented simulations show the robustness of the proposed control approach against unknown parameter and external disturbances.

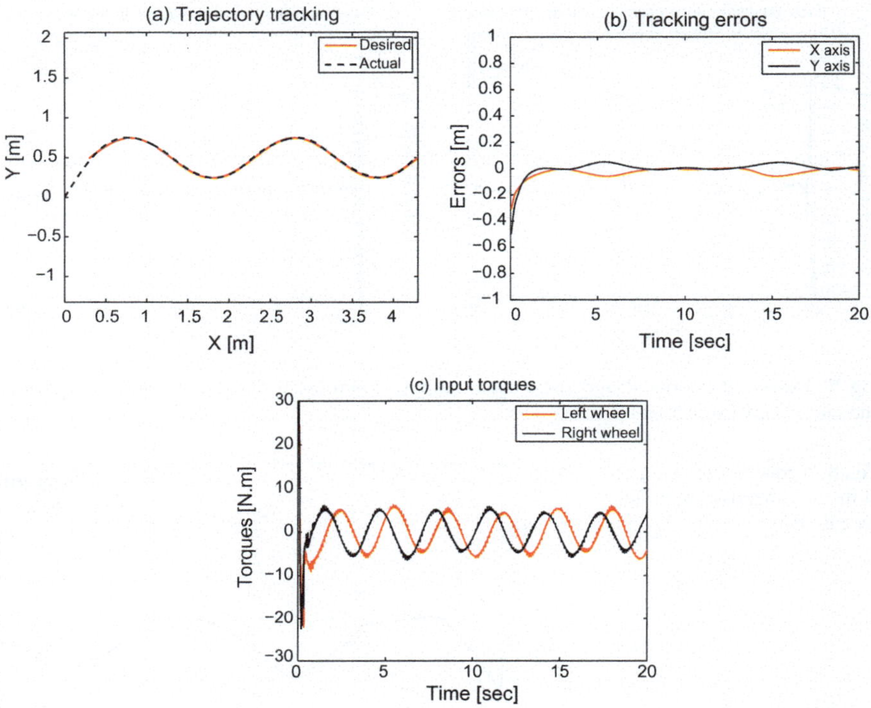


**Fig. 9** Response of the proposed adaptive controller with external torque disturbance and uncertainty estimation for case ii

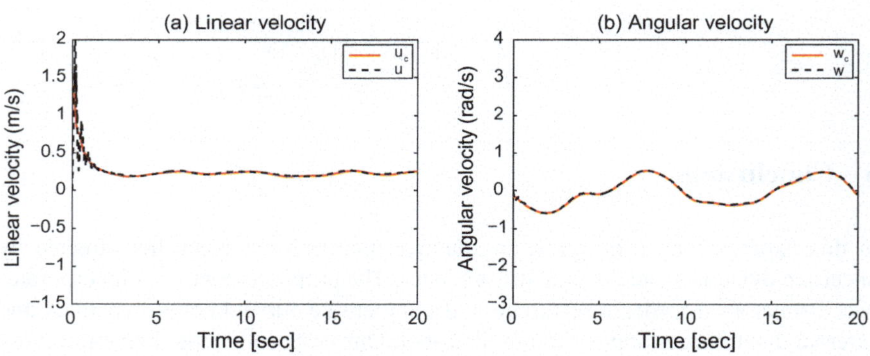


**Fig. 10** Actual velocity and desired velocity using adaptive controller with external torque disturbance and parameters estimation for case ii

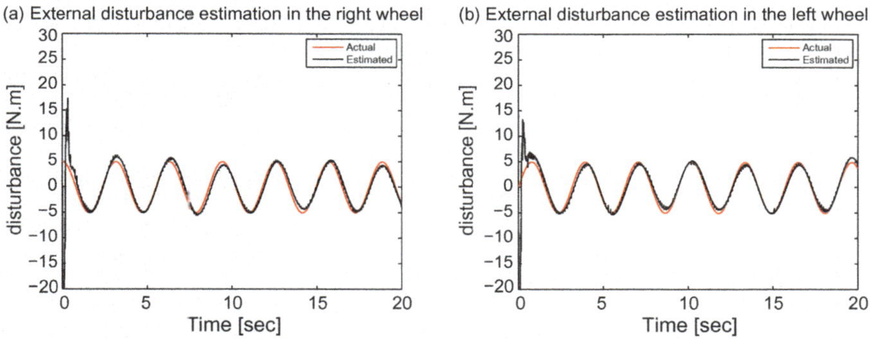


**Fig. 11** Estimated external disturbances using adaptive controller with external torque disturbance and uncertainty estimation for case ii

**Acknowledgements** We thank the ministry of higher education and scientific research of Tunisia for funding this work.

## References

1. Abadi, D.N.M., Khooban, M.H.: Design of optimal mamdani-type fuzzy controller for nonholonomic wheeled mobile robots. J. King Saud Univ-Eng. Sci. **27**(1), 92–100 (2015)
2. Almeida Martins, N., El'youssef, E.S., De Pieri, E.R., Lombardi, W.C., Jungers, M.: An adaptive variable structure controller for the trajectory tracking of a nonholonomic mobile robot with uncertainties and disturbances. J. Comput. Sci. Technol. **11** (2011)
3. Blaič, S.: A novel trajectory-tracking control law for wheeled mobile robots. Robot. Auton. Syst. **59**(11), 1001–1007 (2011)
4. Boukattaya, M., Jallouli, M., Damak, T.: On trajectory tracking control for nonholonomic mobile manipulators with dynamic uncertainties and external torque disturbances. Robot. Auton. Syst. **60**(12), 1640–1647 (2012)
5. Cui, M., Liu, H., Liu, W., Qin, Y.: An adaptive unscented Kalman filter-based controller for simultaneous obstacle avoidance and tracking of wheeled mobile robots with unknown slipping parameters. J. Intell. Robot. Syst. 1–16 (2017)
6. Danesh, M., Sheikholeslam, F., Keshmiri, M.: An adaptive manipulator controller based on force and parameter estimation. IEICE Trans. Fundam. Electron. Commun. Comput. Sci. **89**(10), 2803–2811 (2006)
7. Esmaeili, N., Alfi, A., Khosravi, H.: Balancing and trajectory tracking of two-wheeled mobile robot using backstepping sliding mode control: design and experiments. J. Intell. Robot. Syst. **87**(3–4), 601–613 (2017)
8. Fierro, R., Lewis, F.L.: Control of a nonholonomic mobile robot using neural networks. IEEE Trans. Neural Netw. **9**(4), 589–600 (1998)
9. Ghasemi, M., Nersesov, S.G., Clayton, G.: Finite-time tracking using sliding mode control. J. Frankl. Inst. **351**(5), 2966–2990 (2014)
10. Kim, D.-H., Oh, J.-H.: Tracking control of a two-wheeled mobile robot using input-output linearization. Control Eng. Pract. **7**(3), 369–373 (1999)
11. Liu, Z., Chen, C., Zhang, Y., Chen, C.P.: Coordinated fuzzy control of robotic arms with actuator nonlinearities and motion constraints. Inf. Sci. **296**, 1–13 (2015)

12. Mohareri, O., Dhaouadi, R., Rad, A.B.: Indirect adaptive tracking control of a nonholonomic mobile robot via neural networks. Neurocomputing **88**, 54–66 (2012)
13. Peng, Z., Yang, S., Wen, G., Rahmani, A., Yu, Y.: Adaptive distributed formation control for multiple nonholonomic wheeled mobile robots. Neurocomputing **173**, 1485–1494 (2016)
14. Pourboghrat, F., Karlsson, M.P.: Adaptive control of dynamic mobile robots with nonholonomic constraints. Comput. Electr. Eng. **28**(4), 241–253 (2002)
15. Rani, M., Kumar, N., Singh, H.P.: Efficient position/force control of constrained mobile manipulators. Int. J. Dyn. Control 1–10 (2018)
16. Shojaei, K., Shahri, A.M., Tarakameh, A.: Adaptive feedback linearizing control of nonholonomic wheeled mobile robots in presence of parametric and nonparametric uncertainties. Robot. Comput.-Integr. Manuf. **27**(1), 194–204 (2011)
17. Singh, V.K., Sharma, V., Sharma, B., Nath, R.: Synchronisation of different order chaotic systems with delay and parametric uncertainty. Int. J. Appl. Nonlinear Sci. **2**(3), 184–199 (2016)
18. Su, K.-H.: Robust tracking control design and its application to balance a two-wheeled robot steering on a bumpy road. Proc. Inst. Mech. Eng. Part I: J. Syst. Control Eng. **226**(7), 887–903 (2012)
19. Sun, T., Pei, H., Pan, Y., Zhou, H., Zhang, C.: Neural network-based sliding mode adaptive control for robot manipulators. Neurocomputing **74**(14–15), 2377–2384 (2011)
20. Tsai, C.-C., Cheng, M.-B., Lin, S.-C.: Robust tracking control for a wheeled mobile manipulator with dual arms using hybrid sliding-mode neural network. Asian J. Control **9**(4), 377–389 (2007)
21. Tsai, C.-C., Li, Y.-Y., Tai, F.-C., Lu, C.-H.: Intelligent adaptive motion control using fuzzy basis function networks for electric unicycle. Asian J. Control **17**(3), 977–993 (2015)
22. Wu, J., Xu, G., Yin, Z.: Robust adaptive control for a nonholonomic mobile robot with unknown parameters. J. Control Theory Appl. **7**(2), 212–218 (2009)
23. Ye, J.: Tracking control for nonholonomic mobile robots: integrating the analog neural network into the backstepping technique. Neurocomputing **71**(16–18), 3373–3378 (2008)
24. Yue, M., Wang, S., Zhang, Y.: Adaptive fuzzy logic-based sliding mode control for a nonholonomic mobile robot in the presence of dynamic uncertainties. Proc. Inst. Mech. Eng. Part C: J. Mech. Eng. Sci. **229**(11), 1979–1988 (2015)
25. Yue, M., Wei, X.: Dynamic balance and motion control for wheeled inverted pendulum vehicle via hierarchical sliding mode approach. Proc. Inst. Mech. Eng. Part I: J. Syst. Control Eng. **228**(6), 351–358 (2014)
26. Zhang, Y., Liu, G., Luo, B.: Finite-time cascaded tracking control approach for mobile robots. Inf. Sci. **284**, 31–43 (2014). Special issue on Cloud-assisted Wireless Body Area Networks
27. Zhou, B., Han, J., Dai, X.: Backstepping based global exponential stabilization of a tracked mobile robot with slipping perturbation. J. Bionic Eng. **8**(1), 69–76 (2011)

# Adaptive Event-Triggered Regulation Control of Nonholonomic Mobile Robots

**Haci Mehmet Güzey**

**Abstract** In this chapter, the design of adaptive-regulation control of mobile robots (MR) in the presence of uncertain MR dynamics with event-based feedback is provided. Two layer neural-networks (NN) are utilized to represent the uncertain dynamics of the MR which is subsequently employed to generate the control torque with event-sampled measurement update. By relaxing the perfect velocity tracking assumption, control torque is developed to minimize the velocity tracking errors, by explicitly taking into account the dynamics of the MR. The Lyapunov's method is utilized to develop an event-sampling condition and to demonstrate the regulation error performance of the MR. At the end of the chapter, simulation results are given to verify our theoretical claims.

**Keywords** Event-triggered control · Adaptive control · Mobile robot · Robot dynamics

## 1 Introduction

Traditionally, feedback control systems are developed based on periodic sampling of the sensors to compute appropriate control signal. However, such controllers may be computationally inefficient since the sampling instants are not based on the actual system states and are based on the worst-case analysis. Therefore, event-triggered sampling has become popular [1–5] wherein the execution time of the control inputs are based on the real-time operation of the system. In the event-triggered framework [1–5], the measured state vector is sampled based on certain state dependent condition referred to as event-triggering condition and the controller is executed at those aperiodic sampling instants. The event-triggering conditions are designed by

H. M. Güzey (✉)
Department of Electrical and Electronics Engineering, Erzurum Technical University, Erzurum, Turkey
e-mail: mehmet.guzey@erzurum.edu.tr



J. Ghommam et al. (eds.), *New Trends in Robot Control*, Studies in Systems, Decision and Control 270, https://doi.org/10.1007/978-981-15-1819-5_9

taking into account the stability and closed-loop performance, and, hence, proven to be advantageous over its periodic counterpart.

Initially, the event-triggered techniques from the literature [1, 4, 5] were designed for ensuring stable operation of the closed-loop system by assuming that a stabilizing controller exists for the system under consideration. Developing an event-triggering condition and establishing the existence of positive inter-event time with the proposed event-sampling condition was the main focus in these works [1, 4, 5]. The traditional optimal control problem under the event-sampled framework is studied in [6] while in [2], the event-sampled adaptive controller design was presented for physical systems with uncertain dynamics.

Motivated by the aforementioned benefits of event-triggered control, event-triggered adaptive dynamical regulation controls are provided for MR. The velocity tracking errors those are due to MR dynamics are considered as well. Using the two layer NN-based representations of the MR dynamics, the control inputs are derived to minimize the velocity tracking errors with event-triggered feedback. It should to noted that, in contrast to the existing event-triggered controller of MRs [7, 8] the dynamics of the MR is explicitly taken into account.

For non-holonomic MR, an adaptive event-triggering condition is required to determine the sampling instants to generate the feedback information to update the controllers. Since uncertain parameters are tuned at the event-triggered instants, the computations are reduced when compared to traditional adaptive-control schemes, but it causes aperiodic parameter updates. Therefore, an event-triggering condition is derived using the stability conditions directly to ensure that the performance of the adaptive controller is not deteriorated due to the intermittent feedback. Finally, the extension of the Lyapunov's direct method is used to prove the local uniform ultimate boundedness (UUB) of the tracking errors and the parameter estimation error with event-triggered feedback.

The contributions of the chapter include: (a) the design of novel event-triggered adaptive regulation control of MRs; (b) development of a novel event-triggering condition to determine the sampling instants and generate feedback signals to the controller; and (c) demonstration of stability of the MR with the proposed event-triggered controller using Lyapunov stability theory.

In this chapter, $\mathbb{R}^n$ is used to denote $n$ dimensional Euclidean space. Euclidean-norm is used for vectors while Frobenius norm is used for matrices.

## 2 Background and Problem Formulation

In this section, first, a brief background on the event-triggered control implementation is provided. Then, the dynamics of the nonholonomic MRs are introduced. In this chapter, the state vector is considered measurable.

## 2.1 Event-Triggered Control

The traditional periodic/continuous feedback based control problem is well-established in the literature and several techniques are available to analyze the performance of these controllers. In an event-triggered framework, the system state vector is measured continuously but available to the controllers only at the event-triggered instants. To denote the sampling instants, define an increasing sequence of time $\{t_k\}_{k=0}^{\infty}$, referred to as the event-triggered instants satisfying $t_{k+1} > t_k$, $\forall k = 0, 1 \ldots n$. Let $t_0 = 0$ be the initial sampling instant. At the instant $t_k$, the sampled state $x(t_k)$ is available to the controllers, and the last sampled states at the controllers denoted by $\breve{x}(t)$.

The error $e_{ET}(t)$ introduced due to the event-triggered state can be written as, $\forall k = 0, 1 \ldots n$, $t_k < t < t_{k+1}$ >:

$$e_{ET} = x(t) - \breve{x}(t) \tag{1}$$

where $e_{ET}$ is referred to as event-triggering error. Therefore, the event-triggering errors become zero at every sampling instant and update of the state, that is: $e_{ET}(t_k) = 0$, $\forall k = 0, 1 \ldots n$.

For the event-triggered controllers, as mentioned before, an event-sampling mechanism/condition is required to determine the sampling instants, without jeopardizing the system stability. Also, if the controller parameters are adaptive and updated from the feedback information, the parameter adaptation process is also dependent on the event-based sampling instants. Therefore, the event-sampling mechanism should be carefully designed so that event-based feedback does not impede with the adaptation process of the controller.

Next, the dynamics of the MRs will be presented.

## 2.2 MR Dynamics

Consider the non-holonomic MR shown in Fig. 1, where $x_r$ and $y_r$ denote Cartesian positions with respect to the robot frame, $d$ is the distance between the rear-axis and the center of mass of the robot, $r$ and $R$ are the radius of the rear wheels, and half of the robot width, respectively.

The equations of motion about the center of mass $C$, for the $i$th robot in a networked robot formation are written as:

$$\dot{q} = \begin{bmatrix} \dot{x} \\ \dot{y} \\ \dot{\theta} \end{bmatrix} = \begin{bmatrix} \cos\theta & -d\sin\theta \\ \sin\theta & d\cos\theta \\ 0 & 1 \end{bmatrix} \begin{bmatrix} v \\ \omega \end{bmatrix} S(q)\overline{v} \tag{2}$$

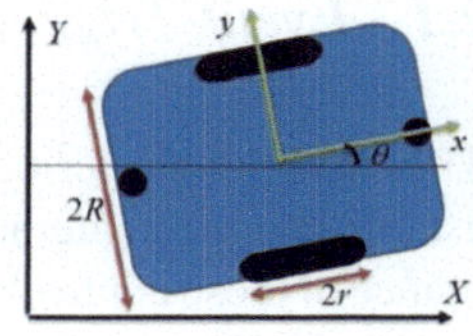


**Fig. 1** Differentially driven mobile robot

with $d$ being the distance from the rear axle to the MR's center of mass; $q = [x,\ y,\ \theta]^T$ denotes the Cartesian positions of the center of mass and orientation of the robot; $\omega$ and $\upsilon$ represent angular and linear velocities, respectively, and $\overline{\upsilon} = [\upsilon,\ \omega]^T$.

MR systems, in general, can be characterized as having an $n$-dimensional configuration space $\mathcal{C}$ with generalized coordinates $(q_1, \ldots, q_n)$ subject to $\ell$ constraints [9]. Applying the transformation, the dynamics of the mobile robot are given by:

$$\overline{M}\dot{\overline{\upsilon}} + \overline{V}_m(q, \dot{q})\overline{\upsilon} + \overline{F}\overline{\upsilon} + \overline{\tau}_d = \overline{\tau} \tag{3}$$

where $\overline{M} \in \mathbb{R}^{\rho\times\rho}$ is a constant positive definite inertia matrix, $\overline{V}_m \in \mathbb{R}^{\rho\times\rho}$ is the bounded centripetal and Coriolis matrix, $\overline{F} \in \mathbb{R}^{\rho}$ is the friction vector, $\overline{\tau}_d \in \mathbb{R}^{\rho}$ represents unknown bounded disturbances such that $||\overline{\tau}_d|| \leq d_m$ for a known constant $d_m$, $\overline{B} \in \mathbb{R}^{\rho\times\rho}$ is a constant, nonsingular input transformation matrix, $\overline{\tau} = \overline{B}\tau \in \mathbb{R}^{\rho}$ is the input vector, and $\overline{\tau} \in \mathbb{R}^{\rho}$ is the control torque vector. For complete details on (3) and the parameters, refer to [9].

**Assumption 1** The MR system satisfies the following properties: $\overline{M}$ is a known positive definite matrix and it is bounded by $B_M$ and $0 < \overline{M}^{-1} < B_m$ the norm of $\overline{V}_m$ and $\overline{\tau}_d$ are all bounded. The matrix $\overline{M} - 2\overline{V}_m$ is the skew-symmetric [9].

## 3 Controller with Periodic Update

Consider the kinematics of the mobile robot. The objective of the regulation controller design synthesis is to stabilize about a desired posture, $q_d = [x_d,\ y_d,\ \theta_d]^T$. However, due to Brockett's theorem [10], smooth stabilizability of the driftless regular system requires the number of inputs to equal to the number of states, a property not satisfied by. The above obstruction has a significant impact on controller design. In fact, to obtain a posture stabilizing controller, it is necessary to use discontinuous and/or time-varying control laws [10].

A technique which allows us to overcome the complication presented by the Brockett theorem is to apply a change of coordinates such that the input vector fields of the transformed equations are singular at the origin. This approach is carried out using a polar coordinate transformation, and the control law, once rewritten in terms of the original state variables, is discontinuous at the origin of the configuration space $\mathcal{C}$.

Consider again the robot shown in Fig. 1. Let $\rho$ be the distance of the point $(x, y)$ of the robot to the goal point $(x_d, y_d)$. Let $\alpha$ be the angle of the pointing vector to the goal with respect to the robot's main axis (labeled as $X_R$ in Fig. 1.), and define $\beta$ to be the angle of the same pointing vector with respect to the orientation error [10]. That is:

$$\begin{aligned}\rho &= \sqrt{(\Delta x)^2 + (\Delta y)^2} \\ \alpha &= \pi - \theta + \operatorname{atan2}(\Delta y, \Delta x) \\ \beta &= \alpha + \theta - \theta_d\end{aligned} \tag{4}$$

where $\Delta x = x - x_d$ and $\Delta y = y - y_d$.

Then, the polar coordinate kinematics of a MR can be given as discussed in [10], and expressed as:

$$\begin{bmatrix} \dot{\rho} \\ \dot{\alpha} \\ \dot{\beta} \end{bmatrix} = \begin{bmatrix} -\cos\alpha & 0 \\ \frac{1}{\rho}\sin\alpha & -1 \\ \frac{1}{\rho}\sin\alpha & 0 \end{bmatrix} \begin{bmatrix} v \\ \omega \end{bmatrix} \tag{5}$$

From Eq. (5), it is observed that the input vector field associated with $v$ is singular for $\rho = 0$, thus satisfying Brockett's Theorem. To drive MRs from any initial position to a goal position, a nonlinear control law is given as [10]:

$$v_d = k_\rho \rho \cos\alpha\,, \quad \omega_d = k_\alpha \alpha + \frac{k_\rho}{\alpha} \sin\alpha \cos\alpha(\alpha + k_\beta \beta) \tag{6}$$

where $k_\rho$, $k_\alpha$ and $k_\beta$ are positive design constants. As shown in [10], the controller (6) provides asymptotic converge to the constant desired posture.

However, the results are obtained by assuming the perfect velocity tracking (assuming that $v_d = v$ and $\omega_d = \omega$ which does not hold in practice. To relax the perfect velocity tracking assumption, the backstepping technique was employed in [11].

Define the velocity tracking error as:

$$e_v^R = \begin{bmatrix} e_{v1}^R \\ e_{v2}^R \end{bmatrix} = \overline{v}_d - \overline{v} \tag{7}$$

where $\overline{v}_d = [v_d, \omega_d]^T$. Rearranging (7) gives $\overline{v} = \overline{v}_d - e_v^R$, and substituting this expression into the open loop system (5) while using (7) reveals:

$$\begin{bmatrix} \dot{\rho} \\ \dot{\alpha} \\ \dot{\beta} \end{bmatrix} = \begin{bmatrix} -k_\rho \rho \cos^2\alpha + e_{v1}^R \cos\alpha \\ -k_\alpha \alpha - \frac{\beta}{\alpha} k_\rho k_\beta \sin\alpha\cos\alpha \\ k_\rho \sin\alpha \cos\alpha - \frac{1}{\rho} e_{v1}^R \sin\alpha \end{bmatrix} \begin{bmatrix} v \\ \omega \end{bmatrix} \tag{8}$$

The closed loop kinematic system (8) explicitly considers the velocity tracking error (7). Therefore, the back-stepping technique ensures the robot tracks the design velocities (6).

Differentiating (6) and using (3), the mobile robot velocity tracking error system as

$$\overline{M}\dot{e}_v^R = -\overline{V}_m(q,\dot{q})e_v^R - \overline{\tau} + f(z) + \overline{\tau}_d \tag{9}$$

with $f(z) = \overline{M}\dot{\overline{v}}_d + \overline{V}_m(q,\dot{q})\overline{v}_d + \overline{F}\overline{v}$ and contains the MR parameters such as masses, moments of inertia, friction coefficients, and so on. When the robot dynamics are known, the control torque applied to the robot system (3), which ensures the desired velocity (6), is achieved and is written as:

$$\overline{\tau} = K_v e_v^R + f(z) + \lambda(\rho,\alpha,\beta) \tag{10}$$

where:

$$\lambda(\rho,\alpha,\beta) = \left[\frac{\left[\rho + \frac{1}{2}(\alpha^2 + k_\beta\beta^2)\right]\cos\alpha - (\alpha + k_\beta\beta)\sin\alpha}{\rho\alpha}\right]$$

is a function of the polar coordinate error system (4) and is required for stability. Substituting (10) into (9) reveals the closed loop velocity tracking error dynamics

$$\overline{M}\dot{e}_v^R = -K_v e_v^R - \overline{V}_m(q,\dot{q})e_v^R + \overline{\tau}_d - \lambda(\rho,\alpha,\beta) \tag{11}$$

**Lemma 1** *Given the MR system described by (3) and, (10) let the velocity tracking error and its dynamics for driving the nonholonomic system to the goal configuration* $q_d$, *be given by (7) and (9), respectively, and let the control torque vector be defined by (10). Then, in the absence of disturbances* ($\overline{q}_d = 0$), *the velocity tracking error system (9) and kinematic system (5) converge to the origin asymptotically, and the robot reaches its desired velocity and converges to its desired posture. That is:* $e_v^R(t,t_0) \longrightarrow 0$ *and* $q \longrightarrow q_d$.

***Proof*** See [11] for the proof.

Next, event sampled regulation controller is developed.

## 4 Event Sampled Controller Design

In this section, the NN controller design with event-sampled feedback, for the nonholonomic MRs will be presented and the aperiodic NN weight adaptation law will be derived from the Lyapunov stability analysis. The event-sampling mechanism is designed using stability analysis such that the event-sampling error (1) satisfies a state-dependent threshold for every inter-event period for each robot, which is of the form:

$$||\overline{E}_{ET}|| \leq \sigma\mu_k||\breve{E}|| \tag{12}$$

for $t_k \leq t < t_{k+1}$, $k = 1, 2, 3$, and where $0 < \sigma < 1$, and $\mu_k$ is a positive design parameter, and $E_{ET}$, $\breve{E}$ are functions of event-sampling error and the formation, and velocity tracking errors respectively. By using the event-sampled feedback, the objective is to reduce the computations from periodic parameter adaptation without compromising the stability while ensuring acceptable velocity tracking performance.

**Definition 1** First, define for $t_k \leq t < t_{k+1}$, and $\forall i = 1, 2, \ldots N$:

$$\begin{cases} \rho(t) = \rho(t_k), \ \alpha(t) = \alpha(t_k), \ \beta(t) = \beta(t_k), \\ \varepsilon_\rho(t) = \varepsilon_\rho(t_k), \ \varepsilon_\alpha(t) = \varepsilon_\alpha(t_k), \ \varepsilon_\beta(t) = \varepsilon_\beta(t_k), \end{cases} \tag{13}$$

Now, to define the regulation error dynamics with event-sampled measurement error, consider (6), during the inter-event period, the desired virtual control equations are obtained as

$$\breve{\overline{v}}_d \begin{bmatrix} \breve{v}_d \\ \breve{\omega}_d \end{bmatrix} = \begin{bmatrix} k_\rho \breve{\rho} \cos \breve{\alpha} \\ k_\alpha \breve{\alpha} + \frac{k_\rho}{\breve{\alpha}} \sin \breve{\alpha} \cos \breve{\alpha} (\breve{\alpha} + k_\beta \breve{\beta}) \end{bmatrix} \tag{14}$$

for $t_k \leq t < t_{k+1}$.

After defining the event-sampled signals, (14) can be rewritten with the measurement errors, for $t_k \leq t < t_{k+1}$, as

$$\dot{\breve{\overline{v}}}_d \begin{bmatrix} \breve{v}_d \\ \breve{\omega}_d \end{bmatrix} = \begin{bmatrix} k_\rho \rho \cos \alpha + \varepsilon_v \\ k_\alpha \alpha + \frac{k_\rho}{\breve{\alpha}} \sin \alpha \cos \alpha (\alpha + k_\beta \beta) + \varepsilon_\omega \end{bmatrix} \tag{15}$$

with $\varepsilon_v$, $\varepsilon_\omega$ are given by:

$$\begin{aligned} \varepsilon_v &= k_\rho \breve{\rho} \cos \breve{\alpha} - k_\rho (\breve{\rho} + \varepsilon_\rho) \cos(\breve{\alpha} + \varepsilon_\alpha) \\ \varepsilon_\omega &= k_\alpha \breve{\alpha} + \frac{k_\rho}{\breve{\alpha}} \sin \breve{\alpha} \cos \breve{\alpha} (\breve{\alpha} + k_\beta \breve{\beta}) \\ & -k_\alpha \alpha - \frac{k_\rho}{\breve{\alpha}} \sin(\breve{\alpha} + \varepsilon_\alpha) \cos(\breve{\alpha} + \varepsilon_\alpha) \big[ (\breve{\alpha} + \varepsilon_\alpha + k_\beta (\breve{\beta} + \varepsilon_\beta) \big] \end{aligned}$$

*Remark 1* Realize that the event triggering errors $\varepsilon_v, \varepsilon_\omega$ are the functions of the event triggering errors of the each state $\varepsilon_\rho$, $\varepsilon_\alpha$, $\varepsilon_\beta$ as well as the last measured state values $\breve{\rho}$, $\breve{\alpha}$, $\breve{\beta}$ which is available for the controller in the interval $t_k \leq t < t_{k+1}$.

To get the closed-loop formation error dynamics in the presence of measurement error, use (15) in (5), which reveals the event-sampled regulation error dynamics as:

$$\begin{bmatrix} \dot{\rho} \\ \dot{\alpha} \\ \dot{\beta} \end{bmatrix} = \begin{bmatrix} -k_\rho \rho \cos^2 \alpha + \left(e_{v1}^R - \varepsilon_v\right) \cos \alpha \\ -k_\alpha \alpha - \dfrac{k_\rho \sin \alpha \cos \alpha}{\alpha} k_\beta \beta + \dfrac{\sin \alpha}{\rho} (\varepsilon_v - e_{v1}^R) + (e_{v2}^R - \varepsilon_\omega) \\ k_\rho \sin \alpha \cos \alpha + \dfrac{\sin \alpha}{\rho} (\varepsilon_v - e_{v1}^R) \end{bmatrix} \quad (16)$$

The closed loop regulation error dynamics in the presence of event trigger errors are obtained in (16). Similar to that, the velocity tracking errors in the event sampled framework will be derived next.

The unknown NN weights can be estimated as $\widehat{\Theta}$ and estimate of the unknown dynamics with event sampled feedback can be obtained, for $t_k \leq t < t_{k+1}$, as:

$$\widehat{f}(z) = \widehat{\Theta}^T \psi(\breve{z}) \quad (17)$$

with $\widehat{z} = z + \varepsilon_z$, being the event-sampled signals of the mobile robot. The unknown NN weight estimation error is defined as $\widetilde{\Theta} = \Theta - \widehat{\Theta}$ and the estimation error dynamics can be given as $\dot{\widetilde{\Theta}} = -\dot{\widehat{\Theta}}$. The event-sampled control torque, using (17), is obtained, for $t_k \leq t < t_{k+1}$, as:

$$\overline{\tau} = -K_v \breve{e}_v + \widehat{f}(z) - \breve{\gamma}(\breve{\rho}, \breve{\alpha}, \breve{\beta}) \quad (18)$$

where:

$$\breve{e}_v = e_v + e_{ET} \quad (19)$$

with the event triggered velocity tracking error $\breve{e}_v(t) = e_v(t_k)$ is defined similar to the event triggered formation errors and $\breve{\gamma}$ is the stabilizing term with measurement error due to event-sampled mechanism.

Substituting (18) into (9) reveals the closed-loop velocity tracking error dynamics for $t_k \leq t < t_{k+1}$:

$$\overline{M}\dot{e}_v = -K_v e_v + \tau_d - \breve{\gamma}(\breve{\rho}, \breve{\alpha}, \breve{\beta}) + \widetilde{f}(\breve{z}) - K_v e_{ET} + [f(z_i) - f(\breve{z}_i)] - V_{mi}(q_i, \dot{q}_i) e_{iv}^F \quad (20)$$

where $\widetilde{f}(\breve{z}) = \widetilde{\Theta}^T \psi(\breve{z}) + \chi$. With the regulation error dynamics (16) and the velocity tracking error dynamics (20) driven by the event-sampling errors, the stability results for MRs are presented in the theorem statements.

Next the definition for UUB is introduced.

**Definition 2** ([12]) An equilibrium point $x_e$ is said to be uniformly ultimately bounded (UUB) if there exists a compact set $S \subset \mathbb{R}^n$ so that for all $x_0 \in S$ there exists a bound $B$ and a time $T(B, x_0)$ such that $||x(t) - x_e|| \leq B$ for all $t \geq t_0 + T$.

The following theorems provide the stability analysis for the control law that does not require the perfect velocity tracking assumption. As noted before, the controller is updated only at the event-sampling instants.

**Theorem 1** (Input-to-state stability) *Given the regulation error dynamics (16) and the velocity tracking error dynamics (20) of the robot, let the regulation controller (18) be applied to the robot. Define the control torque by (18) with:*

$$\breve{\gamma}(\breve{\rho}, \breve{\alpha}, \breve{\beta}) = \begin{bmatrix} \left[\breve{\rho} + \frac{1}{2}(\breve{\alpha}^2 + k_\beta \breve{\beta}^2)\right] \cos \breve{\alpha} - (\breve{\alpha} + k_\beta \breve{\beta}) \sin \breve{\alpha}_i \\ \breve{\rho}\breve{\alpha} \end{bmatrix} \tag{21}$$

Further, tune the unknown NN weights using the adaptation rule:

$$\dot{\widehat{\Theta}} = \Lambda_1 \psi(z) e_v^T - \Lambda_1 \kappa \widehat{\Theta} \tag{22}$$

with $\Lambda_1 > 0$, $\kappa > 0$ are small positive design parameters and with the measurement error satisfying the inequality $||E_{ET}|| \leq \overline{B}_{ETM}$, with $\overline{B}_{ETM}$ being a positive constant. Consider Assumptions 1 holds. The velocity tracking (20) and regulation errors (16) are UUB and (a) the robot reach any desired posture in the presence of bounded measurement error and (b) the closed-loop system is input-to-state stable (ISS), with the input being a function of the measurement error $e_{ET}$.

***Proof*** See [13] for similar proofs.

In the following theorem, the event-triggering mechanism is designed and stability of the MR is analyzed by using the Lyapunov stability theory in the presence of disturbance torque input and NN re-construction error.

**Theorem 2** *Given the regulation error dynamics (20) for the robot with the disturbance torque and the NN approximation error $\overline{\tau}_d \neq 0$, $\chi_i \neq 0$, respectively.*

Consider the Assumptions 1 holds. Let the regulation control input, (18) with (21), be applied to the MR at the event-based sampling instants and the event-sampling condition be defined by (12). Further, consider the unknown NN weights are tuned at the event sampling instants using the aperiodic tuning rule (22). Then the velocity tracking error (20) and regulation error system (16) are UUB and a) the MR reaches any desired posture.

***Proof*** See [13] for similar proofs.

## 5 Results and Discussions

To illustrate the effectiveness of the proposed event-triggered adaptive controller, a non-holonomic MR is considered. The MR is initiated from an arbitrary position and move to a given desired location. Using the proposed approach, the MR reaches to its desired location while ensuring the velocity tracking errors converge to small values. The initial and desired positions, bearing angles of the non-holonomic MR are given by:

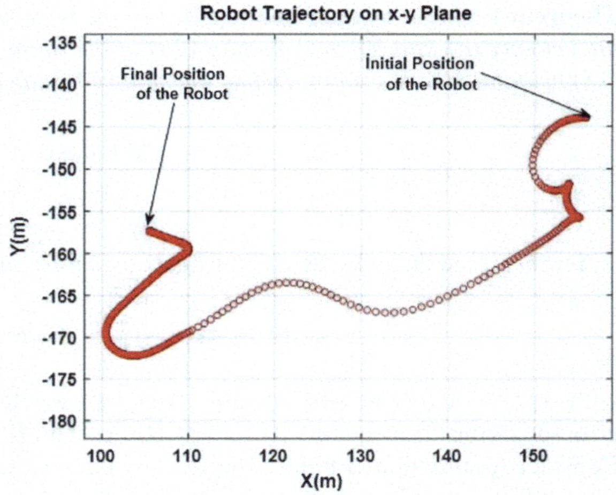


**Fig. 2** Mobile robot moving to its desired location

$$x(t_0) = 156,\ x^d = 106,\ y(t_0) = -144,\ y_d = -157,\ \theta(t_0) = \frac{\pi}{3},\ \theta^d = -\frac{\pi}{3}$$

The controller gains are selected as.

$$K_v = 30,\ k_\rho = 0.9,\ k_\alpha = 0.1,\ k_\beta = 0.5$$

The parameters for the robot dynamics are selected as

$$m = 3\,\text{kg},\ I = 3\,\text{kgm}^2,\ R = 0.25\,\text{m},\ d = 0.6\,\text{m}$$

Figure 2 depicts the motion of the non-holonomic mobile robot.

With the parameter adaptation using $\Lambda = 0.2$, $\kappa = 0.1$ the controllers of the robot learn the unknown parameters. After the uncertain parameters are tuned, the MR starts moving toward its desired location on $x$, $y$.

The difference between the desired angular and linear velocities and the actual angular and linear velocities which are the velocity tracking errors are plotted in Fig. 3.

In the event-sampling mechanism, the event triggering coefficient is selected as $\sigma = 0.5$. With the proposed event-sampled feedback, the MR was able to reach it's desired location. However, due to the aperiodic feedback, the error bounds are slightly larger than their continuous counterparts.

Due to the event-sampling mechanism developed in this work, the total number of parameter update and the communication instants are considerably reduced as seen in Fig. 4.

Next, some concluding remarks are given.

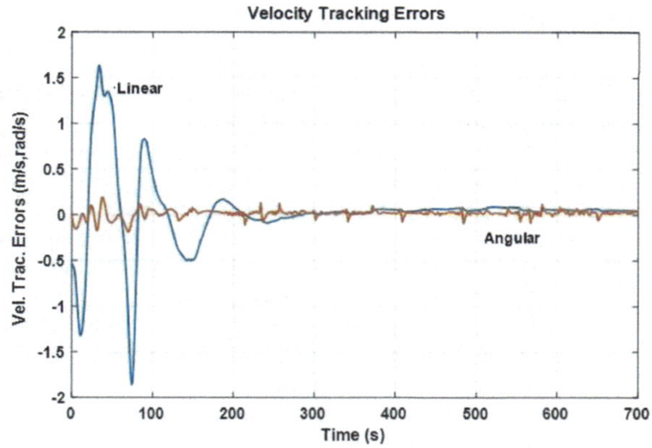


**Fig. 3** Velocity tracking errors

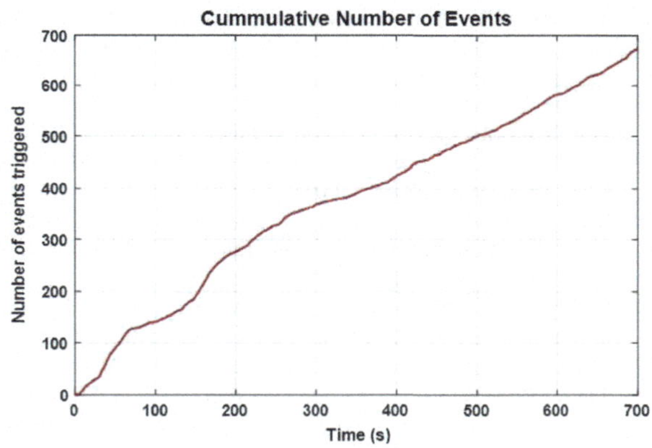


**Fig. 4** Cumulative number of events of the robot

## 6 Conclusion

In this chapter, an event-triggered controller for the MR system was presented. The adaptive event-triggered regulation torque control for the nonholonomic MR was able to drive the system to its goal position with bounded error due to event triggered measurement errors. The event-triggering mechanism was able to generate additional events so that the velocity tracking error remains bounded in the presence of disturbance.

## References


1. Tabuada, P.: Event-triggered real-time scheduling of stabilizing control tasks. IEEE Trans. Autom. Control **52**(9), 1680–1685 (2007)
2. Sahoo, A., Xu, H., Jagannathan, S.: Near optimal event-triggered control of nonlinear discrete-time systems using neuro-dynamic programming. IEEE Trans. Neural Netw. Learn. Syst. **27**(9), 1801–1815 (2016)
3. Zhong, X., Ni, Z., He, H., Xu, X., Zhao, D.: Event-triggered reinforcement learning approach for unknown nonlinear continuous-time system. In: International Joint Conference on Neural Networks (IJCNN), pp. 3677–3684 (2014)
4. Guinaldo, M., Lehmann, D., Sanchez, J., Dormido, S., Johansson, K.H.: Distributed event-triggered control with network delays and packet losses. In: Annual Conference on Decision and Control (CDC), pp. 1–6 (2012)
5. Wang, X., Lemmon, M.D.: Event-triggering in distributed networked control systems. IEEE Trans. Autom. Control **56**(3), 586–601 (2011)
6. Molin, A., Hirche, S.: On the optimality of certainty equivalence for event-triggered control systems. IEEE Trans. Autom. Control **58**(2), 470–474 (2013)
7. Chen, X., Hao, F., Ma, B.: Periodic event-triggered cooperative control of multiple non-holonomic wheeled mobile robots. IET Control Theory Appl. **11**(6), 890–899 (2017)
8. Yilong, Q., Fei, C., Linying, X.: Distributed event-triggered control for coupled nonholonomic mobile robots. In: 34th Chinese Control Conference (CCC), pp. 1268–1273 (2015)
9. Lewis, F.L., Dawson, D.M., Abdallah, C.T.: Robot Manipulator Control: Theory and Practice. CRC Press, Boca Raton (2003)
10. Miah, S., Chaoui, H., Sicard, P.: Linear time-varying control law for stabilization of hopping robot during flight phase. In: 23rd International Symposium on Industrial Electronics (ISIE), pp. 1550–1554 (2014)
11. Guzey, H.M., Dierks, T., Jagannathan, S., Acar, L.: Hybrid consensus-based control of non-holonomic mobile robot formation. J. Intell. Robot. Syst. **88**(1), 181–200 (2017)
12. Dierks, T., Jagannathan, S.: Neural network output feedback control of robot formations. IEEE Trans. Syst. Man Cybern. Part B (Cybern.) **40**(2), 383–399 (2010)
13. Guzey, H.M.: Adaptive consensus based formation control of unmanned vehicles. Doctoral Dissertations. 2537 (2016)

# Optimal Lane Merging for AGV

**Kawther Osman, Jawhar Ghommam and Maarouf Saad**

**Abstract** This paper addresses the generation of an optimal constrained trajectory for lane merging tasks. The developed algorithm ensure a safe and a less fuel consumption for autonomous driving. The security is established with the restriction of both the lateral and the longitudinal AGV's position inside a safe zone while accomplishing a lane change to overtake an obstacle or to follow a lead vehicle. However the path's optimality is carried out by minimizing the lateral and the longitudinal cost functions. The generated trajectory is then tracked accurately with an adaptive computed torque controller. The whole approach is then validated with numerical simulations.

**Keywords** Automated guided vehicle (AGV) · Longitudinal control · Lateral control · Adaptive computed torque · Lane merging

## 1 Introduction

During the last decades, the researchers investigated the automation of driving to make it more comfortable with the implementation of decision support systems and sophisticated controller. Therefrom, the lane keeping assist system are developed to restrict the lateral vehicle's position inside the lane during its travel. On the other hand, the cruise control system acts on the safety distance between the ego vehicle and

K. Osman (✉)
CEM-Lab and the National School of Engineering of Sousse, University of Sousse, Sousse, Tunisia
e-mail: osmankawthar22@gmail.com

J. Ghommam
Departement of Electrical and Computer Engineering, College of Engineering, Sultan Qaboos University, Al Khoudh, Mascut, Oman
e-mail: jawhar.ghommam@gmail.com

M. Saad
Department of Electrical Engineering, Ecole de Technologie Superieure, Montreal, Canada
e-mail: Maarouf.Saad@etsmtl.ca

 
J. Ghommam et al. (eds.), *New Trends in Robot Control*, Studies in Systems, Decision and Control 270, https://doi.org/10.1007/978-981-15-1819-5_10

the preceding one by limiting its longitudinal position to avoid crashes. However for lane change scenario, one may combine both aforementioned systems which make the vehicle increasingly autonomous. In fact, lane merging are performed whether to overtake a front slow vehicle or a static obstacle, it can be concerned also in the case of leader vehicle following namely the platooning as in [27]. To ensure the correct merging, some sensors are needed such as the blind spot to make sure there is no other vehicle overtaking the ego one, a vision sensor to find out the lane marking and limits, a radar to detect the static or the dynamic obstacle to overtake and to calculate the distance between it and the ego vehicle. By validating the security conditions for lane merging, one can change its main lane safely.

In this work, we propose to restrict the AGV's position while generating an optimal trajectory to be followed by the AGV during the lane merging. One way to proceed is by using sensors to specify the safety distance w.r.t the front vehicle and to define the road boundaries in which the optimal path may be bounded.

## 1.1 Related Works

In order to improve road safety and to optimize consumed fuel, a cooperation between various automated driving functionalities are addressed during the last decades. The paper of [4] tackled a cooperative merging for platoons based on vehicle to vehicle communication and a fusion of data sensors to participate in the second grand cooperative driving challenge (GCDC). The cited work used both lateral and longitudinal control for highway roads to perform the lane merging. The used sensors for such a maneuver present an important factor, thus the researchers [7] expressed the importance of the blind spot information system in detecting the coming vehicle rapidly and hence to avoid eventual crashes also, it elaborated the essential conditions for merging such as activating the indicator before lane changing. However, the lane change task was defined as the riskiest driving maneuver in [8], to ensure a safe merging, they defined the minimum allowable longitudinal distance between the front and the ego vehicle.

In other hand, the authors of [19] emphasised the importance of using optimal path planner due to its high efficiency. On this concern, some researchers proposed to optimize the lane merging. In fact, [28] resolved the lane changing conflicts while reducing the total travel time and optimizing the vehicle density across lanes by combining the reactive and proactive lateral control for highway roads. The authors in [2] suggested a proactive optimal merging strategy that optimized the merging time specifically at on-ramp, reduced the bottlenecks and economized the fuel consumption while improving the traffic flow, so it acted basically on the averages of the merging and waiting time, the vehicle's rate and velocity. Otherwise, the work in [9] determined the risks associated with the lane merging in connection with the longitudinal vehicle's acceleration, to cover such risk, the authors anticipated the gap between the ego vehicle and the preceding one to decide whether the lane changing was performed or not.

The researchers in [22] proposed to decompose the lane merging task into three phases, namely, the diverting from the main lane, then, a straight driving in the next lane, the traveled distance here depend on the overtaken vehicle's speed and finally the return to the initial lane which is symmetric to the first phase. To design a smooth and an ergonomic optimal trajectory, the authors got started from a mathematical model based on the minimum jerk trajectory and constrained within a predetermined velocity and acceleration. As well as the work of [26], it planned a traffic adapted path. The proposed system guarantied a good performance in generating an optimal path for lane keeping and merging and collision avoidance in the Frenet frame. This is accomplished by combining the lateral and the longitudinal optimization problem. The authors [21], developed an algorithm that ensure a smooth lane merging with optimizing the vehicle's acceleration and fuel consumption while avoiding obstacle.

To increase safety driving, the work in [24] focused basically on the comfort that an automated lane merging system might offer to passenger and this by developing a probabilistic representation of the vehicle's environment with fusing various data sensors. The developed proactive system allowed to keep the vehicle inside the traveled lane and perform a security distance w.r.t the front vehicle. In other hand, the authors [1] proposed to automate the driving of the AGV to accomplish a collision free motion using two types of controllers based on the initial conditions. [11] tackled a path planning and tracking problem while ensuring a collision free motion to handle different driving routine including lane keeping and merging. Otherwise, [14] developed an innovative algorithm that made rapid decision on lane changing and perform aggressively reaction by collecting sufficient and reliable data form various sensor modalities. The proposed approach allowed to merge lanes as humans.

All the aforementioned works dealated with different types of optimal path planner whether by limiting the vehicle's dynamic or by minimizing the travel time but none of them restrict both the longitudinal and the lateral vehicle's position while planning an optimal path.

In order to ensure an autonomous lane merging, the generated trajectory should be implemented into a controller for AGV. In [12], an optimized road model was parameterized using the non linear integer programming model. The authors in the mentioned paper avowed that the system's stability was affected by the force exchange rate, the velocity's average and the gradient rate.

However in [17], the authors implemented a model predictive control that help driver to take a decision concerning lane merging. While in [6], an analytic approach was proposed for controller developing which allowed a smooth lane change in highway road. The cited paper solved a path tracking problem with optimal longitudinal speed and this was accomplished using a robust switching controller based on the generation of a virtual desired yaw orientation. Others works proceeded the applied longitudinal control for lane change as in [3], the authors first treated a path planning based on LIDAR and stereo vision, secondly they proposed a PID longitudinal controller that acted on AGV's steering wheel, brake and throttle.

Otherwise, some works addressed the lateral control. In [15], among the toughest automation challenge was treated using a lateral fuzzy controller that mimic human driving. This controller permitted to handle two consecutive lane changes, the trajectory generation is based itself on collected data from GPS sensor and a wireless communication between the concerned vehicles. Another lateral control were performed in [5], The mentioned work addressed a predictive control framework based on the concept of a driver in the loop. The proposed controller imposed some constraints on the control input to obtain an optimal lateral control while it handled the AGV's actuator limits. But what if the environment parameters are not measurable or were uncertain, to tackle this problematic, an adaptive sliding mode controller were investigated in [16], besides, a fuzzy boundary layer were added to avoid chattering. The aforementioned works treated weather longitudinal control wether the lateral control separately. However few works combined both types of control. In [13] an intelligent controller was developed in order to overcome the AGV's model uncertainties. The proposed approach addressed the lateral and the longitudinal dynamics while monitoring the vehicle's steering angle, its throttle and brake.

## 1.2 Main Contribution

This work addressed two main contributions. First, an optimized path is planned to perform a constrained lane change routine. In other words, the vehicle is driven autonomously inside a safe zone of the road and this is performed while keeping a security distance between the ego vehicle and the preceding one by bounding its longitudinal position besides, the lateral AGV's position should be restricted inside an interval of two lane. The reason for constraining such a trajectory is to guaranty a safe lane merging free of accidents or drifts.

As a second contribution of this work, the planned path is followed autonomously based on an adaptive computed torque controller(ACT). For sake of clarity, the ACT is chosen because it covers the uncertainties of the parameters of the adopted vehicle model. In [23], ACT proved its efficiency in monitoring a robot arm with high accuracy. Hence we propose to apply it in controlling the lateral and the longitudinal vehicle's position in order to follow the generated optimal constrained trajectory.

## 1.3 Outlines

The remainder of this chapter is structured as follow: In Sect. 2, an algorithm for constrained and optimal path for lane merging is developed wether for vehicle overtaking or for platooning. The Sect. 3 presents the modeling and the resultant dynamic model of the used vehicle in this work. An adaptive computed torque controller is designed in Sect. 4. The whole approach is validated with a numerical simulation in Sect. 5.

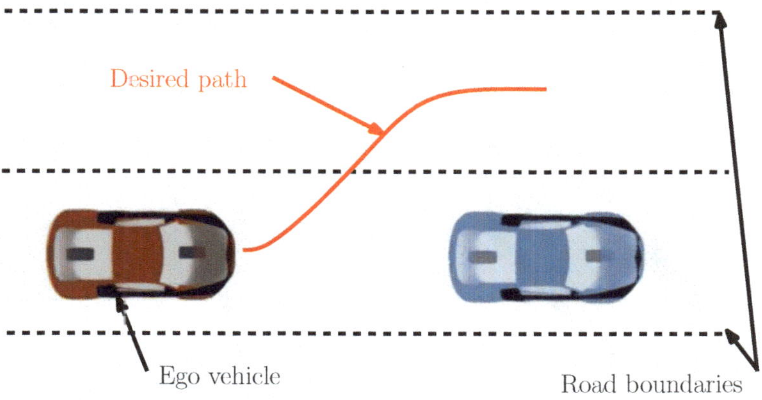


**Fig. 1** The lane change

## 2 Optimal Path Planning

In this section, an optimal trajectory planner is proposed to perform lane merging tasks which cover the case of obstacle avoidance or vehicle following scenarios. It is worth to mention that the faced obstacles can be static or dynamic. In order to overtake it, the ego vehicle should merge to the adjacent lane smoothly, exceed the obstacle and come back to the initial lane. This scenario can be summarized in Fig. 1 as two consecutive lane merging. Obviously, we can divide the planned trajectory into three phases, the first is the first lane merging, the second one is overtaking the obstacle, this phase depends on the velocity of the ego vehicle and the obstacle. The last phase is the return to the initial lane. Hence, it is shown that the lane merging is a symmetric operation. In this paper, we tackle the problem of generating an optimal path for one lane merging and the second one can be derived from a symmetric relationship in the case of overtaking obstacle. However, only one lane change is required for merged road such as on ramp.

### *2.1 Problem Formulation*

In this section, the objective behind developing an optimal path planner is defined according to a simplified road frame. To guaranty a safe lane change, an optimal trajectory is generated while imposing some constraints on the vehicle's movement. Namely, we propose to limit the lateral vehicle's position with respect to the road's boundaries and its longitudinal position within a secure distance from the preceding obstacle. As cited in [26], the street's Frenet frame $[n_c, t_c]$ is adopted as in Fig. 2 where the desired trajectory $x(t)$ is defined by $d(t)$ as the lateral offset from the

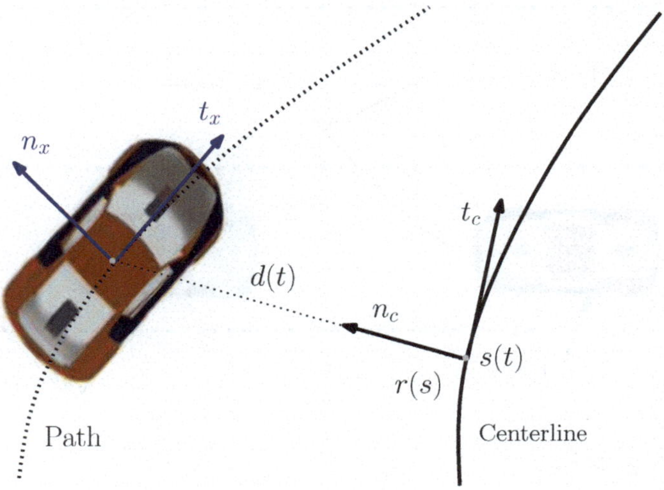


**Fig. 2** The road frenet representation

current lane's center line and $s(t)$ that denotes the covered arc length of a point $r(s)$ from the lane's centerline.

In order to define the optimal trajectory for lane changing routine, we propose to minimize the lateral and the longitudinal AGV's position while imposing strict constraint on both coordinates. So the generated path will be optimal and constrained for more comfortable and secure driving. The resulted cost J to minimise is then the sum of the of lateral cost $J_{lat}$ and longitudinal cost $J_{long}$.

## 2.2 *Optimal Path Planning*

One way to proceed is to minimise the lateral and the longitudinal position's jerk. In fact, a combination of data sensors such as GPS and IMU may be collected and derived to obtain an accurate jerk position. The reason for choosing such a parameter is to have a smooth trajectory. Thus, the optimisation problem is formulated as cited in [26]:

$$\begin{aligned} X_1 &= \dot{X}_2 \\ X_2 &= \dot{X}_3 \\ X_3 &= u \end{aligned} \tag{1}$$

The minimisation of the lateral offset $d(t)$ is first treated, therefore the lateral jerk and the performance index are defined respectively as:

$$u = \dddot{d}(t)$$
$$J = \frac{1}{2}\int_{t_i}^{t_f} u^2(t)dt \quad (2)$$

To guarantee a secure driving, the lateral movement of the vehicle is limited within the road's edges which can be defined using a vision based algorithm, i.e for two lanes roads, the lateral offset is limited inside the road such as: $d(t) \in [0, 2w]$, where $w$ is the lane width. Hence the new constrained cost function of lateral offset is:

$$J_{lat} = k_j \int_{t_i}^{t_f} \dddot{d}(t)dt + k_d[d_i - d_f]^2 \quad (3)$$

such as $k_j$, $k_d$ are positive gains, $t_i$ and $t_f$ are the initial and the final lane change time, $d_i$ and $d_f$ denote the lateral offset's minimal and maximal constraint. By analogy, the cost function of the longitudinal arc length is:

$$J_{long} = k_g \int_{t_i}^{t_f} \dddot{s}(t)dt + k_s[s_i - s_f]^2 \quad (4)$$

while $s(t)$ must not exceed the security distance $ds$, i.e $s(t) \in [0, ds]$. The security distance can be obtained from a radar sensor or vision based system. Otherwise, $k_g$ and $k_s$ are the positive gains $s_i$ and $s_f$ are the applied constraints on the arc length. In this work, the reserved time for the lane change task is assumed to 3 s. Thus, the vehicle is moving from one lane's centerline to the adjacent lane's centerline without exceeding the security distance.

After the definition of the state system in (1) and the constrained lateral (3) and longitudinal (4) costs, the optimal path can be generated on the basis of this parameters. Although we add constraints on the lateral and longitudinal position, new coefficients are needed to find out the optimal path. Therefore, Hamiltonian equations are inserted as cited in [25]. By using the following Eq. (5), we get the optimal coefficients and so on the optimal path.

$$\dot{p}_i^* = -\frac{\partial H}{\partial X_i^*}$$
$$\frac{\partial H}{\partial u^*} = 0 \quad (5)$$

Therefore, to generate the optimal path, we need to solve numerically 6 equations with 6 unknown coefficient as presented:

$$\begin{bmatrix} X_1 \\ X_2 \\ X_3 \\ p_1 \\ p_2 \\ p_3 \end{bmatrix} = \begin{bmatrix} 1 & t & t^2 & t^3 & t^4 & t^5 \\ 0 & 1 & 2t & 3t^2 & 4t^3 & 5t^4 \\ 0 & 0 & 2 & 6t & 12t^2 & 20t^3 \\ 0 & 0 & 0 & 6 & 24t & 60t^2 \\ 0 & 0 & 0 & 0 & -24 & -120 \\ 0 & 0 & 0 & 0 & 0 & 120 \end{bmatrix} \begin{bmatrix} a_0 \\ a_1 \\ a_2 \\ a_3 \\ a_4 \\ a_5 \end{bmatrix} \tag{6}$$

At this stage, we must develop the equation of the lateral offset $d(t)$ and the arc length $s(t)$ in order to transport the vehicle from an initial into a final configuration while limiting its motion in a safe zone. Namely, we need to minimize the trajectory cost while defining $X(t_i) = [X_1(t_i), X_2(t_i), X_3(t_i)]$ and $X(t_f) = [X_1(t_f), X_2(t_f), X_3(t_f)]$. We assume that the vehicle changes its lane smoothly as presented in Fig. 1, the lateral and the longitudinal motion are developed respectively as following

$$\begin{aligned} s(t) &= \alpha \tan\left(\frac{t}{3}\right) \\ d(t) &= \omega \sin\left(\frac{\Pi}{6}\right) + \frac{\omega}{2} \end{aligned}$$

The aforementioned equations allow to calculate the first and the second derivative of lateral and longitudinales AGV's position in order to define the initial and the final path's configuration. As we work on a fixed time interval of 3 s, a numerical solver (i.e f bvp4c of MATLAB) can be used to define the expression of $X_i$ and $p_i$ for $i = 1 \ldots 3$. And therefore, the optimal constrained equation of $s^*(t)$ and $d^*(t)$ are obtained as following:

$$\begin{aligned} s^*(t) &= \Sigma_0^5 a_{si}^* t^i \\ d^*(t) &= \Sigma_0^5 a_{di}^* t^i \end{aligned} \tag{7}$$

where $\alpha =$ Security distance$/1.5574$ and $\omega$ is the road lane width. Considering that the first state $X_1(t) = s(t)$, (by analogy for $d(t)$), the states $X_2(t)$ and $X_3(t)$ are calculated according to the Eq. (1). In order to perform an autonomous lane change either for obstacle overtaking or for vehicle following, an adaptive controller is developed. This system allows to the AGV to follow the generated optimal trajectory.

## 3 Dynamic AGV Modeling

In this work, the single track vehicle model is adopted in order to simulate the longitudinal and the lateral AGV's motion. Some assumptions are considered to simplify the design of the AGV model. In fact, the left and right wheels are regrouped into one central wheel for both front and rear axes, also, the pitch and roll angles are neglected. Hence, the vehicle model is expressed in Fig. 3 as in [18, 20].

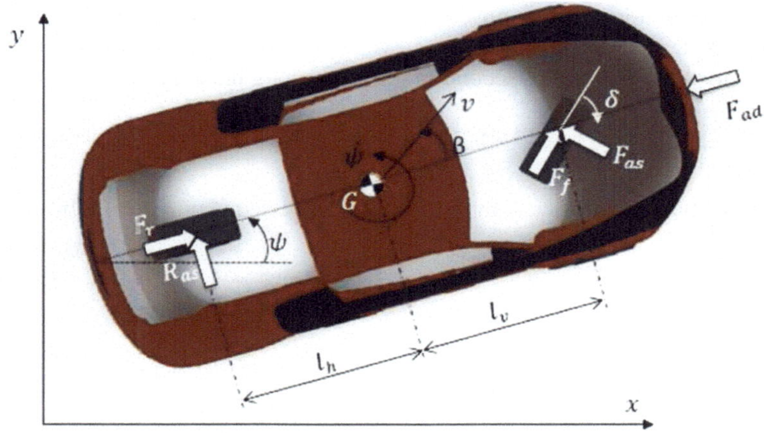


**Fig. 3** AGV's model and applied forces

The equations of the vehicle's motion are defined using geometrical relationships. According to Fig. 3, one can define the AGV's lateral and longitudinal motions $\dot{x}$ and $\dot{y}$ by a projection of the velocity vector on the x and y axis respectively.

$$\begin{aligned} \dot{x} &= v\cos(\psi - \beta) \\ \dot{y} &= v\sin(\psi - \beta) \end{aligned} \tag{8}$$

The rotational motion w.r.t the global z axis namely the yaw rate is parameterised as:

$$\dot{\psi} = r \tag{9}$$

The sideslip angle of the vehicle's wheel can be defined using the Ackerman differential equation as:

$$mV(r - \dot{\beta}) = \sum(Forces) \tag{10}$$

Hence

$$\dot{\beta} = r - \frac{1}{mv}\{(F_r - F_{ad})\sin(\beta) + F_f\sin(\delta + \beta) \tag{11}$$

$$F_{as}\cos(\delta + \beta) + R_{as}\cos(\beta)\} \tag{12}$$

The aforementioned kinematic equations are not enough to control the AGV with dynamic linear and angular velocity, unless by defining its derivative.

By applying the Newton's second motion law along the y axis

$$m\dot{v} = \sum forces \tag{13}$$

Also the moment balance w.r.t the z axis is expressed geometrically by

$$I\ddot{\psi} = I\dot{r} \tag{14}$$

$$= F_{as}l_v \cos(\delta) - R_{as}l_h \tag{15}$$

From the Eqs. (13) and (14), the linear and the angular acceleration are expressed as:

$$\begin{aligned}
\dot{r} &= \frac{1}{I}\{F_{as}l_v \cos(\delta) - R_{as}l_h\} \\
\dot{v} &= \frac{1}{m}\{(F_r - F_{ad})\cos(\beta) + F_f \cos(\delta + \beta) \\
&\quad - F_{as}\sin(\delta + \beta) - R_{as}\sin(\beta)\}
\end{aligned} \tag{16}$$

where $x$, $y$ denote the lateral and the longitudinal vehicle's position, $\psi$, $\delta$ and $\beta$ are the vehicle's yaw angle, the steering angle and the side slip angle respectively. While $r$ and $v$ define the vehicle's yaw rate and linear velocity respectively, $F_{as}$, $R_{as}$ present the vehicle's aerodynamic front and rear forces and $F_{ad}$ is the front aerodynamic drag force.

The control input of this model are the rear longitudinal force $F_r$ that ensure the drag of the vehicle and the steering angle $\delta$ which allows the lateral motion of the vehicle, in other words, to make the vehicle move from one lane to the other one. The expressions of these control input are illustrated as:

$$Fr = \frac{\tau_w}{rad} \tag{17}$$

$$\delta = \arccos\left(\frac{\tau_r}{F_{as}l_v}\right) \tag{18}$$

Hence, the control problem is the definition of aforementioned torques $\tau_w$ and $\tau_r$. A compact AGV's model representation is parameterized as follow

$$\mathbf{M}\ddot{z} + \mathbf{G}(z, \dot{z}) + f = \tau \tag{19}$$

such as the torque vector is

$$\tau = \begin{bmatrix} \tau_w \\ \tau_r \end{bmatrix}$$

First let's define the expression of $\tau_w$

$$\tau_w = F_r r_{ad}$$

$r_{ad}$ is the wheel radius and $F_r$ can be parameterized according to the dynamic AGV's model demonstrated above.

From (11) one can define the expression of $F_r$ as follow

$$F_r = \frac{m\dot{v} + F_{ad} - F_f \cos(\delta + \beta) + F_{as} \sin(\delta + \beta) + R_{as} \sin(\beta)}{\cos(\beta)} \tag{20}$$

One can replace the definition of $F_f$ from (16) in (20)

$$F_f = \frac{mv(\dot{\psi} - \dot{\beta}) - F_r \sin(\beta) + F_{ad} \sin(\beta) - F_{as} \cos(\delta + \beta) - R_{as} \cos(\beta)}{\sin(\delta + \beta)} \tag{21}$$

By assuming that the sideslip angle is very small, we suppose that

$$\sin(\beta) \approx \beta$$
$$\cos(\beta) \approx 1$$

and $\beta$ is neglected in front of $\delta$. From (21), the equation of $F_r$ is expressed as

$$\begin{aligned} F_r &= m\dot{v} + F_{ad} + r_{ad} mv(\dot{\beta} - \dot{\psi}) \cot(\delta) + F_{as} \\ &\quad \times \cot(\delta) \cos(\delta) + R_{as} \cot(\delta) + \frac{1}{r_{ad}^2} \left(I_r + I_f\right) \dot{v} \end{aligned} \tag{22}$$

and Hence the correspondent torque $\tau_w$ is defined as

$$\begin{aligned} \tau_w &= r_{ad} m\dot{v} + r_{ad} F_{ad} + r_{ad} mv(\dot{\beta} - \dot{\psi}) \cot(\delta) + r_{ad} F_{as} \\ &\quad \times \cot(\delta) \cos(\delta) + r_{ad} R_{as} \cot(\delta) + \frac{1}{r_{ad}} (I_r + I_f) \dot{v} \end{aligned} \tag{23}$$

The second torque's expression is modeled according to the equation of $r$ in (16)

$$\tau_r = F_{as} l_v \cos(\delta) \tag{24}$$

## 4 Adaptive Computed Torque Controller

Due to the uncertainties of the parameters of the AGV's model, an adaptive controller is required to compensate the gap between the estimated and the real vehicle's criterions. Among the adaptive controllers, one may find the adaptive computed torque that prove its efficacy in controlling arm robot with high precision and robustness as cited in [23]. Hence, the main objective is to calculate online the necessary torques to control the AGV's steering angle and throttle while compensating the model's uncertainties. Using the predefined torques' expression in (23) and (24), one can define the compact dynamic form of the AGV as follow in (25):

$$\mathbf{M}\ddot{z} + \mathbf{G}(z, \dot{z}) + f = \boldsymbol{\tau} \tag{25}$$

where $z(t)$ presents the state variable such as:

$$\begin{aligned} z(t) &= \left[\int v(\sigma)d\sigma dt\ \psi(t)\right] \\ \dot{z}(t) &= \left[v(t)\ \dot{\psi}(t)\right] \\ \ddot{z}(t) &= \left[\dot{v}(t)\ \ddot{\psi}(t)\right] \end{aligned}$$

$\boldsymbol{\tau}$ denotes the torque vector as $\boldsymbol{\tau} = [\tau_w, \tau_r]^T$, while **M**, **G** and $f$ define the inertia matrix, the nonlinear vector and the torque friction respectively.

$$\begin{aligned} \mathbf{M}(z) &= \begin{bmatrix} mr_{ad} + \frac{1}{r_{ad}}(I_r + I_f) & 0 \\ 0 & I \end{bmatrix} \\ \mathbf{G}(z, \dot{z}) &= \begin{bmatrix} r_{ad}F_{ad} + r_{ad}mv(\dot{\beta} - \dot{\psi})\cot(\delta) \\ +r_{ad}F_{as}\cot(\delta)\cos(\delta) + r_{ad}R_{as}\cot(\delta) \\ -R_{as}l_h \end{bmatrix} \\ f &= \left[\mu F_N\ 0\right] \end{aligned}$$

Clearly, the torque friction $f$ depends on the normal force $F_N$ which is the sum of the front and the rear wheel normal forces. The friction coefficient $\mu$ depends on the road nature, the weather and the vehicle speed as cited in [10]. Therefore, $\mu$ can vary from 0.75 to 1 for dry roads.

The main purpose of such a controller is to achieve the optimal path following by the AGV, so that the lateral $x(t)$ and the longitudinal $y(t)$ vehicle's position follow the optimal trajectory's coordinate $s^*(t)$ and $d^*(t)$ respectively. Thus, the control objective is to make the tracking error tend to zero.

The tracking error and its first and second derivatives can be expressed as follow:

$$\begin{aligned} e &= z_d - z \\ \dot{e} &= \dot{z}_d - \dot{z} \\ \ddot{e} &= \ddot{z}_d - \ddot{z} \end{aligned} \tag{26}$$

where $z_d$ and $z$ denote the desired and the actual vehicle configuration. For sake of clarity, $z_d$ is got from the previous section such as:

$$z_d = \left[\int v_d\ \psi_d\right]$$

As $v_d = \sqrt{\dot{s^*}^2 + \dot{d^*}^2}$ and $\psi = \arctan\left(\frac{\dot{d^*}}{\dot{s^*}}\right)$ where $d^*$ and $s^*$ are defined in (7).

In the ideal case, all the model parameters are known and the friction parameter is neglected i.e $f = 0$, hence, the control input can be considered as a PD feedback control.

By substituting the expression of $\ddot{z}$ from (25) into (26), the vehicle torque can be parameterized as:

$$\tau = \mathbf{M}(\ddot{z}_d - u) + \mathbf{G} + f \tag{27}$$

One way to define the control input $u$ is to perform the proportional derivative feedback.

$$u = -\boldsymbol{K_d}\dot{e} - \boldsymbol{K_p}e \tag{28}$$

where $\boldsymbol{K_p}$ and $\boldsymbol{K_d}$ are $2 \times 2$ diagonal matrix. $\boldsymbol{K_p} = \mathbf{diag}\{k_p\}$, $\boldsymbol{K_d} = \mathbf{diag}\{k_d\}$ So the torque expression is obtained as follow

$$\tau = \mathbf{M}(\ddot{z}_d + \boldsymbol{K_d}\dot{e} + \boldsymbol{K_p}e) + \mathbf{G}(z, \dot{z}) + f \tag{29}$$

Then the chosen PD gains should ensure the convergence of the tracking error to zero.

Otherwise, in the real case, some uncertainties can be added to the parameters of the dynamic vehicle model, then the estimated dynamic model can be expressed as follow:

$$\tau = \hat{\mathbf{M}}(z)\ddot{z} + \hat{\mathbf{G}}(z, \dot{z}) + \hat{f} \tag{30}$$

where $\hat{\mathbf{M}}$ and $\hat{\mathbf{G}}$ are the same as $\mathbf{M}$ and $\mathbf{G}$ are the inertia and the regression matrices estimates and $\hat{f}$ is obtained by replacing $\mu$ with $\hat{\mu} = 0.8$. In this case, the aforementioned control input in (28) is not valid anymore. Therefore, the new adaptive computed torque control law is defined such as:

$$\tau = \hat{\mathbf{M}}(\ddot{z}_d + \boldsymbol{K_d}\dot{e} + \boldsymbol{K_p}e) + \hat{\mathbf{G}}(z, \dot{z}) + \hat{f} \tag{31}$$

Based on (26), the aforementioned Eq. (31) can be written as:

$$\tau = \hat{\mathbf{M}}(\ddot{e} + \boldsymbol{K_d}\dot{e} + \boldsymbol{K_p}e) + \hat{\mathbf{M}}\ddot{z} + \hat{\mathbf{G}}(z, \dot{z}) + \hat{f} \tag{32}$$

By combining the Eqs. (32) and (25) we obtain:

$$\ddot{e} + \boldsymbol{K_d}\dot{e} + \boldsymbol{K_p}e = \hat{\mathbf{M}}^{-1}(\mathbf{M}\ddot{z} + \mathbf{G}(z, \dot{z}) + f - (\hat{\mathbf{M}}\ddot{z} + \hat{\mathbf{G}}(z, \dot{z}) + \hat{f})) \tag{33}$$

The error dynamics can be parameterized as

$$\ddot{e} + \boldsymbol{K}_d\dot{e} + \boldsymbol{K}_p e = \hat{\mathbf{M}}^{-1}\mathbf{R}(h - \hat{h}) \tag{34}$$

Such as $h$ denotes the vehicle model parameter vector presented as:

$$h = \begin{bmatrix} r_{ad} \; m \; I \; I_h \; I_v \; l_v \; \mu \end{bmatrix}$$

and $\hat{h}$ is its estimate. $\hat{\mathbf{M}}^{-1}$ is the inverse of the estimated inertia matrix, $\mathbf{R}$ is the regression matrix and $\tilde{h}$ is the error of the model parameter vector.

One can present the error parameter vector by $\tilde{h} = h - \hat{h}$ which yields to the following error representation:

$$\dot{\mathbf{E}} = A\mathbf{E} + B\hat{\mathbf{M}}^{-1}\mathbf{R}\tilde{h} \tag{35}$$

where:

$$\mathbf{E} = \begin{bmatrix} e \; \dot{e} \end{bmatrix}^T$$

$$\mathbf{A} = \begin{bmatrix} \mathbf{0} & I \\ -K_p & -K_d \end{bmatrix}$$

$$B = \begin{bmatrix} \mathbf{0} \; I \end{bmatrix}^T$$

**Theorem 1** *For an autonomous non linear system $\dot{x} = f(x)$, there exists a function $V$ positive definite and continuously differentiable, having as derivative $\dot{V} \leq 0$. The system is asymptotically stable if and only if $\dot{V}(s) = 0$ for $s(t) = 0 \forall t$*

*Proof* In order to define the new adaptive control law for an asymptotically stable system, the Lyapunov quadratic function is used.

$$V = \frac{1}{2}\mathbf{E}^T K\mathbf{E} + \frac{1}{2}\tilde{h}^T L\tilde{h} \tag{36}$$

where $K$ and $L$ are two symmetric positive definite matrices. The derivative of the Lyapunov candidate is

$$\dot{V} = \mathbf{E}^T K\dot{\mathbf{E}} + \dot{\mathbf{E}}^T K\mathbf{E} + 2\tilde{h}^T L\dot{\tilde{h}} \tag{37}$$

By substituting $\dot{\mathbf{E}}$ by its expression in (35), we obtain:

$$\begin{aligned} \dot{V} &= \mathbf{E}^T K(A\mathbf{E} + B\hat{\mathbf{M}}^{-1}\mathbf{R}\tilde{h}) + (A\mathbf{E} + B\hat{\mathbf{M}}^{-1}\mathbf{R}\tilde{h})^T K\mathbf{E} \\ &+ 2\tilde{h}^T L\dot{h} \\ &= -\mathbf{E}^T W\mathbf{E} + 2\tilde{h}^T(L\dot{\tilde{h}} + \mathbf{R}^T\hat{\mathbf{M}}^{-1}B^T K\mathbf{E}) \end{aligned} \tag{38}$$

where $W$ is a symmetric positive matrix defined as:

$$A^T K + K A = -W \tag{39}$$

To obtain an asymptotically stable system i.e. $\dot{V} \leq 0$, the derivative of the error parameter vector is chosen as:

$$\dot{\tilde{h}} = -L^{-1}\mathbf{R}^T\hat{\mathbf{M}}^{-1}B\mathbf{E} \tag{40}$$

Since:

$$\begin{aligned} \tilde{h} &= h - \hat{h} \\ \dot{\tilde{h}} &= \dot{h} - \dot{\hat{h}} \\ \dot{h} &\approx 0 \end{aligned}$$

We choose the adaptation update rule as

$$\begin{aligned} \dot{\hat{h}} &\approx -\dot{\tilde{h}} \\ &= L^{-1}\mathbf{R}^T\hat{\mathbf{M}}^{-1}B\mathbf{E} \end{aligned} \tag{41}$$

**Theorem 2** *The control law (31) and the adaptation law (41) of the ACT controller guaranty the AGV's stability while merging the adjacent lane. Besides, the tracking error $e$ and its derive $\dot{e}$ go to zero while the time $t$ goes to the infinity.*

*Proof* By replacing the expression (40) in (38), we obtain

$$\dot{V} = -\mathbf{E}^T W\mathbf{E} \tag{42}$$

While the derivative of (42) is

$$\ddot{V} = -2\mathbf{E}^T W\dot{\mathbf{E}} \tag{43}$$

From (42) and (36), one can conclude that $\mathbf{E}$ and $\tilde{h}$ are bounded and since the desired trajectory and the desired vehicle's velocity are limited, $\mathbf{R}$ is bounded either. Based on this aforementioned bounded parameters, it is clear that $\dot{\mathbf{E}}$ in (35) is also bounded. Thus, (43) is bounded. By applying the Barbarat's lemma, $\dot{V} \to 0$ as $t \to \infty$ which yields the tracking error $e$ and its derive $\dot{e}$ to tend to zero as $t \to \infty$.

The proposed ACT controller is summarized in Fig. 4 as the desired configuration $z_d$ and $\dot{z}_d$ are defined using the developed optimal path generator in Sect. 1.1. However the actual AGV's configuration $z$ and $\dot{z}$ are obtained from a post processed data sensors.

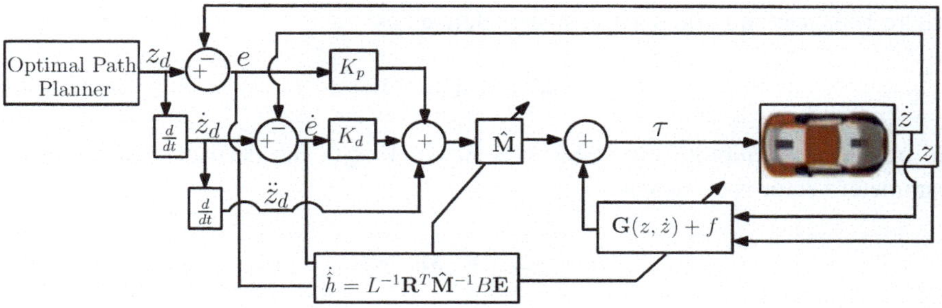


**Fig. 4** The control scheme of the AGV's controller

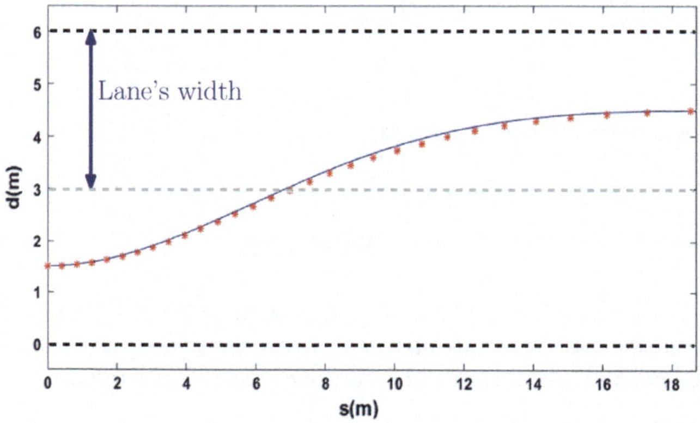


**Fig. 5** The planned optimal trajectory

# 5 Numerical Simulation

To validate the developed approach, a numerical simulation is established first to define the optimal lateral $s^*(t)$ and longitudinal $d^*(t)$ desired positions that yields to define the optimal trajectory $x^*(s^*(t), d^*(t))$ for lane merging, and second to prove the efficacy of the adaptive computed torque controller. The approximated and the optimal path are presented by the red asterisk and the blue line respectively in Fig. 5.

It is shown that the optimal trajectory is bounded inside the lane boundaries.

To follow the planned optimal trajectory, an adaptive computed torque controller is developed. The used vehicle parameters are $m = 1000Kg\ I = 1210, lv = 0.87, lh = 1.29$. In addition, the proportional and the derivative gain for $\tau_w$ and $\tau_r$ are respectively $k_{pw} = 0.64\ k_{dw} = 1.6\ k_{pr} = 1.69\ k_{dr} = 2.6$. The result of the simulation shows that

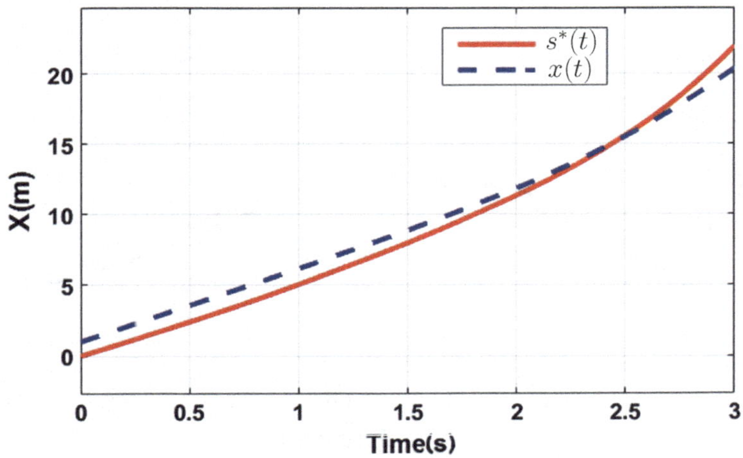


**Fig. 6** The optimal vehicle's lateral position

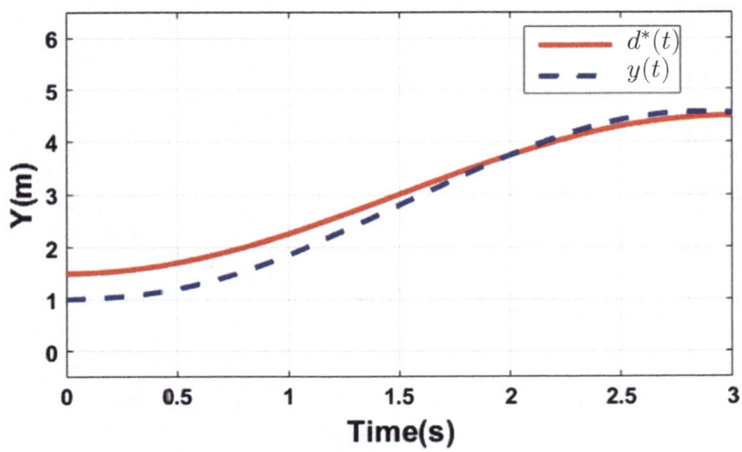


**Fig. 7** The optimal vehicle's longitudinal position

the longitudinal and the lateral vehicle's position track the optimal ones as presented in Fig. 7 and in Fig. 6, the vehicle's velocity is increased in Fig. 9 due to the need of overtaking the preceding vehicle. Otherwise, the vehicle change its lane, so the yaw angle augmented then the vehicle's orientation returns to the initial state which is expressed by the decrease of the yaw angle in Fig. 8.

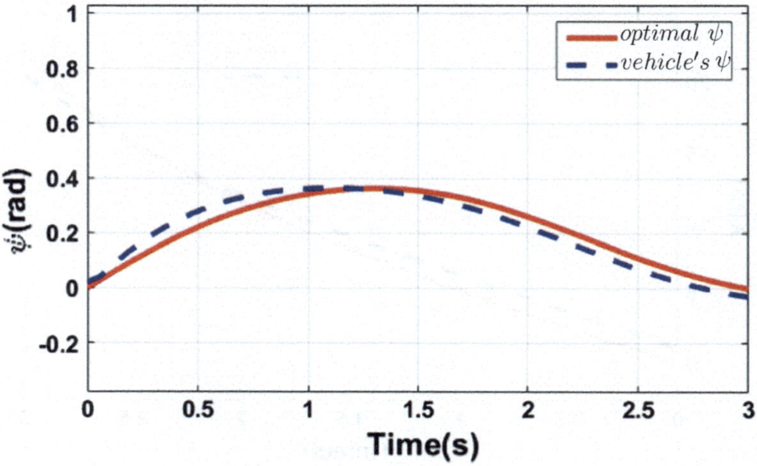


**Fig. 8** The optimal vehicle's orientation

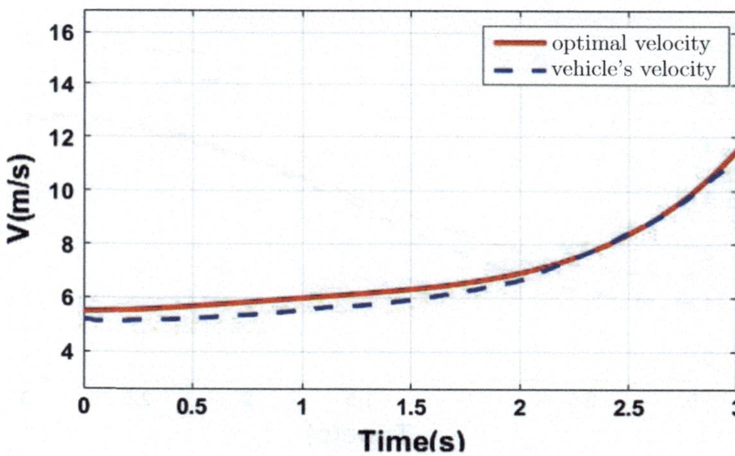


**Fig. 9** The optimal vehicle's velocity

## 6 Conclusion

This paper introduce the development of a trajectory planner for lane merging tasks. The main purpose of such planner is to generate at once an optimal and a constrained path while merging the other lane by an automated guided vehicle. This is accomplished to ensure a safer driving. Secondly, an adaptive computed torque controller is implemented in the vehicle to follow autonomously the generated path without deviating from the road boundaries and with guarding a safe distance w.r.t the preceding vehicle. The proposed approach is then validated by a numerical simulation.

## References


1. Arora, S., Raina, A.K., Mittal, A.K.: Collision avoidance among agvs at junctions. In: Proceedings of the IEEE on Intelligent Vehicles Symposium IV, pp. 585–589 (2000)
2. Awal, T., Kulik, L., Ramamohanrao, K.: Optimal traffic merging strategy for communication- and sensor-enabled vehicles. In: 16th International IEEE Conference on Intelligent Transportation Systems-(ITSC), pp. 1468–1474 (2013)
3. Broggi, A., Medici, P., Zani, P., Coati, A., Panciroli, M.: Autonomous vehicles control in the vislab intercontinental autonomous challenge. Annu. Rev. Control **36**(1), 161–171 (2012)
4. Dolk, V., Ouden, J.D., Steeghs, S., Devanesan, J.G., Badshah, I., Sudhakaran, A., Elferink, K., Chakraborty, D.: Cooperative automated driving for various traffic scenarios: experimental validation in the GCDC 2016. IEEE Trans. Intell. Transp. Syst. **19**(4), 1308–1321 (2018)
5. Ercan, Z., Carvalho, A., Tseng, H.E., Gökaşan, M., Borrelli, F.: A predictive control framework for torque-based steering assistance to improve safety in highway driving. Veh. Syst. Dyn. **56**(5), 810–831 (2018)
6. Hatipoglu, C., Ozguner, U., Redmill, K.A.: Automated lane change controller design. IEEE Trans. Intell. Transp. Syst. **4**(1), 13–22 (2003)
7. Isaksson-Hellman, I., Lindman, M.: An evaluation of the real-world safety effect of a lane change driver support system and characteristics of lane change crashes based on insurance claims data. Traffic Inj. Prev. **19**(sup1), S104–S111 (2018)
8. Jula, H., Kosmatopoulos, E.B., Ioannou, P.A.: Collision avoidance analysis for lane changing and merging. IEEE Trans. Veh. Technol. **49**(6), 2295–2308 (2000)
9. Kesting, A., Treiber, M., Helbing, D.: General lane-changing model mobil for car-following models. Transp. Res. Rec.: J. Transp. Res. Board **1999**, 86–94 (2007)
10. Kost, F.: Basic principles of vehicle dynamics. In: Fundamentals of Automotive and Engine Technology, pp. 114–129. Springer (2014)
11. Li, X., Sun, Z., He, Z., Zhu, Q., Liu, D.: A practical trajectory planning framework for autonomous ground vehicles driving in urban environments. In: IEEE Intelligent Vehicles Symposium (IV), pp. 1160–1166 (2015)
12. Lin, Li, Liang, Su, Chen, Shaozhen, Zhang, Yuanbiao: Optimization of the merging area after toll. Eng. Manag. Res. **7**(1), 17 (2018)
13. Menhour, L., d'Andréa Novel, B., Fliess, M., Gruyer, D., Mounier, H.: An efficient model-free setting for longitudinal and lateral vehicle control: validation through the interconnected pro-sivic/rtmaps prototyping platform. IEEE Trans. Intell. Transp. Syst. **19**(2), 461–475 (2018)
14. Miley, J.: New algorithm allows autonomous cars to change lanes more like humans. Technical Report, MIT, https://interestingengineering.com/new-algorithm-allows-autonomous-cars-to-change-lanes-more-like-humans (2018)
15. Naranjo, J.E., Gonzalez, C., Garcia, R., De Pedro, T.: Lane-change fuzzy control in autonomous vehicles for the overtaking maneuver. IEEE Trans. Intell. Transp. Syst. **9**(3), 438–450 (2008)
16. Norouzi, A., Kazemi, R., Azadi, S.: Vehicle lateral control in the presence of uncertainty for lane change maneuver using adaptive sliding mode control with fuzzy boundary layer. Proc. Inst. Mech. Eng. Part I: J. Syst. Control. Eng. **232**(1), 12–28 (2018)
17. Okuda, H., Harada, K., Suzuki, T., Saigo, S., Inoue, S.: Design of automated merging control by minimizing decision entropy of drivers on main lane. In: IEEE Intelligent Vehicles Symposium (IV), pp. 640–646 (2017)
18. Osman, K., Ghommam, J., Saad, M.: Combined road following control and automatic lane keeping for automated guided vehicles. In: 14th IEEE International Conference on Control, Automation, Robotics and Vision (ICARCV), pp. 1–6 (2016)

19. Raja, P., Pugazhenthi, S.: Optimal path planning of mobile robots: a review. Int. J. Phys. Sci. **7**(9), 1314–1320 (2012)
20. Rajamani, R.: Vehicle Dynamics and Control. Springer Science & Business Media (2011)
21. Rios-Torres, J., Malikopoulos, A., Pisu, P.: Online optimal control of connected vehicles for efficient traffic flow at merging roads. In: 18th IEEE International Conference on Intelligent Transportation Systems (ITSC), pp. 2432–2437 (2015)
22. Shamir, T.: How should an autonomous vehicle overtake a slower moving vehicle: design and analysis of an optimal trajectory. IEEE Trans. Autom. Control **49**(4), 607–610 (2004)
23. Shang, W.-W., Cong, S., Ge, Y.: Adaptive computed torque control for a parallel manipulator with redundant actuation. Robotica **30**(3), 457–466 (2012)
24. Tawari, A., Sivaraman, S., Trivedi, M.M., Shannon, T., Tippelhofer, M.: Looking-in and looking-out vision for urban intelligent assistance: Estimation of driver attentive state and dynamic surround for safe merging and braking. In: IEEE Intelligent Vehicles Symposium, pp. 115–120 (2014)
25. Wang, X.: Solving optimal control problems with matlab: indirect methods. https://www.researchgate.net/publication/239575747_Solving_optimal_control_problems_with_MATLAB_Indirect_methods (2009)
26. Werling, M., Kammel, S., Ziegler, J., Gröll, L.: Optimal trajectories for time-critical street scenarios using discretized terminal manifolds. Int. J. Robot. Res. **31**(3), 346–359 (2012)
27. Yao, W., Zhao, X., Yu, Y., Fang, Y., Wang, C., Yang, T.: A leader-path-following formation system for agvs with multi-sensor data fusion based vehicle tracking. In: IOP Conference Series: Materials Science and Engineering, vol. 235, pp. 012006. IOP Publishing (2017)
28. Ye, E., Ramezani, M.: Lane distribution optimisation of autonomous vehicles for highway congestion control. Technical Report (2018)

# Finite Time Consensus for Higher Order Multi Agent Systems with Mismatched Uncertainties

**Sanjoy Mondal, Jawhar Ghommam and Maarouf Saad**

**Abstract** This chapter presents a finite time distributed consensus problem in the context of higher order multi-agent systems (MAS), when the agent dynamics are potentially affected by mismatched uncertainties. With the development of integral sliding mode (ISM), using finite time disturbance observer(FTDO), allows to neglect the effect of mismatched uncertainties during the sliding mode. Based on homogeneous continuous finite time protocol, the nominal consensus control is designed to track the desired target trajectories. The finite time convergence is presented using strict Lyapunov function for the higher order MAS. Simulation results validates the effectiveness of the proposed controller.

**Keywords** Finite time convergence · Higher order sliding mode · Multi-agent systems · Disturbance observer

## 1 Introduction

In Network Consensus, a group of interacting agents reaches a common agreement by local interaction. As compared to the traditional coordination approaches, the distributed control approaches for multi-agent systems (MAS) exhibits many advantages, such as stronger robustness, higher efficiency and less communication requirements. Among the many formations of MAS, leader-follower MAS explicitly studied due to its vast applications in engineering background, physical meaning and

S. Mondal (✉)
School of EEE, Nanyang Technologcal, Jurong West, Singapore
e-mail: mondal.sanjoy@gmail.com

J. Ghommam
Department of Electrical and Computer Engineering, Sultan Qaboos University Muscat, Al Khoudh, Oman
e-mail: jawher@squ.edu.om

M. Saad
Department of Electrical Engineering, Ecole de technologie superieure, Montreal, Canada
e-mail: maarouf.saad@etsmtl.ca



J. Ghommam et al. (eds.), *New Trends in Robot Control*, Studies in Systems, Decision and Control 270, https://doi.org/10.1007/978-981-15-1819-5_11

simplicity [11]. In this formation, the leader is usually independent of its followers, but can affect the follower's behaviours, while the leader's behaviour can easily be controlled, in order to achieve the control objective. In the analysis of consensus problem, convergence rate is an important issue. Hence compared with the asymptotical control method [19, 20], a finite-time control approach demonstrates higher control accuracy, stronger robustness and disturbance rejection properties besides a high speed convergence rate [5, 7, 9, 14].

Therefore, many results related to finite time consensus of MAS, i.e., the states of all agents reach an agreement in finite time and remain there for all further times, have been developed in recent years from different perspectives [5, 7, 9]. In [21] a fixed time consensus tracking problem for second-order MAS is developed to estimate the convergence time off-line.

An integral sliding mode control proposed to counteract the effect of uncertainties is recently proposed for second order MAS [18]. All these new consensus control techniques can only be used when the agent dynamics are of second order i.e. double integrators and it is only affected by matched uncertainties. When the uncertainties come though the input channel it holds the matching condition and thus termed as matched uncertainty. However uncertainties existing in many physical system may not always satisfy the matching condition. It is quite natural for a system that it can be affected by both matched and mismatched or unmatched uncertainties. Nevertheless, traditional sliding mode design methods of [2–4, 15, 16] may fail to give stabilization faced with mismatched uncertainty. Thus it can be very restrictive, as the agent dynamics can be of any order can be affected by match and mismatched uncertainties. So designing a more general finite time controller for higher order agent dynamics with mismatched uncertainty is still a considerable challenge. Therefore the original contribution of this paper is to propose a finite-time control technique for a more general higher order agent dynamics subjected to matched and mismatched uncertainties both via FTDO. In the context of MAS, a new sliding surface based on HOSM, with disturbance estimation is designed to guarantee the finite time sliding motion in the presence of mismatched uncertainties. The use of HOSM guarantees the higher accuracy and diminishing the chattering phenomenon.

The rest of the chapter is organized as follows. Section 2 and 3 includes the mathematical preliminaries and problem formulation. Design of disturbance observer and controller are given in Sect. 4 and 5. Section 6 provides the illustrative examples. The conclusions are given in Sect. 7.

## 2 Mathematical Preliminaries, Graph Theory

Consider a MAS consisting of one leader and $N$ followers. The communication topology between agents, is modeled by a weighted directed graph directed graph $\mathcal{G} = \{\succeq, \upharpoonleft, \mathcal{A}\}$ , where $\nu = 0, 1, 2, \ldots, N$ is the vertex set of the agents, node i represents the ith agent, $\varepsilon = \{(i, j) \subseteq \nu \times \nu\}$, where $\nu$ is the set of vertices while $\varepsilon$ represents the set of edges respectively. The weight adjacent matrix $\boldsymbol{\mathcal{A}} = (a_{ij} \geq 0) \in$

$\Re^{(N\times 1)\times(N\times 1)}$. $(i, j) \in \varepsilon \Leftrightarrow a_{ij} = a_{ji} = 1$, otherwise $a_{ij} = a_{ji} = 0, a_{ii} = 0$ for all $i \in \nu$ because $(i, i) \notin \varepsilon$. Therefore $\boldsymbol{\mathcal{A}}$ is symmetric.

For the leader-follower MAS, another graph $\bar{\mathcal{G}}$ can be considered to associate the system consisting of $N$ followers with the leader. The leader adjacency matrix is defined as $\bar{\boldsymbol{\mathcal{B}}} = [b_1, b_2, \ldots, b_N]^T \in \Re^N$ with the adjacency element $b_i > 0$ if agent $i$ is a neighbor of the leader, otherwise $b_i = 0$. The followers can receive information from the leader, but cannot send information to the leader [8, 9, 20].

**Lemma** [13] *Consider a system described with $\dot{x} = f(x)$ with $f(0) = 0$ and there exists a continuous differential positive-definite function $V(x) : D \to \Re$, and $\beta > 0$, $0 < \eta < 1$, $\eta \in \Re$ such that*

$$\dot{V}(x) + \beta V^{\eta}(x) \leq 0 \;\; \forall x \in D \tag{1}$$

*Then, the origin of the system is a locally finite-time stable, with the settling time depending on the initial state $x(0) = x_0$, satisfies*

$$T(x_0) \leq \frac{V^{1-\eta}(x_0)}{\beta(1-\eta)} \tag{2}$$

## 3 Problem Formulation

Let the followers are governed by nth order integrator dynamics

$$\begin{aligned} \dot{x}_{ik} &= x_{i(k+1)} + d_{ik}, \quad 1 \leq k \leq n-1 \\ &\vdots \\ \dot{x}_{in} &= u_i + d_{in}, \;\; 1 \leq i \leq N \end{aligned} \tag{3}$$

where $y_i = x_{i1} \in \Re$ is the output, $x_i = [x_{i1}\; x_{i2}, \ldots x_{in}]^T \in \Re^n$ are the state variables, $u_i \in \Re$ is the control input. $d_{ik}$ and $d_{in}$ are the unknown mismatched and matched uncertainties.

The leader dynamics can be modeled as

$$\begin{aligned} \dot{x}_{0k} &= x_{0(k+1)}, \; 1 \leq k \leq n-1 \\ &\vdots \\ \dot{x}_{0n} &= u_0 \end{aligned} \tag{4}$$

where $x_0 = [x_{01}, \ldots, x_{0n}]^T \in \Re^n$ are the state vectors, $y_0 = x_{01} \in \Re$ is the output and $u_0 \in \Re$ is the control input.

For further analysis we assume the disturbance is bounded and continuously differentiable [4, 14, 17].

# 4 Effect of Mismatched Uncertainty

Let us consider the first state equation of the follower dynamics (3) as,

$$\dot{x}_{i1} = x_{i2} + d_{i1} = \bar{x}_{i2} \tag{5}$$

Taking the derivative of (5) one can easily write

$$\dot{\bar{x}}_{i2} = x_{i3} + d_{i2} + \dot{d}_{i1} = \bar{x}_{i3} \tag{6}$$

Thus the overall agent dynamics can be expressed as

$$\begin{aligned} \dot{\bar{x}}_{ik} &= \bar{x}_{i(k+1)} \;\; 1 \leq k \leq n-1 \\ &\vdots \\ \dot{\bar{x}}_{in} &= u_i + d_{in} + \sum_{p=1}^{n-1} d_{i(n-p)}^{p} \;\; i = 1, 2, \ldots, N \end{aligned} \tag{7}$$

Now for the ith and jth agent, the new states are:

$$\begin{aligned} \bar{x}_{i1} &= x_{i1}, \; \bar{x}_{j1} = x_{j1} \\ \bar{x}_{il} &= x_{il} + \sum_{p=1}^{l-1} d_{ip}^{l-p-1}, \; \bar{x}_{jl} = x_{jl} + \sum_{p=1}^{l-1} d_{jp}^{l-p-1} \end{aligned} \tag{8}$$

where $l = 2, \ldots n$.

In the presence of mismatched uncertainties, the existing consensus algorithms [6, 8, 9, 18, 21] do not guarantee finite time tracking [17]. As the states are contaminated with uncertainties. Now a finite time disturbance observer(FTDO) [14] is employed to estimate the unknown uncertainties and to make the tacking of MAS in finite time.

## 4.1 Design of MAS Finite Time Disturbance Observer

Suppose that the disturbance $d_{ik}$, $\forall k = 1, \ldots, n$, in (3) is $(r_i - 1)$th order differentiable and $d_{ik}^{[r_i-1]}$ has a Lipschitz constant $L_{ik} \forall i = 1, \ldots, N$. A finite-time disturbance observer [14] is used for the MAS (3) to estimate the uncertainties, given by

$$
\begin{aligned}
\dot{\xi}_{i0}^k &= v_{i0}^k + f_{ik}(x,u), \dot{\xi}_{i1}^k = v_{i1}^k, \ldots, \dot{\xi}_{ir_i}^k = v_{ir_i}^k. \\
v_{i0}^k &= -\lambda_{i0}^k L_{ik}^{\frac{1}{r_i+1}} |\xi_{i0}^k - x_{ik}|^{\frac{r_i}{r_i+1}} \operatorname{sign}(\xi_{i0}^{ik} - x_{ik}) + \xi_{i1}^k \\
v_{i1}^k &= -\lambda_{i1}^k L_{ik}^{\frac{1}{r_i}} |\xi_{i1}^k - v_{i0}^k|^{\frac{r_i-1}{r_i}} \operatorname{sign}(\xi_{i1}^k - v_{i0}^k) + \xi_{i2}^k \\
&\vdots \\
v_{i(r_i-1)}^k &= -\lambda_{i(r_i-1)}^k L_{ik}^{\frac{1}{2}} |(\xi_{i(r_i-1)}^k - v_{i(r_i-2)}^k)|^{\frac{1}{2}} \operatorname{sign}(\xi_{i(r_i-1)}^k - v_{i(r_i-2)}^k) + \xi_{ir_i}^k \\
v_{ir_i}^k &= -\lambda_{ir_i}^k L_{ik} |\xi_{ir_k}^k - v_{r_i-1}^k| \\
\hat{x}_{ik} &= \xi_{i0}^k, \hat{d}_{ik} = \xi_{i1}^k, \hat{\dot{d}}_{ik} = \xi_{i2}^k, \ldots, \hat{d}_{ik}^{[r_i-1]} = \xi_{ir_i}^k
\end{aligned}
\tag{9}
$$

for $i = 1, \ldots, N$ and $f_{ik}(x,u) = x_{i(k+1)}$ for $k = 1, \ldots, n-1$, $f_{in}(x,u) = u_i$, $\lambda_{im}^k > 0$ $m = 0, 1, \ldots, r_i$ ,are the observer coefficients to be designed, $\hat{x}_{ik}, \hat{d}_{ik}, \hat{\dot{d}}_{ik}, \hat{d}_{ik}^{[r_i-1]}$ are the estimates of $x_{ik}, d_{ik}, \dot{d}_{ik}, d_{ik}^{[r_i-1]}$ respectively. The auxiliary estimation error variables are defined as $\mathfrak{e}_{i0}^k = \xi_{i0}^k - x_{ik}$, $\mathfrak{e}_{i1}^k = \xi_{i1}^k - d_{ik}, \ldots, \mathfrak{e}_{ir_i}^k = \xi_{ir_i}^k - d_{ik}^{[r_i-1]}$. Combining (3) and (9) the estimation error dynamics can be expressed as,

$$
\begin{aligned}
\mathfrak{e}_{i0}^k &= -\lambda_{i0}^k L_{ik}^{\frac{1}{r_i+1}} |\mathfrak{e}_{i0}^k|^{\frac{r_i}{r_i+1}} \operatorname{sign}(\mathfrak{e}_{i0}^k) + \mathfrak{e}_{i1}^k \\
\dot{\mathfrak{e}}_{i1}^k &= -\lambda_{i1}^k L_{ik}^{\frac{1}{r_i}} |\mathfrak{e}_{i1}^k - \dot{\mathfrak{e}}_{i0}^k|^{\frac{r_i-1}{r_i}} \operatorname{sign}(\mathfrak{e}_{i1}^k - \dot{\mathfrak{e}}_{i0}^k) + \mathfrak{e}_{i2}^k \\
&\vdots \\
\dot{\mathfrak{e}}_{i(r_i-1)}^k &= -\lambda_{i(r_i-1)}^k L_{ik}^{\frac{1}{2}} |\mathfrak{e}_{i(r_i-1)}^k - \dot{\mathfrak{e}}_{i(r_i-2)}^k|^{\frac{1}{2}} \operatorname{sign}(\mathfrak{e}_{i(r_i-1)}^k - \dot{\mathfrak{e}}_{i(r_i-2)}^k) + \mathfrak{e}_{ir_i}^k \\
\dot{\mathfrak{e}}_{ir_i}^k &\in -\lambda_{ir_i}^k L_{ik} \operatorname{sign}(\mathfrak{e}_{ir_i}^k - \dot{\mathfrak{e}}_{i(r_i-1)}^k) + [-L_{ik}, L_{ik}]
\end{aligned}
$$

The stability of the FTDO can be achieved by selecting appropriate parameters $\lambda_{im}^k$ $m = 0, 1, \ldots, r_i$ and $L_{ik}$ such that the errors $\mathfrak{e}_{im}^k$ can converge to zero in finite time. The proof of this stability can be followed from [14].

## 5 Homogeneous Finite Time Sliding Mode Control

In this section with (8), the finite time consensus algorithm for nth order MAS with one leader under directed network topology is proposed. The dynamics of the followers and the single leader are given by (3) and (4). Here for each follower, define the consensus tracking errors as follows:

$$
\begin{aligned}
e_{ik} &= \sum_{j=1}^{N} a_{ij}(\bar{x}_{ik} - \bar{x}_{jk}) + b_i(\bar{x}_{ik} - x_{0k}),\ 1 \le k \le n-1 \\
&\vdots \\
e_{in} &= \sum_{j=1}^{N} a_{ij}(\bar{x}_{in} - \bar{x}_{jn}) + b_i(\bar{x}_{in} - x_{0n})
\end{aligned}
\tag{10}
$$

Hence the error dynamics can be written as,

$$
\begin{aligned}
\dot{e}_{ik} &= e_{ik},\ 1 \le k \le n-1 \\
&\vdots \\
\dot{e}_{in} &= \sum_{j=1}^{N}(a_{ij} + b_i)\left(u_i + d_{in} + \sum_{p=1}^{n-1} d^p_{i(n-p)}\right) \\
&\quad - \sum_{j=1}^{N} a_{ij}\left(u_j + d_{jn} + \sum_{p=1}^{n-1} d^p_{j(n-p)}\right) - b_i u_0 \\
&= h_i u_i - \sum_{j=1}^{N} a_{ij} u_j + h_i\left(d_{in} + \sum_{p=1}^{n-1} d^p_{i(n-p)}\right) \\
&\quad - \sum_{j=1}^{N} a_{ij}\left(d_{jn} + \sum_{p=1}^{n-1} d^p_{j(n-p)}\right) - b_i u_0
\end{aligned}
\tag{11}
$$

where $h_i = \sum_{j=1}^{N}(a_{ij} + b_i)$ is invertible. The design of the control law$(u_i)$ requires the information of the uncertainties. In practical cases, which are not always available. So by measuring the uncertainties with FTDO (9), the estimated states are designed and then a finite time controller$(u_i)$ is developed to achieve desired consensus performance.

## 5.1 FTDO Based Integral Sliding Mode Control for MAS

Defining the new FTDO based states of the MAS as:

$$\begin{aligned}
\tilde{x}_{i1} &= x_{i1}, \quad \tilde{x}_{j1} = x_{j1} \\
&\vdots \\
\tilde{x}_{il} &= x_{il} + \sum_{p=1}^{l-1} \hat{d}_{ip}^{l-p-1}, \quad \tilde{x}_{jl} = x_{jl} + \sum_{p=1}^{l-1} \hat{d}_{jp}^{l-p-1}
\end{aligned} \tag{12}$$

$\forall l = 2, \ldots, n$. $\hat{d}$ is the estimated uncertainty.

Now the consensus error with FTDO can be expressed as,

$$\begin{aligned}
e_{ik}^{o} &= \sum_{j=1}^{N} a_{ij}(\tilde{x}_{ik} - \tilde{x}_{jk}) + b_i(\tilde{x}_{ik} - x_{0k}),\ 1 \le k \le n-1 \\
&\vdots \\
e_{in}^{o} &= \sum_{j=1}^{N} a_{ij}(\tilde{x}_{in} - \tilde{x}_{jn}) + b_i(\tilde{x}_{in} - x_{0n})
\end{aligned} \tag{13}$$

To converge the consensus error (13) in finite, the integral sliding surface for the ith agent is considered as:

$$s_i = e_{in}^{o} + \int_0^{T} \sum_{q=1}^{n} [k_{iq} |e_{iq}^{o}|^{\alpha_{iq}} \operatorname{sign}(e_{iq}^{o})] d\tau \tag{14}$$

where $k_{iq}$ and $\alpha_{iq}(q = 1, \ldots, n)$ are constants. $k_{iq}$ can be found as, such that the polynomial $F(\psi) = \psi^n + k_{in}\psi^{n-1} + \cdots + k_{i2}\psi + k_{i1}$ is Hurwitz. $\alpha_{in}$ can be obtained as follows [1]:

$$\begin{aligned}
\alpha_{iq} &= \alpha, \quad q = 1 \\
\alpha_{i(q-1)} &= \frac{\alpha_{iq}\alpha_{i(q+1)}}{2\alpha_{i(q+1)} - \alpha_{iq}}, \quad \forall q \ge 2 \ldots n
\end{aligned} \tag{15}$$

where $\alpha_{i(n+1)} = 1$, $\alpha_{iq} = \alpha$, $\alpha \in (1 - \epsilon, 1)$, $\epsilon \in (0, 1)$.

When the ideal sliding-mode $s_i = 0$ is reached, one can easily get

$$\begin{aligned}
&\dot{e}_{in} + k_{in}|\dot{e}_{i(n-1)}|^{\alpha_{in}} sign(\dot{e}_{i(n-1)}) + k_{i(n-1)}|\dot{e}_{i(n-2)}|^{\alpha_{i(n-1)}} sign(\dot{e}_{i(n-2)}) \\
&+ \cdots + k_{i1}|e_{i1}|^{\alpha_{i1}} sign(e_{i1}) = 0
\end{aligned} \tag{16}$$

Hence the error dynamics (11) can be expressed as,

$$\begin{aligned}
\dot{e}_{i1} &= e_{i2} \\
\dot{e}_{in} &= -k_{in}|\dot{e}_{i(n-1)}|^{\alpha_{in}} sign(\dot{e}_{i(n-1)}) - k_{i(n-1)}|\dot{e}_{i(n-2)}|^{\alpha_{i(n-1)}} sign(\dot{e}_{i(n-2)}) \\
&\quad - \cdots - k_{i1}|e_{i1}|^{\alpha_{i1}} sign(e_{i1})
\end{aligned} \tag{17}$$

which represents the establishment of the ideal sliding-mode $s_i = 0$, for system (11). It can converge to its equilibrium point from any initial condition in finite-time.

**Theorem** *The consensus error (13) for the ith agent will converge to equilibrium in finite time, if the sliding surface $s_i$ is selected as (14) and the control is designed as follows:*

$$
\begin{aligned}
u_i &= h_i^{-1}[u_i^{nom} + u_i^{sw} - \hat{d}_{oi}] \\
u_i^{nom} &= b_i u_0 + \sum_{j=1}^{N} a_{ij} u_j - \sum_{q=1}^{n} [k_{iq} |e_{iq}^o|^{\alpha_{iq}} sign(e_{iq}^o)] \\
u_i^{sw} &= -\lambda_i |s_i|^{0.5} sign(s_i) - \alpha_i \int sign(s_i) d\tau
\end{aligned} \tag{18}
$$

*where* $h_i = \sum_{j=1}^{N}(a_{ij} + b_i)$, $\lambda_i > 0$ *and* $\alpha_i > 0$ *and* $\hat{d}_{oi} = h_i(\hat{d}_{in} + \sum_{k=1}^{n-1} v_{i(n-k)}^k) - \sum_{j=1}^{N} a_{ij}(\hat{d}_{jn} + \sum_{k=1}^{n-1} v_{j(n-k)}^k)$ *are the estimation of the overall uncertainties.*

***Proof*** For the consensus error (13) and using the control law (18), the derivative of the sliding surface can be written as

$$
\begin{aligned}
\dot{s}_i = \dot{e}_{in}^o + \sum_{q=1}^{n} [k_{iq} |e_{iq}^o|^{\alpha_{iq}} \text{sign}(e_{iq}^o)] &= -\lambda_i |s_i|^{0.5} \text{sign}(s_i) \\
&\quad - \alpha_i \int \text{sign}(s_i) d\tau - h_i \mathfrak{e}_{i1}^n + \sum_{j=1}^{N} a_{ij} \mathfrak{e}_{j1}^n
\end{aligned} \tag{19}
$$

For the MAS using (12) and (14) the consensus tracking error dynamics can be obtained as:

$$
\begin{aligned}
\dot{e}_{ik}^o &= e_{i(k+1)}^o + h_i \tilde{\mathfrak{e}}_{ik} - \sum_{j=1}^{N} a_{ij} \tilde{\mathfrak{e}}_{jk} \\
&\vdots \\
\dot{e}_{in}^o &= - \sum_{q=1}^{n} [k_{iq} \text{sign}(e_{iq}^o) |e_{iq}^o|^{\alpha_{iq}}] + \dot{s}_i
\end{aligned} \tag{20}
$$

where (for $k = 2, \ldots, n-1$):

$$
h_i \tilde{\mathfrak{e}}_{ik} - \sum_{j=1}^{N} a_{ij} \tilde{\mathfrak{e}}_{jk} = \mathfrak{e}_{ik}
$$

$$
\tilde{\mathfrak{e}}_{i1} = -\mathfrak{e}_{i1}^1
$$

$$\tilde{\mathfrak{e}}_{ik} = \sum_{m=1}^{k-1}(\dot{\mathfrak{e}}_{i(k-m)}^{m} - \mathfrak{e}_{i(k-m+1)}^{m}) - \mathfrak{e}_{i1}^{k}$$

$$\tilde{\mathfrak{e}}_{j1} = -\mathfrak{e}_{j1}^{1}$$

$$\tilde{\mathfrak{e}}_{jk} = \sum_{m=1}^{k-1}(\dot{\mathfrak{e}}_{j(k-m)}^{m} - \mathfrak{e}_{j(k-m+1)}^{m}) - \mathfrak{e}_{j1}^{k}$$

From the dynamics (20), it depicts that the error state variables are effected by sliding surface dynamics (19) and observer dynamics (9). Next we will show that with propose control law the observer dynamics (9) will not drive the sliding surface dynamics (19) and the error state dynamics (20) to infinity.

Let us define a Lyapunov function as,

$$V_i = \frac{1}{2}s_i^2 + \sum_{k=1}^{n}\frac{1}{2}[e_{ik}^{o}]^2 \tag{21}$$

Note for $\alpha \in (0, 1)$, $|\tilde{x}|^{\alpha} < 1 + |\tilde{x}|$ holds. Now taking the derivative of (21) and using (19) and (20) we obtain,

$$\dot{V}_i = s_i\dot{s}_i + \sum_{k=1}^{n} e_{ik}^{o}\dot{e}_{ik}^{o} = s_i\dot{s}_i + \sum_{k=1}^{n-1} e_{ik}^{o}(e_{i(k+1)}^{o} + \mathfrak{e}_{ik})$$

$$- e_{in}^{o}\sum_{q=1}^{n} k_{iq}\text{sign}(e_{iq}^{o})|e_{iq}^{o}|^{\alpha_{iq}} + e_{in}^{o}\dot{s}_i$$

$$\leq h_i|s_i\mathfrak{e}_{i1}^{n}| + \sum_{j=1}^{N} a_{ij}|s_i\mathfrak{e}_{j1}^{n}| + \sum_{k=1}^{n-1}|e_{ik}^{o}|(|e_{i(k+1)}^{o}| + |\mathfrak{e}_{ik}|)$$

$$+ |e_{in}^{o}|\sum_{q=1}^{n} k_{iq}(1 + |e_{iq}^{o}|) + |e_{in}^{o}|(\lambda_i(1 + |s_i|)$$

$$+ \alpha_i + h_i|\mathfrak{e}_{i1}^{n}| + \sum_{j=1}^{N} a_{ij}|\mathfrak{e}_{j1}^{n}|)$$

$$\leq h_i\frac{s_i^2 + (\mathfrak{e}_{i1}^{n})^2}{2} + \sum_{j=1}^{N} a_{ij}\frac{s_i^2 + (\mathfrak{e}_{j1}^{n})^2}{2}$$

$$+\sum_{k=1}^{n-1}\frac{(e^o_{ik})^2+(e^o_{i(k+1)})^2}{2}+\sum_{k=1}^{n-1}\frac{(e^o_{ik})^2+\mathfrak{e}^2_{ik}}{2}$$

$$+\frac{(e^o_{in})^2+\sum_{q=1}^{n}\left(k_{iq}\right)^2}{2}+\sum_{q=1}^{n-1}k_{iq}\frac{(e^o_{in})^2+(e^o_{iq})^2}{2}$$

$$+(\lambda_i+\alpha_i)\frac{1+(e^o_{in})^2}{2}+\lambda_i\frac{s_i^2+(e^o_{in})^2}{2}$$

$$+h_i\frac{|\mathfrak{e}^n_{i1}|^2+(e^o_{in})^2}{2}+\sum_{j=1}^{N}a_{ij}\frac{s_i^2+(\mathfrak{e}^n_{j1})^2}{2}$$

$$\leq K_iV_i+L_i$$

$$K_i=\left[3+k_{in}+\sum_{q=1}^{n}k_{iq}+2\lambda_i+\alpha_i+h_i+2\sum_{j=1}^{N}a_{ij}\right]$$

$$L_i=\frac{1}{2}max\left[h_i(\mathfrak{e}^n_{i1})^2+\sum_{j=1}^{n}a_{ij}(\mathfrak{e}^n_{j1})^2+\sum_{k=1}^{n-1}\mathfrak{e}^2_{ik}\right.$$

$$\left.+\sum_{q=1}^{n}\left(k_{iq}\right)^2+(\lambda_i+\alpha_i)+h_i|\mathfrak{e}^n_{i1}|^2+\sum_{j=1}^{N}a_{ij}(\mathfrak{e}^n_{j1})^2\right]$$

Since the estimation error are bounded and vanishes in finite time thus, $V_i>0,\ K_i>0,\ L_i>0$.

$$\frac{\dot{V}_i}{K_iV_i+L_i}\leq 1 \tag{22}$$

Integrating both sides,

$$ln(K_iV_i+L_i)-ln(K_iV_i(0)+L_i)\leq K_it \tag{23}$$

where $V_i(0)$ is the initial condition of the system. From (23) it can be derived

$$V_i\leq\frac{(K_iV_i(0)+L_i)e^{K_it}-L_i}{K_i} \tag{24}$$

Inequality (24) implies that, $V_i$ is a bounded function in finite time and does not tend to infinity in finite time, which implies that the state of system will not escape in finite time [10, 17]. Now with the assumptions the disturbances estimation errors are bounded and its derivative exists, then the sliding surface (19) for the ith agent can be written as,

$$\dot{s}_i = -\lambda_i |s_i|^{\frac{1}{2}} \text{sign}(s_i) - \alpha_i \int \text{sign}(s_i) d\tau + d_{\epsilon_i} \tag{25}$$

where $d_{\epsilon_i} = -h_i e^n_{i1} + \sum_{j=1}^N a_{ij} e^n_{j1}$. Thus (25) can be simplified as,

$$\begin{aligned} \dot{s}_i &= -\lambda_i |s_i|^{\frac{1}{2}} \text{sign}(s_i) + \rho_i \\ \dot{\rho}_i &= -\alpha_i \text{sign}(s_i) \end{aligned} \tag{26}$$

Let $z_i = \rho_i + d_{\epsilon_i}$, then (26) can be rewritten as,

$$\begin{aligned} \dot{s}_i &= -\lambda_i |s_i|^{\frac{1}{2}} \text{sign}(s_i) + z_i \\ \dot{z}_i &= \dot{\rho}_i + \dot{d}_{\epsilon_i} \end{aligned} \tag{27}$$

To analyze the overall stability of the system (27) a Lyapunov function chosen as, $V_o = \sum_{i=1}^N V_{oi} = \sum_{i=1}^N \gamma_i^T P_i \gamma_i$, where $\gamma_i = [\lambda_i |s_i|^{\frac{1}{2}} \text{sign}(s_i) \; s_i]^T$ and $P_i$ is a positive definite matrix. Which is absolutely continuous but not locally Lipschitz on the set $s, z \in \Re^2 | s = 0$, but still can be used for the stability analysis. The overall detailed can be followed from [18].

Thus the Lyapunov function is not continuously differentiable nor locally Lipschitz. The work shown in [12] states, with certain initial conditions the only requirement of a Lyapunov function is need to be continuous for a stability analysis. More details analysis can be found in [12].

Now the derivative of $\gamma_i$

$$\begin{aligned} \dot{\gamma}_i &= \begin{bmatrix} \frac{1}{2}|s_i|^{-\frac{1}{2}} \dot{s}_i \\ \dot{z}_i \end{bmatrix} = \begin{bmatrix} \frac{1}{2}|s_i|^{-\frac{1}{2}} [-\lambda_i |s_i|^{\frac{1}{2}} \text{sign}(s_i) + z_i] \\ -\alpha_i \text{sign}(s_i) + \dot{d}_{\epsilon_i} \end{bmatrix} \\ &= \frac{1}{2}|s_i|^{-\frac{1}{2}} \begin{bmatrix} [-\lambda_i |s_i|^{\frac{1}{2}} \text{sign}(s_i) + z_i] \\ -2[\alpha_i - \dot{d}_{\epsilon_i} \text{sign}(s_i)] |s_i|^{\frac{1}{2}} \text{sign}(s_i) \end{bmatrix} \\ &= |s_i|^{-\frac{1}{2}} A_i \gamma_i \end{aligned}$$

where $A_i = \begin{bmatrix} -\frac{1}{2}\lambda_i & \frac{1}{2} \\ -[\alpha_i - \dot{d}_{\epsilon_i}] & 0 \end{bmatrix}$. Then the first time derivative of $V_o = \sum_{i=1}^N V_{oi} = \sum_{i=1}^N \gamma_i^T P_i \gamma_i$

$$\dot{V}_o = \sum_{i=1}^N \dot{V}_{oi} = \sum_{i=1}^N |s_i|^{-\frac{1}{2}} \gamma_i^T (A_i^T P_i + P_i A_i) \gamma_i \tag{28}$$

since $A_i$ martix is Hurwitz as $\lambda_i > 0$, $\alpha_i > 0$ and $P_i$ is positive definite. Thus it can be simplified as,

$$\dot{V}_o = \sum_{i=1}^{N} \dot{V}_{oi} = -\sum_{i=1}^{N} |s_i|^{-\frac{1}{2}} \gamma_i^T Q_i \gamma_i < 0 \tag{29}$$

where $P_i$ and $Q_i$ are related by

$$A_i^T P_i + P_i A_i = -Q_i \tag{30}$$

Now for every $Q_i > 0$, there exists a unique solution $P_i > 0$. Since $V_{oi}$ is absolutely continuous and $\dot{V}_{oi}$ is negative definite. Now if $|s_i|^{\frac{1}{2}} = ||s_i|^{\frac{1}{2}} \text{sign}(s_i)| \le ||\gamma_i||_2 < \xi_{i_{min}}^{\frac{1}{2}} P_i V_{oi}^{\frac{1}{2}}$ then the inequality (28)

$$\dot{V}_o = \sum_{i=1}^{N} \dot{V}_{oi} \le -\sum_{i=1}^{N} \xi_{i_{min}}^{\frac{1}{2}}(P_i) V_i^{-\frac{1}{2}} \frac{\xi_{i_{min}}(Q_i)}{\xi_{i_{max}}(P_i)} V_{oi}$$

$$\le -\beta \sum_{i=1}^{N} V_{oi}^{\frac{1}{2}}, \beta = min_{i=1,\ldots N}\{\frac{\xi_{i_{min}}^{\frac{1}{2}}(P_i)\xi_{i_{min}}(Q_i)}{\xi_{i_{max}}(P_i)}\} > 0$$

Based on the above equality and $\sum_{i=1}^{N} V_{oi}^{\frac{1}{2}} \ge \left(\sum_{i=1}^{N} V_{oi}\right)^{\frac{1}{2}}$ we have,

$$\dot{V}_o = -\beta V_o^{\frac{1}{2}} \tag{31}$$

It follows that $\dot{V}_o(t, s(0), z(0)) \le -\beta V_o^{\frac{1}{2}}(t, s(0), z(0))$. This implies the overall stability of the MAS. The above result holds with $d_{\epsilon_i} = 0$ also.

Note that the observer error system for the MAS (10) is finite-time stable, which implies that the estimation error will converge to zero in a finite time. In addition, it has been shown by that the sliding variable $(s_i)$ for the ith agent will not be driven to infinity in the finite-time disturbance observer.

Thus, when the finite-time stability of observer error dynamics (10) and the sliding surface $(s_i)$ for the MAS is achieved, the error dynamics (32) will reduce to

$$\begin{aligned} \dot{e}_{ik}^o &= e_{i(k+1)}^o, 1 \le k \le n-1 \\ &\vdots \\ \dot{e}_{in}^o &= -\sum_{q=1}^{n} [k_{iq} \text{sign}(e_{iq}^o) |e_{iq}^o|^{\alpha_{iq}}] \end{aligned} \tag{32}$$

which implies the finite time tracking of the MAS in presence of mismatched uncertainty [1].

*Remark* In the absence of uncertainties and disturbances, the observer error dynamics $\dot{\mathbf{e}}_{ir_i}^k = 0$ and $v_{i(n-k)}^k = 0$ if the initial values of the observer states are selected same as that to the agent. In that case the control law can be used for $nth$ order MAS

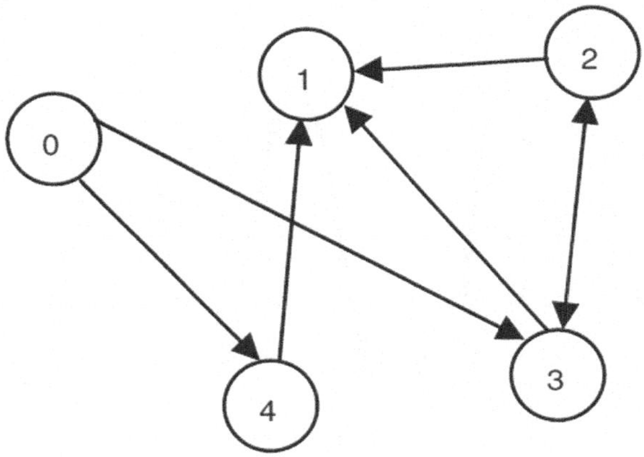


**Fig. 1** The communication graph

in presence of matched uncertainty. In the derivation throughout the paper $\tilde{x}_{ik}$ will be considered as $x_{ik}$. Which implies the nominal control method of the proposed method is retained.

# 6 Results

In this section, two numerical examples will be presented to illustrate the flexibility and effectiveness of the proposed methods.

*Example 1* In this example leader following MAS with four follower and one leader are selected. Where the leader is indexed 0, and the follower are indexed 1, 2, 3, 4 respectively. Figure 1 shows that directed topology used for the simulation [9, 20].

Here we consider the agent dynamics are of second order

$$\begin{aligned} \dot{x}_{i1} &= x_{i2} + d_{i1} \\ \dot{x}_{i2} &= u_i + d_{i2} \end{aligned} \tag{33}$$

where $i = 1, \ldots 4$. The term $d_{i2} = 0.01\,\sin(x_{i1})$ represents the matched uncertainty, and $d_{i1}$ is the mismatched uncertainty. The mismatched values are taken as $d_{11} = 0.15cos(t), d_{12} = 0.2sin(t), d_{13} = 0.1sin(t)$ and $d_{14} = 0.1sin(t)$ respectively.

The initial conditions of the agents are taken as $[x^0_{11}\ \ x^0_{12}]^T = [1\ \ 0]^T$,$[x^0_{21}\ \ x^0_{22}]^T = [1.2\ \ 0]^T$, $[x^0_{31}\ \ x^0_{32}]^T = [2\ \ 0]^T$, $[x^0_{41}\ \ x^0_{41}]^T = [-1.2\ \ 0]^T$. The simulation example is taken from [9, 20], the only difference is that the uncertainties are present in all channels as compared to [9, 20].

The control objective is to track the leader whose dynamic equation is given by [9, 20]

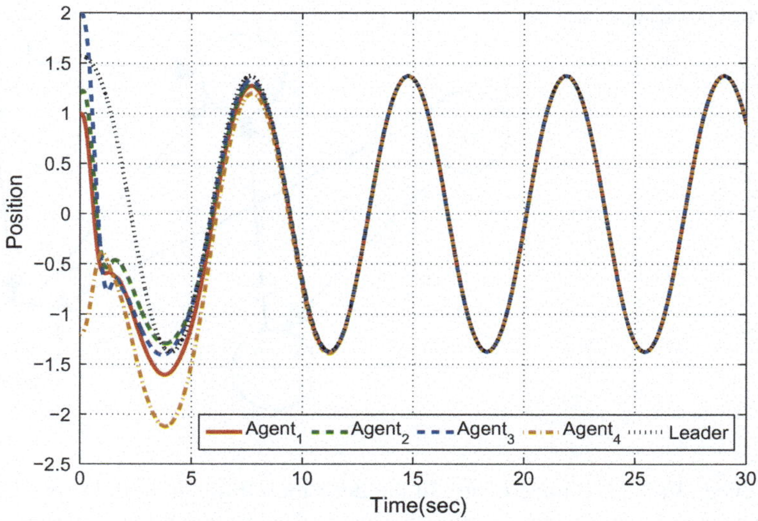


**Fig. 2** Consensus tracking

$$\dot{x}_{01} = \dot{x}_{02}$$
$$\dot{x}_{02} = -\sin(x_{01})/(1 + e^{-t}) \tag{34}$$

The initial conditions of the leader is $[x_{01}^0 \;\; x_{02}^0]^T = [\pi/2 \;\; 0]^T$. The sliding surface for the ith MAS is considered as,

$$s_i = e_{i2}^o + \int_0^t \{k_{i2}|e_{i2}^o|^{\alpha_{i2}}\text{sign}(e_{i2}^o) + k_{i1}|e_{i1}^o|^{\alpha_{i1}}\text{sign}(e_{i1}^o)\}d\tau \tag{35}$$

where $\alpha_{i2} = 9/16$, $\alpha_{i1} = 9/23$, $k_{i1} = 1.2$, $k_{i2} = 1.5$.

The consensus control law (18) for the ith MAS is designed with parameters given by $\lambda_i = 1$ and $\alpha_i = 0.5$. The finite time disturbance observer (9) is used to estimate the disturbances. The designed parameters are chosen as $\lambda_{i0}^k = 2$, $\lambda_{i1}^k = 2$, $\lambda_{i0}^k = 0.01$ and $L_{ik} = 10$. Here for second order MAS, $k = 1, 2$.

Simulation results are shown in the Figs. 2, 3 and 4. The consensus position tracking performance of the multi-agent systems under the proposed controller is shown in Fig. 2. It can be observed that the output of the followers accurately tracks the leader after a short transit.

The integral sliding surface can be resumed in finite time and kept invariant in the sequential time as shown in Fig. 3. Therefore, the accurate consensus position tracking are achieved in finite-time by the proposed protocols.

The control input of each follower is provided in Fig. 4. Since super twisting algorithm is used to converge the sliding surface of MAS in finite time, thus it is more superior in reducing chattering problem from the control input.

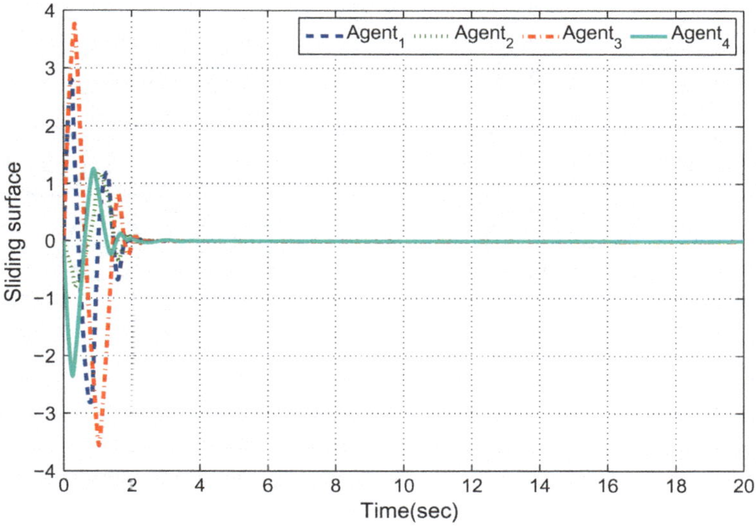


**Fig. 3** Sliding surface

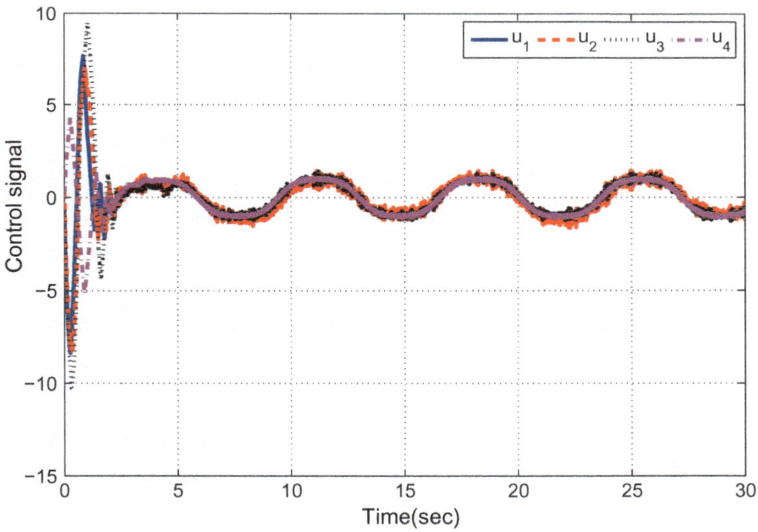


**Fig. 4** Control signal

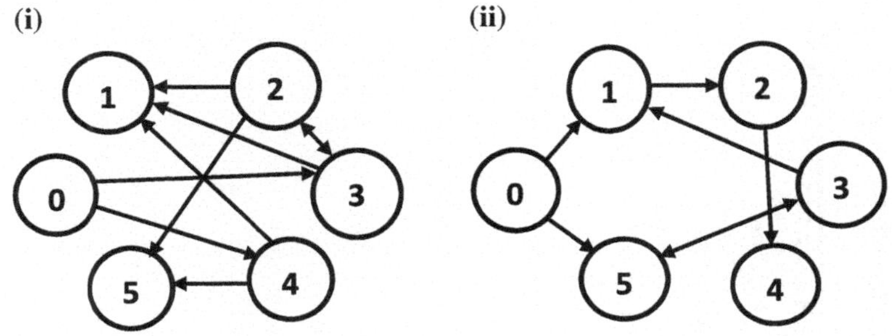


**Fig. 5** The communication graph

*Example 2* In this example two different directed topology of leader following MAS, with five follower and one leader are considered. Where the leader is indexed 0, and the follower are indexed 1, 2, 3, 4, 5 respectively. Figure 5 shows that directed topology used for the simulation.

The dynamic equation of the agent is chosen as of third order:

$$\begin{aligned} \dot{x}_{i1} &= x_{i2} + d_{i1} \\ \dot{x}_{i2} &= x_{i3} + d_{i2} \\ \dot{x}_{i3} &= u_i + d_{i3} \end{aligned} \tag{36}$$

where $y_{i0} = x_{i1}$ is the output.

The dynamic equation of the leader is taken as,

$$\dot{x}_{01} = x_{02},\ \dot{x}_{02} = x_{03},\ \dot{x}_{03} = u_0 \tag{37}$$

$y_0 = x_{01}$ is the output. The control objective is to track the leader whose output equation is given by $y_0 = x_{01} = t$. The initial conditions of the agents are taken as $[x_{11}^0\ x_{12}^0\ x_{13}^0]^T = [5,\ -5,\ -5]^T$, $[x_{21}^0\ x_{22}^0\ x_{23}^0]^T = [-5,\ 0,\ 5]^T$, $[x_{31}^0\ x_{32}^0\ x_{33}^0]^T = [10,\ 0,\ 0]^T$, $[x_{41}^0\ x_{42}^0\ x_{43}^0]^T = [-10,\ 0,\ 5]^T$, $[x_{51}^0\ x_{52}^0\ x_{53}^0]^T = [10,\ 5,\ 0]^T$.

**Case 1** We firstly consider the topology-I. The controller parameters (18) are chosen as, $\alpha_{i1} = 0.5, \alpha_{i2} = 3/5, \alpha_{i3} = 3/4, \alpha_i = 0.15, \lambda_i = 1.5$ and $k_{i1} = 0.5, k_{i2} = 0.75, k_{i3} = 1.00 \forall i = 1, \ldots, 5$.

The observer parameters (9) are $\lambda_{im}^1 = [2\ 1.5\ 1\ 0.1]^T, \forall m = 0, \ldots, 3$. $\lambda_{im}^2 = [2\ 1.5\ 0.1]^T \forall m = 0, \ldots 2$. $\lambda_{im}^3 = [0.2\ 0.1]^T \forall m = 0..1$, and $L_{ik} = 10 \forall i = 1, \ldots, 5$. The uncertainties are $d_{i1} = 0.2\sin(0.1t),\ d_{i2} = 0.3\cos(0.1t), d_{i3} = 0.5\sin(\frac{1}{6}\pi + 0.1t), \forall i = 1, \ldots, 5$.

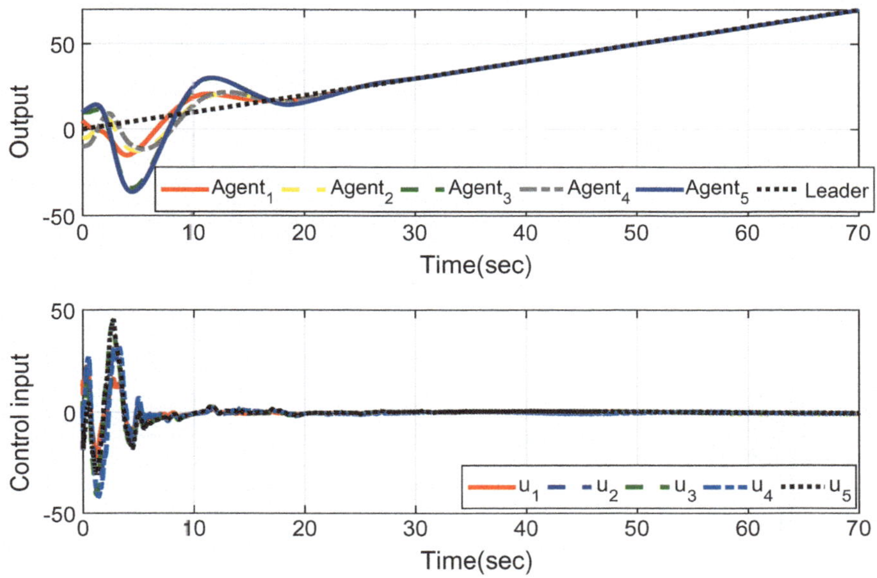


**Fig. 6** Consensus tracking and control input with topology I

The consensus position tracking and control input of the MAS with triple integrator dynamics are shown in Fig. 6. It can be observed that the followers accurately tracks the leader after a short transit in spite the presence of mismatched uncertainty. The use of HOSM guarantees the smooth control actions, which shows its suitability for electro mechanical actuators.

**Case 2** For topology-II, with the same design parameters one can still achieve accurate consensus tracking in finite time as shown in Fig. 7 in spite of the mismatch disturbances.

## 7 Conclusion

The finite-time consensus problem of higher order MAS with mismatched uncertainties is investigated by designing an integral sliding mode based on higher order sliding mode technique. The consensus sliding surface has been proposed for the mismatched disturbance attenuation via finite-time disturbance observer. For the MAS the finite time convergence of the sliding surface is obtained by the use of strict Lyapunov function. Simulation results shows the effectiveness of cooperative control scheme for higher order mismatched multi agent network.

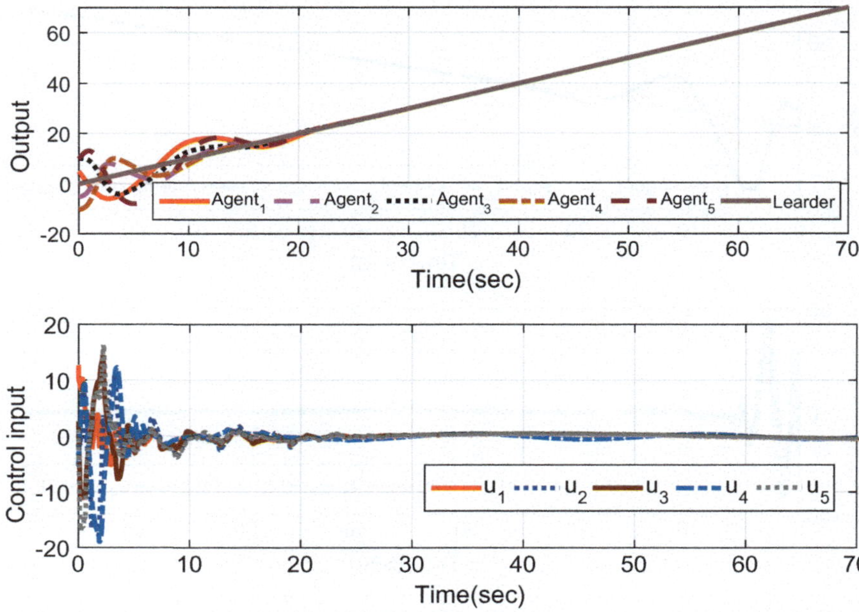


**Fig. 7** Consensus tracking and control input with topology II

## References


1. Bhat, S.P., Bernstein, D.S.: Geometric homogeneity with applications to finite-time stability. Math. Control. Signals Syst. **17**, 101–127 (2005)
2. Cao, W.-J., Xu, J.-X.: Nonlinear integral-type sliding surface for both matched and unmatched uncertain systems. IEEE Trans. Autom. Control **49**(8), 1355–1360 (2004)
3. Choi, H.H.: An explicit formula of linear sliding surfaces for a class of uncertain dynamic systems with mismatched uncertainties. Automatica **34**(8), 1015–1020 (1998)
4. Estrada, A., Fridman, L.M.: Integral hosm semiglobal controller for finite-time exact compensation of unmatched perturbations. IEEE Trans. Autom. Control **55**(11), 2645–2649 (2011)
5. Ghasemi, M., Nersesov, S.G.: Finite-time coordination in multiagent systems using sliding mode control approach. Automatica **50**, 1209–1216 (2014)
6. Ghasemi, M., Nersesov, S.G., Clayton, G., Ashrafiuon, H.: Sliding mode coordination control for multiagent systems with underactuated agent dynamics. Int. J. Control **87**(12), 2612–2633 (2014a)
7. Ghasemi, M., Nersesovn, S.G., Clayton, G.: Finite-time tracking using sliding mode control. J. Frankl. Inst. **351**(5), 2966–2990 (2014b)
8. He, X., Wang, Q., Yu, W.: Finite-time containment control for second-order multiagent systems under directed topology. IEEE Trans. Circuit Syst.-II: Express Briefs **61**(8), 619–623 (2014)
9. Khoo, S., Xie, L., Man, Z.: Robust finite-time consensus tracking algorithm for multirobot systems. IEEE/ASME Trans. Mechatron **14**(2), 219–228 (2009)
10. Li, S., Tian, Y.P.: Finite-time stability of cascaded time-varying systems. Int. J. Control **80**(4), 646–657 (2007)
11. Lin, P., Jia, Y.: Consensus of second-order discrete-time multiagent systems with nonuniform time-delays and dynamically changing topologies. Automatica **45**(9), 2154–2158 (2009)

12. Moreno, J.A., Osorio, M.: Strict lyapunov functions for the super-twisting algorithm. IEEE Trans. Autom. Control **57**(4), 1035–1040 (2012)
13. Moulay, E., Perruquetti, W.: Finite time stability and stabilization of a class of continuous systems. J. Math. Anal. Appl. **323**(2), 1430–1443 (2006b)
14. Shtessel, Y.B., Shkolnikov, I.A., Levant, A.: Smooth second-order sliding modes: missile guidance application. Automatica **43**(8), 1470–1476 (2007)
15. Spurgeon, S.K., Davies, R.: A nonlinear control strategy for robust sliding mode performance in the presence of unmatched uncertainty. Int. J. Control **57**(5), 1107–1123 (1993)
16. Tao, C.W., Chan, M.-L., Lee, T.-T.: Adaptive fuzzy sliding mode controller for linear systems with mismatched time-varying uncertainties. IEEE Trans. Syst. Man Cybern.: Part B: Cybern. **33**(2), 283–294 (2003)
17. Yang, J., Li, S., Su, J., Yu, X.: Continuous nonsingular terminal sliding mode control for systems with mismatched disturbances. Automatica **49**, 2287–2291 (2013)
18. Yu, S., Long, X.: Finite-time consensus for second-order multi-agent systems with disturbances by integral sliding mode. Automatica **54**, 158–165 (2015)
19. Zhang, W., Wang, Z.: Adaptive output consensus tracking of uncertain multi-agent systems. Int. J. Syst. Sci. **46**(13), 2367–2379 (2015)
20. Zhao, D., Zou, T., Li, S., Zhu, Q.: Adaptive backstepping sliding mode control for leader follower multi-agent systems. IET Control Theory Appl. **6**(8), 1109–1117 (2011)
21. Zuo, Z.: Nonsingular fixed-time consensus tracking for second-order multi-agent networks. Automatica **54**, 305–309 (2015)

# Path Planning for a Multi-robot System with Decentralized Control Architecture

**Fethi Metoui, Boumedyen Boussaid and Mohamed Naceur Abdelkrim**

**Abstract** In this chapter we have studied the path planning problem of a group of autonomous Wheeled Mobile Robots in a very dynamic workspace. The work focuses on the combination of artificial potential field approach with the decentralized architecture to coordinate the movements of robots. The technique implemented is adapted to solve the path planning problem for a multi-robot system. So, the problem of coordination of robots at the intersection zone is solved by the use of the priority allocation method. We also used the neighborhood detection technique to reduce the influence area of each robot and to optimize the time of calculation. The solution is tested and simulated with Matlab/Simulink.

**Keywords** Wheeled mobile robot · Multi-robot system · Decentralized control architecture · Artificial potential field · Priority conflict management

## 1 Introduction

The robot is a complex mechanical device equipped with sensors and actuators designed to perform complex tasks either autonomously or under supervision. These tasks are often painful, repetitive, dangerous and too precise for human beings [31]. The new systems are now intended to move in a dynamic and uncertain environment, where can be found both the static and mobile obstacles. It is therefore highly important and difficult to ensure the safety of the robot's movement in such unsecure environments. Thus, path planning is one of tricky issues in the autonomy of mobile robots. It aims to generate a free trajectory without collisions between a robot's initial configuration and a final one.

F. Metoui · B. Boussaid (✉) · M. N. Abdelkrim
National School of Engineers of Gabes, University of Gabes, Gabes, Tunisia
e-mail: boumedyen.boussaid@gmail.com

M. N. Abdelkrim
e-mail: naceur.abldekrim@enig.rnu.tn

 
J. Ghommam et al. (eds.), *New Trends in Robot Control*, Studies in Systems, Decision and Control 270, https://doi.org/10.1007/978-981-15-1819-5_12

In order to guarantee its safety, it is necessary for the robot to move and navigate without colliding or clashing with any kind of obstacle. This safe navigation requires intelligent control strategy capable of overcoming the uncertainties presented by the real world. So, the robot must be able to move from one point to another by finding efficient and safe path to avoid collision with the obstacles.

One of the key points in mobile robotics is to define and implement a strategy of movement towards the desired goal. For example, if one wants to ensure the movement of the robot between two proposed positions in the plane, the process requires two stages:

1. The planning of a trajectory between the initial position and the goal point while ensuring the safety of the robot.
2. The development of an adequate control law in order to track this trajectory by the robot.

To solve the problem of trajectory planning of mobile robots, many approaches have been developed in the literature [21]. One method is proposed by 'Glavaski' [10]. Its principle is based on the decomposition of environment of the robot into cells. The decomposition principle is to decompose the robot free space into a set of non-overlapping cells. This cell represents exactly the space of the configurations. Then, the method of Dynamic Window Approach (DWA) is suggested by 'Fox' [8]. This method is directly attached to the dynamic model of robot and takes the limit of the speed and the acceleration of the robot into consideration. Accordingly, the robot is controlled and oriented towards the target by adjusting the value of speed and the acceleration. In addition to that, a path planning based on a Polynomial Approach is presented in the work of 'Defoort' [6]. This technique consists of two types of trajectories: Bezier curves and spline curves. Besides, it is worth mentioning the Deterministic Kinodynamic Planning (DKP) approach [9]. DKP provides a solution that collects a local path planner based on spline function and global planner with $A^\star$ algorithm. Recently 'Montiel' [22] presented a new robotic navigation algorithm, called 'Bacterial Potential Field', using the APF method reinforced by Bacterial Evolutionary Algorithm. In the same context of mobile robot trajectory planning, 'Tan' [30] proposes an algorithm based on APF and the algorithm $A^\star$.

It is also important to refer to the trajectory planning solution proposed in [14] which combines the three techniques: Particle Swarm Optimization algorithm (PSO), APF approach and the fuzzy logic controller in order to provide a good performance to the robot in trajectory planning. Moreover, [15] proposed a new mechanism of path planning based on "self-adaptive learning particle swarm optimization". Finally, a remarkable research conducted by [12] implemented the Bees algorithm to solve the issue of path planning.

The problem is even more difficult if we plan to move multiple robots in the same dynamic environment. There is a need to coordinate their movements in relation to the movement of the other robots in one hand, and avoid collision between them on the other hand. In recent years, several research efforts have been directed towards controlling a group of autonomous robots. The interest in this area is well justified by the many advantages that these systems have over single autonomous robots and is well supported by the improvements in technologies that allow interaction and integration between multiple systems. The reasons for using multi-robot systems are very different. However, one of the main motivations is that multi-robot systems can be used to increase the efficiency of the system [5]. In other words, a multi-robot system can better accomplish a mission in terms of time and quality. It can also perform non-executable tasks by a single robot (for example, moving a large object). In addition, instead of building and using a single powerful robot, the use of multi-robot system can be simpler and cheaper. It can also offer flexibility in the execution of tasks and make the system tolerant to possible defects. Generally, the use of a group of robots instead of one is motivated by three major factors:

- Either the task to be executed imperatively requires the cooperation between a minimal number of robots, that is to say that the task can be accomplished only by the simultaneous intervention of the critical number of robots.
- Either to improve certain performances related to the execution of the task to be performed, for example: speed, robustness, reliability and flexibility.
- or to execute several tasks in parallel by several robots. Generally in the latter case, the tasks are performed by the robots independently.

In literature, we can distinguish between different works that revolve around the path planning problem of multiple mobile robots. One noteworthy work was conducted by [3] that focused on the algorithm $A^{\star}$ and the application of the notion of priority in area of passage. In their study, the authors applied the algorithm $A^{\star}$ to each robot to determine the optimal trajectory. However, in order to process the cooperation condition between the robots, the technique of priority has been applied. Yet, the application of these two techniques in real time is difficult as it requires the discretization of the environment in a graph at each moment. Another work was carried out by [33] who used and implemented the visibility graph. The method consists of reducing the workspace to the sums of polygonal obstacles. All of these sums constitute the visibility graph and are considered as the set of possible crossing points for robot. The visibility graph approach has the advantage of calculating the optimal trajectory quickly and easily. However, the trajectory generated by this technique includes the sums of obstacles, which means that the robots touch these sums when following the trajectory. In addition to that, the researchers [4] proposed a control architecture for maintaining formation of mobile robots moving in a dynamic environment. Furthermore, [6] relied on the use of polynomial approach to generate the trajectories of cooperative robots. Reference [28] combined the Generalized Connectivity Maintenance with Rapidly exploring Random Tree. This approach rests on implementing a global design planner that generates paths suitable for the fleet of mobile robots.

The potential field is widely used for planning the trajectory of mobile robot. But, it did not receive a great attention in the case of multiple mobile robot. There are various significant works proposed in literature. For instance, the study of [29] was based on the application of potential field approach for the control and manipulation of the multi-robot system. In addition, [2] implemented a new approach based on artificial potential field and described the application of multi-robot system in rescue and search. Besides, [11] focuses on the potential fields method to plan conflict-free trajectories for a large number of aircrafts. It is also important to note the study of [17] who addressed the problem of local minimum in multi-robot system. In the same vein, [13] combined both the APF and the Virtual Obstacles approaches in order to solve the problem of cooperation between multi-robots in the same workspace. Finally, [24] examined the idea of using of the probabilistic approach as a deterministic planner in distributed complex multi-agent system. This approach allows each robot to generate path planning avoiding collisions. However, the problem of these methods is related to scalability.

Despite the impressive progress in the field of autonomous robotics in recent years, a number of problems require even more research. The mobile robot trajectory planning in static and dynamic environments is one of the most widely studied areas of research.

In this work, we seek to focus on solving the issue of path planning for multi-robot system. The environment of simulation is very dynamic and involves different types and forms of obstacles. Our contribution is mainly based on the path planning of group of wheeled mobile robots located in a same dynamic workspace. The major aim of this contribution is to make each robot perfectly autonomous and able to avoid the collision with the static obstacles included in its workspace and with the other robots. The use of multiple robots in a same workspace requires the necessity to coordinate between them. In our contribution, we use the distributed architecture to organize the cooperation between robots. So, each robot uses the local information of workspace to find out the eligible trajectory depending on the path followed by the other robots. To this end, we commune and implement the three following techniques: the application of APF method, the neighborhood system and the application of technique of priority between robots.

In this chapter, at first an overview on the kinematic and dynamic modeling of mobile robot is presented. Subsequently, the main focus is on presenting the proposed multi-robot system trajectory planning algorithms. At this level, we present the decentralized trajectory planning architecture implemented for the multi-robot system. Finally, the simulation results of the proposed techniques will be presented.

## 2 The Mobile Robot Modeling

We were interested in this work with mobile robots with wheels and precisely the robot Pioneer 3DX. It is a wheeled mobile robot manufactured by the company "Adept Mobile Robots". It is dedicated to scientific research in robotics. Pioneer is composed of two independent drive wheels at the rear of the main body of the robot and a wheel not controlled at the front.

Pioneer is composed of two independent drive wheels at the rear of the main body of the robot and a wheel not controlled at the front. The latter is intended only to ensure the stability of the robot. The Pioneer robot is classified in the category of non-holonomic mobile robots of the unicycle type [25].

## 2.1 Kinematic Model

Kinematic modeling is the fundamental study of the motion of a mechanical system. It allows us to describe how mechanical systems behave. At this point, we are only interested in the robot's velocity vectors without considering the forces influencing its motion [27].

In the case of a mobile robot, to achieve navigation, the state of the robot is totally defined by the following vector [27]: $q = [x,\ y,\ \theta]^T$, with $(x, y)$ is the position of robot, $\theta$ is the orientation of robot and $v$ is the linear speed.

In this study, we base our study on the model of Fig. 1 to determine the kinematic and dynamic model of the mobile robot.

The used robot is characterized by:

- $G$: Center of gravity,
- $\theta$: orientation of robot,
- $r$: radius of wheel,
- $2b$: distance between the two-wheel drive,
- $m$: mass of robot,
- $I$: moment of inertia of robot,
- $d$: distance between the center of gravity and the axis of two wheels,
- $\dot{\varphi}_r$ and $\dot{\varphi}_l$ are the angular velocities of right wheel and left wheel.

The kinematic model of the system is written by the following equation [31]:

$$\begin{cases} \dot{x} = v\cos\theta \\ \dot{y} = v\sin\theta \\ \dot{\theta} = \omega \end{cases} \tag{1}$$

If we set $u = [v,\ w]^T$ as control vector we obtain:

$$\begin{bmatrix} \dot{x} \\ \dot{y} \\ \dot{\theta} \end{bmatrix} = \begin{bmatrix} \cos\theta & 0 \\ \sin\theta & 0 \\ 0 & 1 \end{bmatrix} \begin{bmatrix} v \\ w \end{bmatrix} \tag{2}$$

with $v$ is the linear velocity of the robot and $\omega = \dot{\theta}$ is the angular velocity.

In the case of a differential drive system, where two wheels are controlled by two motors separately, the linear and angular speed are calculated by:

$$v = \frac{(v_r + v_l)}{2} = \frac{r\left(\dot{\varphi}_r + \dot{\varphi}_l\right)}{2} \tag{3}$$

$$w = \frac{(v_r - v_l)}{2.b} = \frac{r\left(\dot{\varphi}_r - \dot{\varphi}_l\right)}{2b} \tag{4}$$

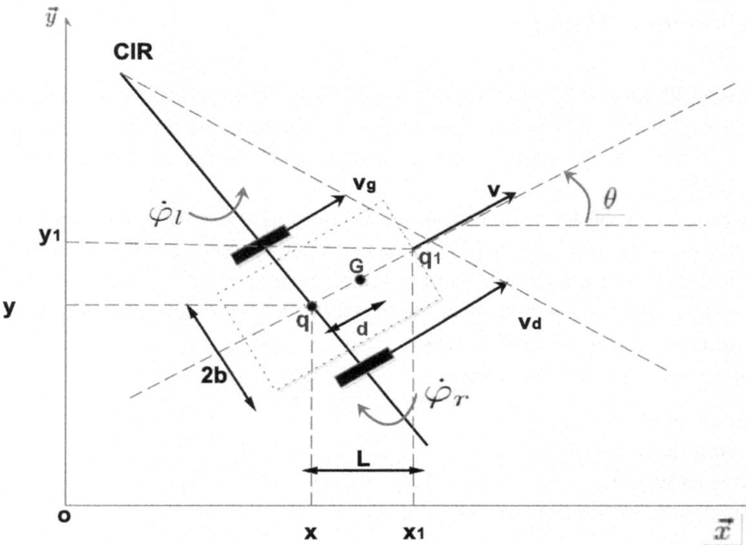


**Fig. 1** Differential drive robot model

Therefore:

$$\begin{cases} \dot{\varphi}_r = \dfrac{v + b\omega}{r} \\ \dot{\varphi}_l = \dfrac{v - b\omega}{r} \end{cases} \tag{5}$$

with $\dot{\varphi}_r$ and $\dot{\varphi}_l$ are the two control variables. The Eqs. (4) and (5) can be resumed in (6).

$$\begin{bmatrix} v \\ w \end{bmatrix} = \begin{bmatrix} \frac{r}{2} & \frac{r}{2} \\ \frac{r}{2.b} & -\frac{r}{2b} \end{bmatrix} . \begin{bmatrix} \dot{\varphi}_r \\ \dot{\varphi}_l \end{bmatrix} \tag{6}$$

By introducing the Eq. (1) in (6), the kinematic model can be written as follows:

$$\dot{q} = \begin{bmatrix} \dot{x} \\ \dot{y} \\ \dot{\theta} \end{bmatrix} = \begin{bmatrix} \frac{r}{2}\cos\theta & \frac{r}{2}\cos\theta \\ \frac{r}{2}\sin\theta & \frac{r}{2}\sin\theta \\ \frac{r}{2b} & -\frac{r}{2b} \end{bmatrix} . \begin{bmatrix} \dot{\varphi}_r \\ \dot{\varphi}_l \end{bmatrix} \tag{7}$$

where the control vector is $[\dot{\varphi}_r, \dot{\varphi}_l]^T$.

## 2.2 Dynamic Model

The dynamic model is the study of the motion of a mechanical system, taking into consideration the different forces that affect its motion. It defines the equations of motion of the robot. This model makes it possible to establish the relations between the forces exerted by the actuators on the one hand, and the location of the robot and the speeds of movement of the robot on the other hand. The identification of this model is important for simulating the motion of the robot, analyzing the structure of the system and developing control algorithms [7, 31].

The most used methods to obtain the dynamic model in robotics: the method based on the Lagrange formulation, and the method based on the Newton–Euler formulation [31].

The dynamic model determination by the Euler-Lagrange technique is based on the following equation:

$$\frac{d}{dt}\left(\frac{\partial L}{\partial \dot{q}}\right) - \frac{\partial L}{\partial q} = S\left(q\right).\tau - A\left(q\right)\lambda \tag{8}$$

with:

- $L$: lagrangian of the system equal to $(E - U)$,
- $E$: total kinetic energy of the system,

$$E = \frac{1}{2}mv^2 + \frac{1}{2}Iw^2 \tag{9}$$

- $U$: total potential energy of the system $(U = 0)$,
- $S\left(q\right)$: transformation matrix,
- $\tau$: Forces/Couples,
- $\lambda$: The multiplier vector of Lagrange,
- $A\left(q\right)$: Matrix of non-holonomic constraints.

The generalized coordinates of the system are written by the following vector:

$$q = \begin{bmatrix} x & y & \theta & \varphi_r & \varphi_l \end{bmatrix}^T \tag{10}$$

The components are thus obtained $x_i$ and $y_i$ of the center of mass and the wheels according to the generalized coordinates as follows:

- For the platform

$$\begin{cases} x_{cr} = x + d\cos\theta \\ y_{cr} = y + d\sin\theta \end{cases} \tag{11}$$

- For the right wheel

$$\begin{cases} x_{rd} = x + b\sin\theta \\ y_{rd} = y + b\cos\theta \end{cases} \tag{12}$$

- For the left wheel

$$\begin{cases} x_{rg} = x - b\sin\theta \\ y_{rg} = y + b\cos\theta \end{cases} \tag{13}$$

The kinetic energies of the wheeled mobile robot are determined by: the kinetic energy of the robot platform $E_{cr}$ (14), the kinetic energy applied to the right wheel of the robot $E_{rd}$ (15) and the kinetic energy at the left wheel of the robot $E_{rg}$ (16).

$$E_{cr} = \frac{1}{2}m_{cr}\left(\dot{x}_{cr}^2 + \dot{y}_{cr}^2\right) + \frac{1}{2}I_{cr}\dot{\theta}^2 \tag{14}$$

$$E_{rd} = \frac{1}{2}m_r\left(\dot{x}_{rd}^2 + \dot{y}_{rd}^2\right) + \frac{1}{2}I_r\dot{\varphi}_r^2 \tag{15}$$

$$E_{rg} = \frac{1}{2}m_r\left(\dot{x}_{rg}^2 + \dot{y}_{rg}^2\right) + \frac{1}{2}I_r\dot{\varphi}_l^2 \tag{16}$$

The total kinetic energy of mobile robot is calculated as follows:

$$L = E_{cr} + E_{rd} + E_{rg} \tag{17}$$

So

$$\begin{aligned} L = {} & \frac{1}{2}\left({x_0}^2 + {y_0}^2\right)(m_c + 2m_r) + \frac{1}{2}\dot{\theta}^2\left(m_c d^2 + 2m_r b^2 + I_c\right) \\ & + m_c\dot{\theta}d\left(\dot{y}_0\cos\theta - \dot{x}_0\sin\theta\right) + \frac{1}{2}I_r\left(\dot{\varphi}_r^2 + \dot{\varphi}_l^2\right) \end{aligned} \tag{18}$$

if we take $m = m_c + 2m_r$ et $I = I_c + m_c d^2 + 2m_r b^2$.

The new expression of $L$ is expressed as follows:

$$L = \frac{1}{2}m\left(\dot{x}_0^2 + \dot{y}_0^2\right) + \frac{1}{2}I\dot{\theta}^2 + m_c\dot{\theta}d\left(\dot{y}_0\cos\theta - \dot{x}_0\sin\theta\right) + \frac{1}{2}I_r\left(\dot{\varphi}_r^2 + \dot{\varphi}_l^2\right) \tag{19}$$

The derivatives of $L$ with respect to the components of the vector $\dot{q}$ sont:

$$\frac{\partial L}{\partial \dot{x}_0} = m\dot{x}_0 - m_c d\dot{\theta}\sin\theta \tag{20}$$

$$\frac{\partial L}{\partial \dot{y}_0} = m\dot{y}_0 - m_c d\dot{\theta}\cos\theta \tag{21}$$

$$\frac{\partial L}{\partial \dot{\theta}} = I\dot{\theta} - m_c d\left(\dot{y}_0\cos\theta - \dot{x}_0\sin\theta\right) \tag{22}$$

$$\frac{\partial L}{\partial \dot{\varphi}_r} = I_r\dot{\varphi}_r \tag{23}$$

$$\frac{\partial L}{\partial \dot{\varphi}_l} = I_r\dot{\varphi}_l \tag{24}$$

The derivatives of these expressions with respect to time are:

$$\frac{d}{dt}\left(\frac{\partial L}{\partial \dot{x}_0}\right) = m\ddot{x}_0 - m_c d\ddot{\theta}\sin\theta - m_c d\dot{\theta}\cos\theta \tag{25}$$

$$\frac{d}{dt}\left(\frac{\partial L}{\partial \dot{y}_0}\right) = m\ddot{y}_0 + m_c d\ddot{\theta}\cos\theta - m_c d\dot{\theta}\sin\theta \tag{26}$$

$$\frac{d}{dt}\left(\frac{\partial L}{\partial \dot{\theta}}\right) = I\ddot{\theta} + m_c d\ddot{y}_0\cos\theta - m_c d\dot{y}_0\sin\theta - m_c d\ddot{x}_0\sin\theta - m_c d\dot{x}_0\cos\theta \tag{27}$$

$$\frac{d}{dt}\left(\frac{\partial L}{\partial \dot{\varphi}_r}\right) = I_r\ddot{\varphi}_r \tag{28}$$

$$\frac{d}{dt}\left(\frac{\partial L}{\partial \dot{\varphi}_l}\right) = I_r\ddot{\varphi}_l \tag{29}$$

The matrix writing of this system of equations (25), (26), (28) et (29) is given by the expression:

$$M(q)\ddot{q} + V(q,\dot{q})\dot{q} = B(q)\tau + A^T(q)\lambda \tag{30}$$

with $M(q)$ is the inertia matrix, $V(q,\dot{q})$ is the Centrifugal Matrix, $B(q)$ is the input matrix and $A^T(q)$ is the matrix associated with cinematic constraints.

$$M(q) = \begin{pmatrix} m & 0 & m_c d\sin\theta & 0 & 0 \\ 0 & m & m_c d\cos\theta & 0 & 0 \\ -m_c d\sin\theta & -m_c d\cos\theta & I & 0 & 0 \\ 0 & 0 & 0 & I_r & 0 \\ 0 & 0 & 0 & 0 & I_r \end{pmatrix} \tag{31}$$

$$V(q,\dot{q}) = \begin{pmatrix} 0 & 0 & -m_c d\cos\theta & 0 & 0 \\ 0 & 0 & -m_c d\sin\theta & 0 & 0 \\ 0 & -m_c d(\sin\theta + \cos\theta) & 0 & 0 & 0 \\ 0 & 0 & 0 & 0 & 0 \\ 0 & 0 & 0 & 0 & 0 \end{pmatrix} \tag{32}$$

$$B(q) = \begin{bmatrix} 0 & 0 \\ 0 & 0 \\ 0 & 0 \\ 1 & 0 \\ 0 & 1 \end{bmatrix} \tag{33}$$

$$A^T(q) = \begin{bmatrix} -\sin\theta & \cos\theta & \cos\theta \\ \cos\theta & \sin\theta & \sin\theta \\ 0 & b & -b \\ 0 & -r & 0 \\ 0 & 0 & -r \end{bmatrix} \tag{34}$$

the Eq. (7) can be written in the form:

$$\dot{q} = S(q)\,\eta \tag{35}$$

with $\eta = \begin{pmatrix} \dot{\varphi}_r \\ \dot{\varphi}_l \end{pmatrix}$

Then, taking the time derivative of the Eq. (35) we obtain:

$$\ddot{q} = \dot{S}(q)\,\eta + S(q)\,\dot{\eta} \tag{36}$$

By substituting the Eqs. (35) and (36) in the main Eq. (30) we obtain:

$$M(q)\left[\dot{S}(q)\,\eta + S(q)\,\dot{\eta}\right] + V(q,\dot{q})\left[S(q)\,\eta\right] = B(q)\,\tau - A^T(q)\,\lambda \tag{37}$$

hence:

$$\begin{aligned} &S^T(q)\,M(q)\,S(q)\,\dot{\eta} + S^T(q)\left[M(q)\,\dot{S}(q) + V(q,\dot{q})\,S(q)\right]\eta = \\ &S^T(q)\,B(q)\,\tau - S^T(q)\,A^T(q)\,\lambda \end{aligned} \tag{38}$$

if we take:

$$\begin{aligned} &S^T(q)\,A^T(q)\,\lambda = 0 \\ &M_c = S^T(q)\,M(q)\,S(q) \\ &V_c = S^T(q)\left[M(q)\,\dot{S}(q) + V(q,\dot{q})\,S(q)\right] \\ &I_c = S^T(q)\,B(q) \end{aligned}$$

Hence the dynamic model can be written in the following form:

$$M_c\,\dot{\eta} + V_c\,\eta = I_c\,\tau \tag{39}$$

Table 1 expresses the physical designation of each parameter used as well as its numerical value.

## 3 Control Architectures

The control architecture represents a central part of the multi-robot system because it determines the success of the missions and strongly influences the overall performance of the system. Coordination is done through regular and systematic acts

**Table 1** The parameters of the robot used

| | Designation | Value |
|---|---|---|
| $m_{cr}$ | Mass of robot (Kg) | 0.5 |
| $m_r$ | Mass of the wheel (Kg) | 0.1 |
| $b$ | Half distance between the wheels (m) | 0.33 |
| $r$ | Radius of the wheel (m) | 0.095 |
| $I_r$ | Rotational inertia of the robot (Kg/m$^2$) | 0.05 |
| $I$ | Moment of inertia of the robot (Kg/m$^2$) | 0.5 |

to ensure that all parties involved in a given activity or project work together in an organized manner to be effective [26]. In robotics sometimes the execution of a given task requires the presence of several robots in the same environment. The coordination between these robots is a fundamental process to improve profitability and avoid collision in the case of trajectory planning [1]. We present below the main control mechanisms used for a multi-robot system.

## 3.1 Centralized Approach

Centralized approaches allow to synthesize for a group of robots, a set of trajectories via a supervisor. This supervisor can be a robot from the set or a computer outside the group [1].

The use of the centralized architecture is motivated by several advantages. First, the robots used do not need embedded sensors or a powerful computing unit. This greatly reduces their cost. In addition, the concentration of sensors at a supervisor level provides a better understanding of the overall environment. However this approach is very complex from a computational point of view. Moreover the complexity of this technique increases exponentially with the increase of number of robots. Note also the great dependence of robots on the supervisor.

## 3.2 Decentralized Approach

Decentralized approaches do not require the use of a supervisor to synthesize the trajectories of the robot group. Thus, each robot individually plans its own trajectory taking into account the activities of other robots. In addition, unlike the centralized control architecture, the acquisition and calculation resources (the sensors, the calculation units, the locators, etc.) are distributed over the elements of the multi-robot system.

This approach also has the advantage of being easy to adapt to different numbers of robots. It responds better to the problematic of unknown or dynamic environments. It also guarantees better results in terms of reliability, flexibility and adaptability. In

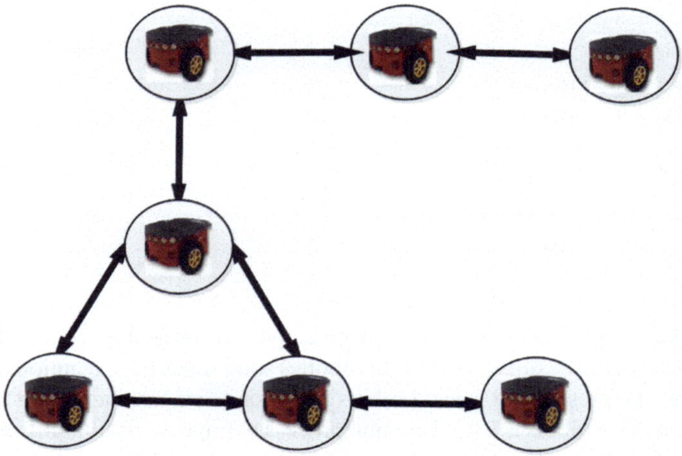

**Fig. 2** Illustration of the decentralized coordination of a team of mobile robots

addition, this structure is robust because if some of the robots stop working, this failure does not affect the other flotilla robots. In addition, the complexities of calculations are solved since each robot decides its actions by oneself. However, the solutions proposed by this type of architecture are sub-optimal because at each moment, each robot has only limited and incomplete information on the other robots [1].

## 4 Proposed Decentralized Trajectory Planning Architecture

The path planning methods for multi-robot systems are different from those used for a single robot. Although some algorithms are feasible and effective for single robot trajectory planning, they can not be applied directly to multi-robot system trajectory planning due to a number of requirements and constraints introduced by the multi-robot system.

In the following, we propose a decentralized trajectory planning approach for the multi-robot system [20].

The local approaches address trajectory planning issues with decentralized methods by allocating planning responsibilities among all team members.

Thus, each robot plans its action according to its local observations. Typically, few calculations are necessary, since each robot only plans and executes its own activities. In addition, less stringent constraints on communication are required, since the robots communicate only with the robots in their vicinity, as shown in Fig. 2.

Decentralized architecture is better suited to unknown or dynamic environments, since robots detect and react locally to the environment [19]. In addition, the system is more robust since the performance of the entire team no longer depends on the

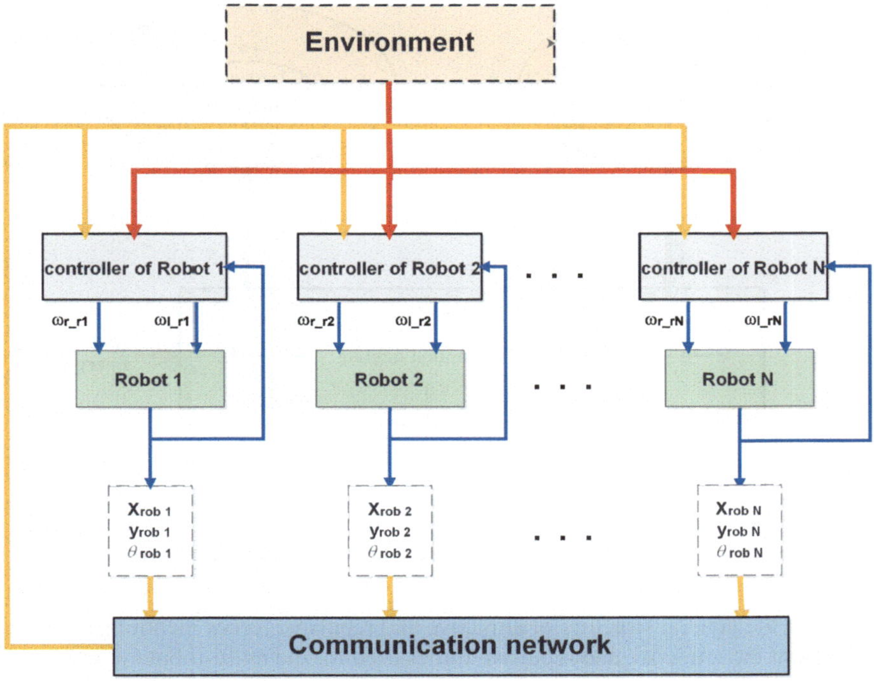


**Fig. 3** Architecture of decentralized approach proposed

direction of a single supervisor. In general, there is no single point of failure for distributed systems and the approach can easily accommodate a large number of robots. The architecture to be used by the robots is presented by Fig. 3.

With the decentralized architecture, each robot has local information from the workspace to find its most optimized path. Receiving this data for the robot is systematic in order to make the best move decision. The internal architecture of mobile robot is presented in Fig. 4.

The architecture shown in Fig. 4 illustrates three modules:

- Locomotion module: this module contains the mechanical robot structure as well as the energy used for moving the mobile robot.
- Perception module: The perception module includes the ability of the mobile robot to collect, process and implement data useful to the system to act in its environment. Thus to extract the data useful for its operation, the robot has several types of sensors.
- Decision module: The data coming from the sensors must be interpreted since they are the elements useful for the decision making. So using this module we can locate the robot, or apply control methods and planning for the robot. Robot movement requires a planning method that takes into account the specificity of robot structure and the applied control strategy.

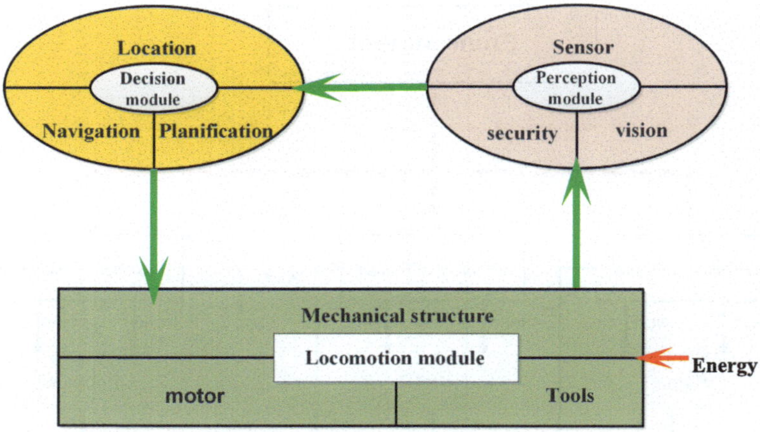


**Fig. 4** Structure of a mobile robot

# 5 Potential Field Modeling for the Multi-robot System

Based on the general structure of attractive and repulsive forces mentioned in [17], we present the attractive and repulsive forces used for the multi-robot system.

## 5.1 *Force Attractive*

The attractive force is used to guide the robot towards the target. Its expression is calculated according to the position of the robot and the position of the target [16]. There are some choices of attractive potential forms used such as exponential potential, conical potential, and parabolic potential [32]. In this work we use the parabolic potential.

The profile of the parabolic attractive potential is defined by the relation (40) [23].

$$U_a(q_r, q_{cible}) = \frac{1}{2}\xi p^2(q_r, q_{cible}) \tag{40}$$

The expression of the parabolic attractive force $F_a(q, g_{cible})$ derivative of potential (40) is given by:

$$F_a\left(q_r, q_{cible}\right) = -\nabla U_a\left(q_r, q_{cible}\right) = -\xi\left(q_r - q_{cible}\right) \tag{41}$$

Consider a group of $N$ robots. The equation of the attractive force generated by the target point $q_i$ applied to the robot $i$ is:

$$F_a(q_i, g_i) = -\xi p(q_i, q_i) \tag{42}$$

where $i = 1 \ldots N$ and each robot $i$ has a target $q_i$ different from other robots.

## 5.2 Repulsive Force

The repulsive force is used to ensure the safety of the robot. Its expression is calculated according to the position of the robot at every moment and the positions of the obstacles found in its vicinity. The repulsive force must create a potential whose value increases with decreasing distance to obstacles. Thus, this function is used to push the robot away from the obstacle area. However, if the robot is far from the obstacles, the path of the robot is not changed and the repulsive force has no effect on the robot [18].

Consider $M$ obstacles, $qo_j$ is the obstacle coordinate $j$ $(j = 1, 2, \ldots, M)$ et $p(q_i, qo_j)$ is the distance between the robot $i$ and the obstacle $j$. The expression of the repulsive force produced by the obstacle $j$ and applied on the robot $i$ is presented by (43).

$$F_{ro}(q_i, qo_j) = \left| \begin{array}{l} \xi \left[ \dfrac{1}{p(q, qo_j)} - \dfrac{1}{po_j} \right]^2 \dfrac{1}{p^2(q, qo_j)} \quad \nabla p(q, qo_j) \text{ if } p(q, qo_j) \le po_j \\ 0 \text{ if } p(q, qo_j) > po_j \end{array} \right. \tag{43}$$

where $po_j$ is a positive constant representing the distance of the influence of obstacles $j$. $P\left(q, qo_j\right) = \left\| q - qo_j \right\|$ is the distance between the robot $i$ and the obstacles $j$ and $\nabla$ is the gradient operator.

$$\nabla P\left(q, qo_j\right) = \frac{q - qo_j}{\left\| q - qo_j \right\|} \tag{44}$$

The repulsive force (43) can be written in a Cartesian form by the Eqs. (45) and (46).

$$F_{ro_{(x)}}(q_i, qo_j) = \left| \begin{array}{l} \xi \left[ \dfrac{1}{p(q, qo_j)} - \dfrac{1}{po_j} \right]^2 \dfrac{1}{p^2(q, qo_j)} \dfrac{x_i - x_j}{\| q - qo_j \|} \quad \text{if } p(q, qo_j) \le po_j \\ 0 \qquad \text{if } p(q, qo_j) > po_j \end{array} \right. \tag{45}$$

$$F_{ro_{(y)}}(q_i, qo_j) = \left| \begin{array}{l} \xi \left[ \dfrac{1}{p(q, qo_j)} - \dfrac{1}{po_j} \right]^2 \dfrac{1}{p^2(q, qo_j)} \dfrac{y_i - y_j}{\| q - qo_j \|} \quad \text{if } p(q, qo_j) \le po_j \\ 0 \qquad \text{if } p(q, qo_j) > po_j \end{array} \right. \tag{46}$$

where $(x_i, y_i)$ is the coordinate of the robot $i$ and $(x_j, y_j)$ is the coordinate of the obstacle $j$, $F_{ro_{(x)}}$ is the repulsive force projected on the axis $x$ and $F_{ro_{(y)}}$ is the repulsive force projected on the axis $y$.

In the workspace can have several obstacles. To implement this situation, the Eq. (43) will be modified in (47).

$$F_{ro}(q_i) = \sum_{j=1}^{M} F_{ro}(q_i, qo_j) \tag{47}$$

If the robot $k$ is the neighbor of the robot $i$. The expression of the repulsive force generated between the robot $i$ and $k$ is noted by the Eq. (48). This force is generated based on the actual position of the robot $i$, the position of robot $k$ and the distance between the robot $i$ and $k$.

$$F_{rr}(q_i, q_k) = \begin{cases} \xi \left[ \dfrac{1}{p(q_i, q_k)} - \dfrac{1}{pr_k} \right]^2 \dfrac{1}{p^2(q_i, q_k)} \nabla p(q_i, q_k) & \text{if } p(q_i, q_k) \leq pr_k \\ 0 \quad \text{if } p(q_i, q_k) > pr_k \end{cases} \tag{48}$$

where $pr_k$ is the minimum threshold of the distance between the robots $i$ and $k$, $p(q_i, q_k)$ is the distance between the robot $i$ and $k$.

In the general case, the neighborhood area of the robot $i$ can contain $T$ robots. Where $T = 1 \ldots N - 1$ and $k = 1 \ldots T$, the Eq. (48) is written in the general case as follows:

$$F_{rr}(q_i) = \sum_{k=1, k \neq i}^{T} F_{rr}(q_i, q_k) \tag{49}$$

In general, the workspace contains several robots and several obstacles. In order to take this situation into account, the repulsive force can be written by the Eq. (50).

$$F_r(q_i) = F_{ro}(q_i) + F_{rr}(q_i) \tag{50}$$

where $F_{ro}$ is the repulsive force induced by static obstacles and $F_{rr}$ is the repulsive force induced by the robots.

The robot is considered as a system subject to the influence of an attractive force induced by the target to attract the robot to the final destination and another repulsive force generated by the obstacles to avoid collision and ensure the safety of the robot.

The resulting force associated with the robots $i$ is calculated by the Eq. (51). This force depends on three factors [20]:

- The attractive force $F_a$ exerted by the target point on the robot $i$,
- The repulsive force $F_{ro}$ exerted by the obstacles on the robot $i$,
- The repulsive force $F_{rr}$ produced by the robots $k$ neighbors of robot $i$.

$$F(q_i) = F_a(q_i) + F_r(q_i) = F_a(q_i) + F_{ro}(q_i) + F_{rr}(q_i) \tag{51}$$

# 6 Proposed Algorithms for Path Planning: Multi-robot Framework

Now, we assume that the simulation environment is composed of several obstacles and several mobile robots. Each robot does its own task that is independent of other group robots. The environment is considered very dynamic where all the robots move at the same time and each robot considers the other robots as dynamic obstacles. The path planning is based on the decentralized architecture previously proposed. The strategy according to which the potential field is applied and illustrated by the algorithm 1 [20].

---

**Algorithm 1:** Path planning

---

- **Step 1:** initialization of Parameters
  - **Inputs:** $i$, $j$, $q_i$, $g_i$, $qo_j$, $po_j$, $p_{ik}$
  - **Outputs:** the robots are at the points of arrival.
- **Step 2: attraction towards the targets**
  - **For** $i = 1$ **to** $N$
    - **if** $(q_i \neq g_i)$ **then**
      - Application of the Eq. (42)
    - **End**
  - **End**
- **Step 3: repulsion between the robot** $i$ **and the obstacles** $j$
  - **For** $i = 1$ **to** $N$
    - **For** $j = 1$ **to** $M$
      - $p(q_i, qo_j) = \|qo_j - q_i\|$
      - **if** $p(q_i, qo_j) \leq po_j$ **then**
        - Application of the Eq. (47)
      - **End**
    - **End**
  - **End**
- **Step 4: repulsion between the robot** $i$ **and the robot** $k$
  - **For** $i = 1$ **to** $N$
    - **For** $k = 1$ **to** $N$ and $i \neq k$
      - $p(q_i, q_k) = \|q_k - q_i\|$
      - **if** $p(q_i, q_k) \leq p_{ik}$ **then**
        - Application of the Eq. (49)
      - **End**
    - **End**
  - **End**

---

where $N$ is the number of robots included in the same workspace, $M$ is the number of obstacles in this workspace, $i$ is a counter on the number of robots $i \in [1 \ldots N]$, $k$ is another counter on the number of robots $k \in [1 \ldots N]$, $j$ is a counter on the number of obstacles $j \in [1 \ldots M]$ in this workspace, $q_i$ is the position of robot $i$,

$g_i$ is the position of the target of robot $i$, $qo_j$ is the position of obstacle $j$, $po_j$ is the danger zone of the obstacle $j$, $p_{ik}$ is the danger radius of the robot $i$, $p(q_i, qo_j)$ is the distance between the robot $i$ and the obstacle $j$, $p(q_i, q_k)$ is the distance between the robot $i$ and the robot $k$.

Below, we present the results of the implementation of three robots working in the same workspace. We treat cases where each robot has its own destination, compared to other robots, and the workspace contains static obstacles. Figure 5a shows the attractive forces applied to the three robots. Similarly, the repulsive forces applied to the three robots are shown in Fig. 5b.

The angular velocities applied to the wheels are different according to the situation of movement: forward movement, turn right and turn left. The profile of these control speeds applied to the robot 1 is presented in Fig. 6.

The robots are controlled by the angular velocities associated with the wheels. These speeds directly describe the rotational speeds applied to the right and left wheels of the robot. Similarly, the angular velocities applied to robots 2 and 3 are determined.

## 6.1 Coordination Between Robots

The problem of coordination between the different robots working in the same workspace is a very interesting scientific problem. Whenever, the mobile robots operate in the same environment, their movements must be coordinated to avoid blockages and congestion (Fig. 7).

We start by implementing the potential field technique to plan the trajectory of three mobile robots working in the same workspace [20]. The simulation results are shown in Fig. 8.

At the intersection point shown in Fig. 8, the problem of collision is not solved. The use of the potential field method does not solve the coordination problem between the robots. Thus, the risk of collision of robots at intersections is very likely.

In order to make the trajectory planning problem safer and more reliable, we combine the notion of priority between robots and the potential field technique. The idea is to use a mixed approach to solve the problem of cooperation for a multi-robot system. In this work, we use the notion of priority between robots to solve the problem of the passage in the congestion zone [20].

We consider a system of $N$ robots where $i$ is a counter on the number of robots, $i \in [1 \dots N]$, $k$ is the neighborhood of the robot $i$, $k \in [1 \dots T]$, $k \neq i$ and $T$ is the number of neighbors, $d_{ik}$ is the distance between the robot $i$ and $k$, $p_{ik}$ is the neighborhood area, $d_{ig}$ is the distance between the robot $i$ and his goal, $d_{kg}$ is the distance between the robot $k$ and his goal.

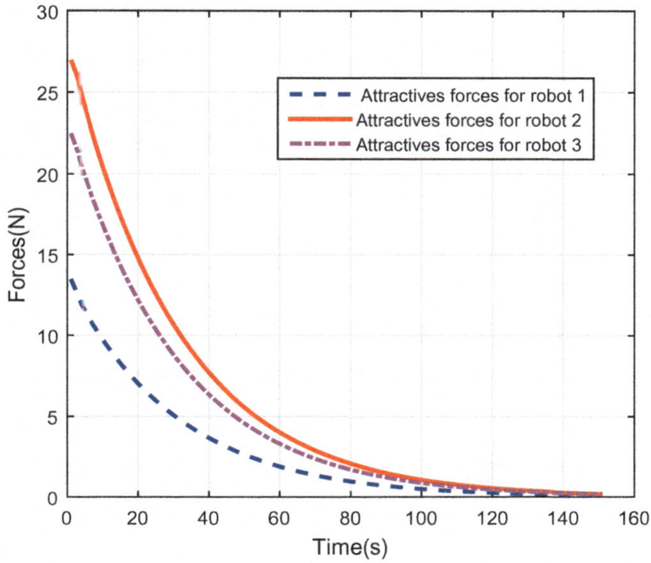


(a) Attractive forces applied to robots

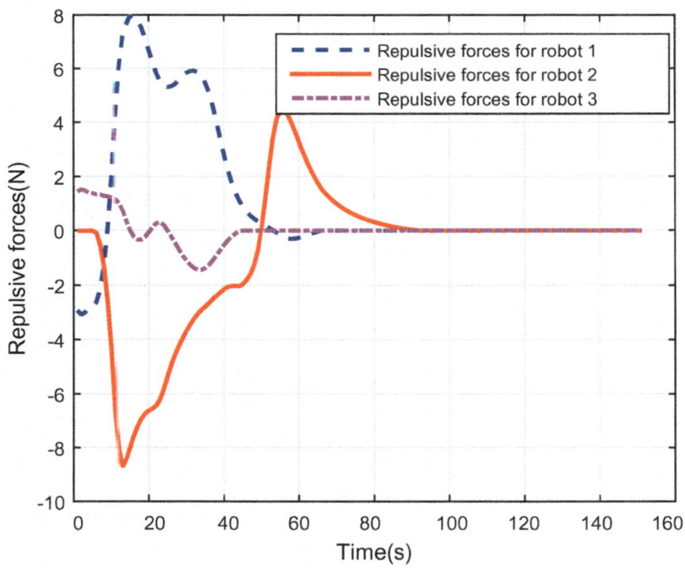


(b) Repulsive forces applied to robots

**Fig. 5** Attractive and Repulsive forces applied to robots

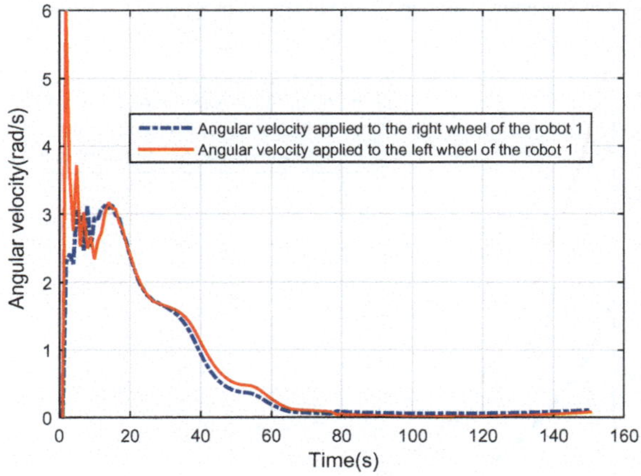


**Fig. 6** Angular velocity of robot 1

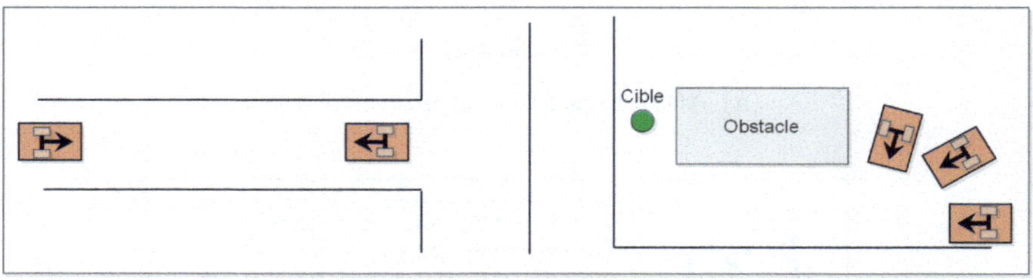


**Fig. 7** Problem of coordination between the robots

**Fig. 8** Collision between the robots

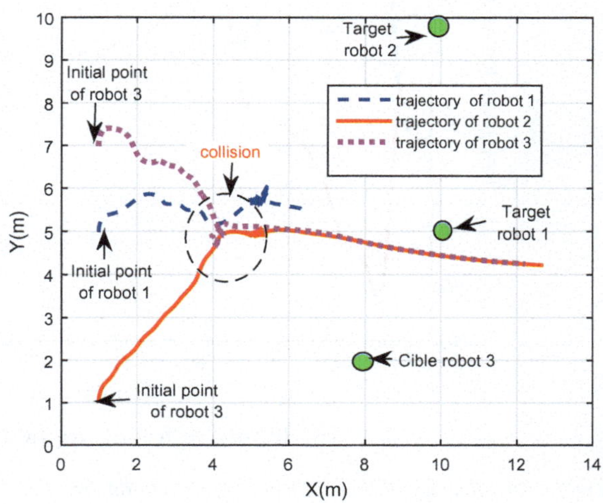

**Algorithm 2:** Algorithm of priority

- **Step 1:** Initialization of Parameters:
  **Input:** The robot $i \in [1 \ldots N]$ and the robot $k \in [1 \ldots T]$,
  **Output:** Assign the priority
- **Step 2:** Detect the neighborhoods of robot:
  $d_{ik} = p(q_k, q_i) = \|q_k - q_i\|$,
  $d_{ig} = p(g_i, q_i) = \|g_i - q_i\|$,
  $d_{kg} = p(g_k, q_k) = \|g_k - q_k\|$ where $i \in [1 \ldots N]$, $k \in [1 \ldots N]$ and $i \neq k$.
  **if** $d_{ik} <= 1$ **then**
  the robot $i$ and $k$ are neighbors
  Assign the priority between the robot $i$ and the robot $k$:
  **else**
  the robot $i$ and $k$ are not neighbors
  **End**
- **Step 3:** assign the priority
  **if** $d_{ig} \leq d_{kg}$ **then**
  the robot $i$ has the priority over the robot $k$,
  **else**
  the robot $k$ has the priority over the robot $i$,
  **End**

### *6.2 Neighborhood Detection Method*

In order to simplify the task of obstacle avoidance, we propose the use of the concept of robot neighborhood detection. Thus, the robot detects all objects coming into its neighborhood. The neighborhood principle is illustrated in Fig. 9. Thus, each robot has a neighborhood circle. We consider a system of $N$ robots where: $i$ and $j$ two values $\in [1 \ldots N]$, $dis$ is the distance threshold, $d_{ij}$ is the distance between the robot $i$ and $j$ [20]. After the use of algorithm 3, the neighborhood vector takes the value 1 if the robot $i$ and $j$ are neighbors but he takes the value 0 if the robot $i$ and $j$ are not neighbors. In order to clarify the proposed approach, we present the flow diagram of the global system (Fig. 10):

## 7 Simulation and Presentation of the Results

In this part, we use the results presented previously to plan the trajectory of three mobile robots working in the same workspace. The environment is composed of static obstacles and three mobile robots. Each robot performs its own task independently of

**Algorithm 3:** Algorithm of Neighborhood

- **Step 1:** Initialization of Parameters
  **Input:** $N$: number of robots used, $i$ and $j \in [1 \ldots N]$, $d_0$: the threshold of distance
  **Output:** neighborhood vector
- **Step 2:**
  $x = [x_{r1}, x_{r2}, \ldots, x_{ri}, \ldots, x_{rN}]$
  $y = [y_{r1}, y_{r2}, \ldots, y_{ri}, \ldots, y_{rN}]$
  N=length(x);
  Vec=zeros(N,N);
  **For i=1 to N**
      **For j=1 to N**
          $dis(i, j) = sqrt((x(i) - x(j))^2 + (y(i) - y(j))^2)$
          **if** $(dis(i, j) > d_0)||(i == j)$ **then**
              $Vec(i, j) = 0;$
          **else**
              $Vec(i, j) = 1;$
          **end**
      **end**
  **end**

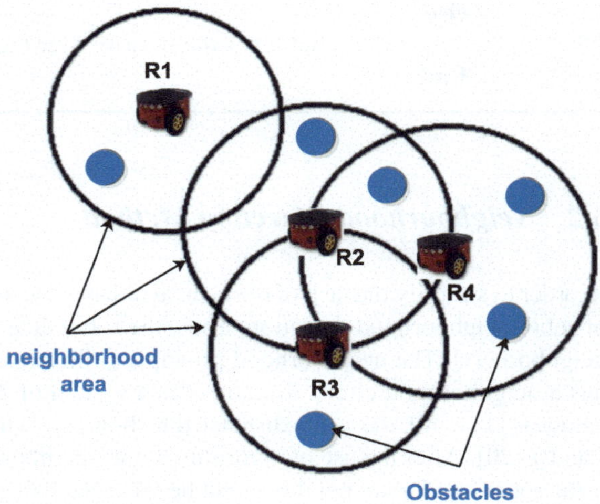


**Fig. 9** The principle of neighborhood

the others robots. The environment is considered very dynamic and the robots move at the same time. We present some situations of path planning to see the reaction of the implemented approach (Table 2).

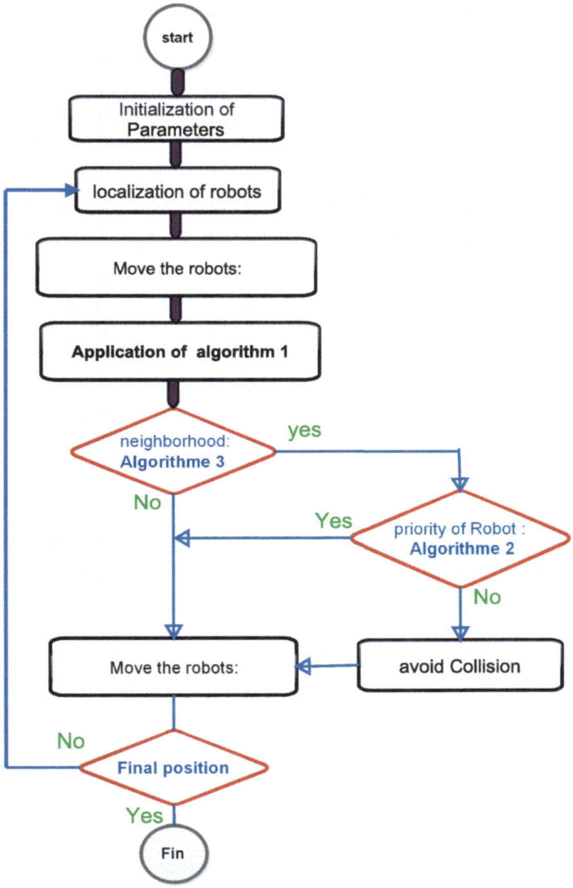


**Fig. 10** Organizational chart of decentralized approach proposed

**Table 2** The initial positions of the robots and the positions of their targets

| | Initial point | Final point |
|---|---|---|
| Robot 1 | (1, 5) | (10, 5) |
| Robot 2 | (1, 1) | (10, 10) |
| Robot 3 | (1, 7) | (8, 2) |

## *7.1 Scenario 1: Case of Cooperation Between Robots Without Presence of Obstacles*

We start the simulation by initializing the location of the robots as follows:

**Case 1:** In the first case the robots are attracted only by the attractive forces. The simulation results are shown in Fig. 11.

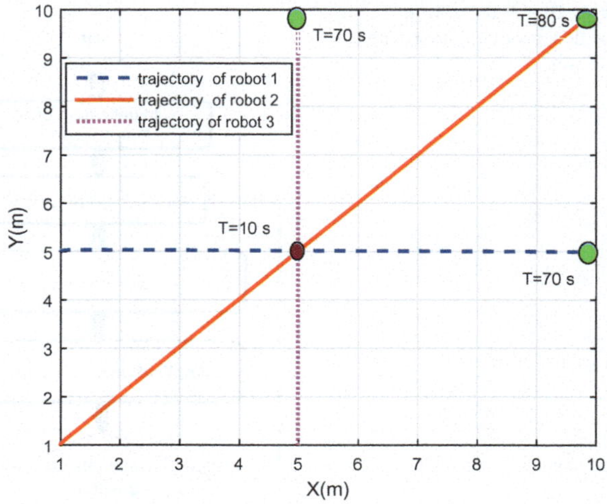


**Fig. 11** Trajectories of three mobile robots without collision

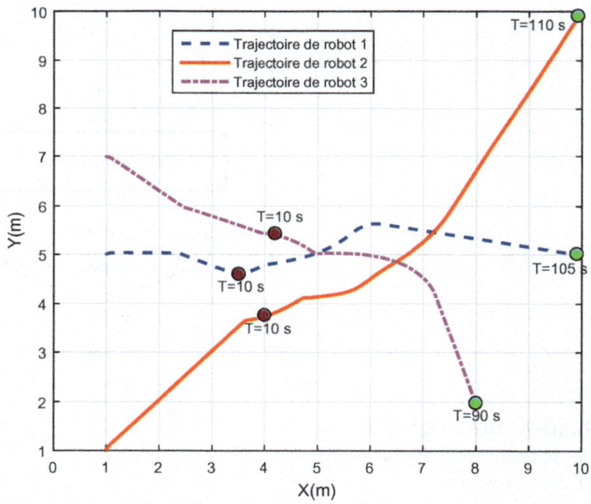


**Fig. 12** Trajectories of three mobile robots with a safety distance of 1 m

The trajectories of the robots in this case are linear trajectories. They present the paths followed by robots towards their targets. At $t = 15$ s, the trajectories of robots intersect at the coordinate point [5 5].

**Case 2:** Now the robots are moving in a field of attractive and repulsive forces. The attractive forces are generated by the target to guide the robots to their destinations and the repulsive forces are generated by robots to avoid collisions between them. The Simulation results are presented in Fig. 12.

The trajectories of the three robots intersect at the crossing points, hence the need to use security concepts to coordinate their movement at these points. The concepts used combine the previously presented algorithms according to the flowchart shown in Fig. 10.

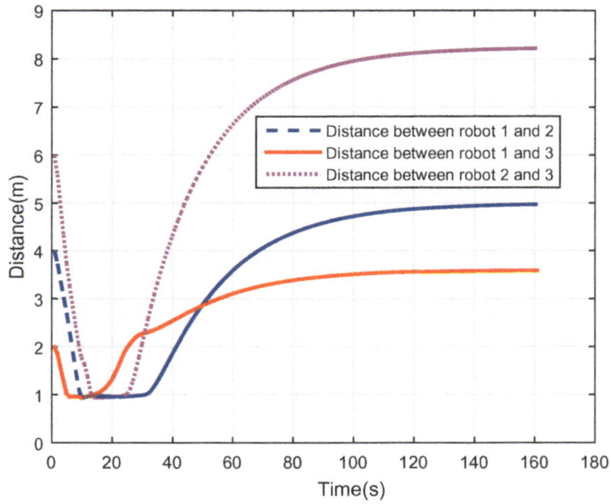


**Fig. 13** Distances between the three robots

The simulation results show that each robot reaches its desired goal within a reasonable time. Also, the problem of cooperation is solved and the security of each robot is assured (Fig. 12). The neighborhood area programmed between the robots in this case is 1 m. This distance is the area of influence of each robot. The distances between the robots are shown in Fig. 13.

The potential field (Algorithm 1), the priority between the robots (Algorithm 2) and the notion of neighborhood (Algorithm 3) are combined to provide the path planning of robots.

## *7.2 Scenario 2: Case of Cooperation Between Robots with the Presence of Obstacles*

**Case 1:** In order to produce a more complicated case compared to scenario 1, we include some static obstacles in the workspace of robots. The configuration settings are not changed from the 1 scenario. But, we add three static obstacles to complicate the case of path planning. The positions of obstacles are: (3, 2), (4, 7) and (7, 8). The results are shown in Fig. 14.

**Case 2:** In this case, the initial position of robot 3 is changed to the position (5, 1) and its target is changed to the position (5, 10). The objective is to evaluate the reaction of the change of position of a robot on the trajectory of the other robots. The simulation results are shown in Fig. 15.

In this case, we saw through the simulation result in Fig. 15 that the trajectories of the three robots are admissible trajectories and that the safety of each robot is ensured. Thus, the system reacts with the new situations by updating the trajectories assigned to the robots.

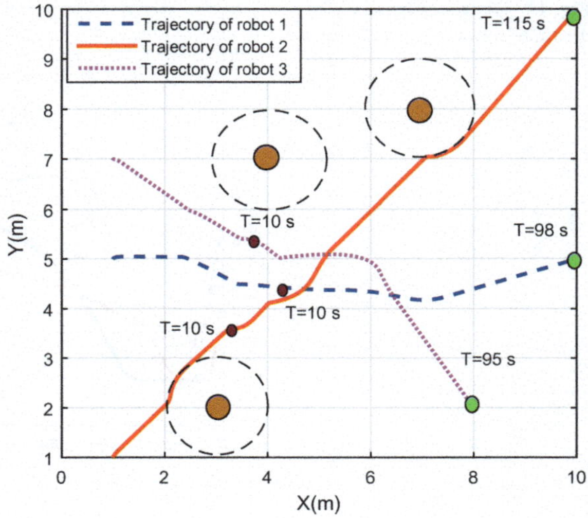


**Fig. 14** Trajectories of three mobile robots in the scenario 2 (case 1)

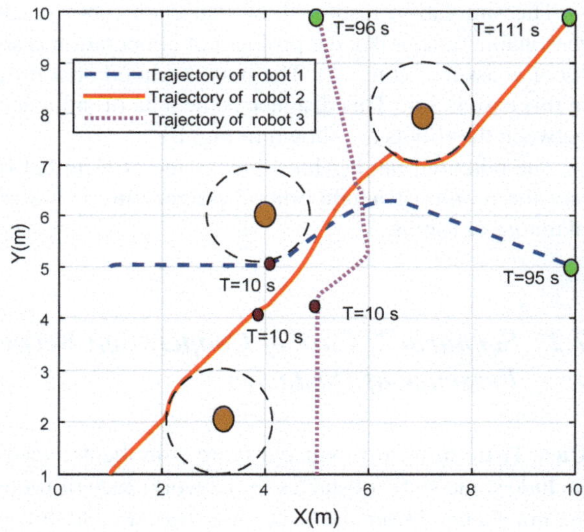


**Fig. 15** Trajectories of three mobile robots in the scenario 2 (case 2)

**Case 3:** In this case, we change the target of the robot 3 at the position (10, 2) to see the reaction of this action on the trajectory of other robots. The results of the simulation are shown in Fig. 16.

Now we change the destination of robot 3 at position (7, 7). The results of the simulation are shown in Fig. 17. For this case, the presence of the robot 3 at its destination causes a flow barrier for the robot 2. So, the robot 2 is obliged to change his trajectory.

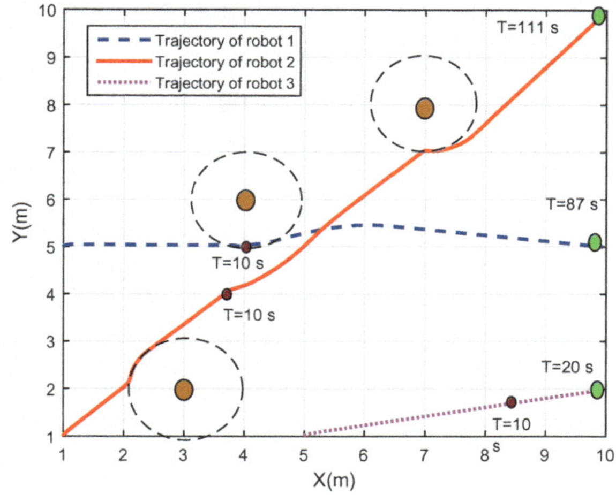


**Fig. 16** Trajectories of three mobile robots in the scenario 2 (case 3)

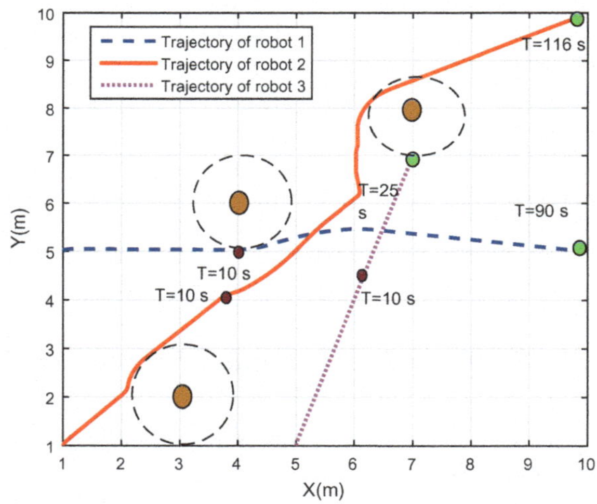


**Fig. 17** Trajectories of three mobile robots in the scenario 2 (case 3)

## 7.3 *Scenario 3: Simulation with Multiple Obstacles*

The goal of this case is to evaluate the proposed approach in more complex cases. For this, we added several obstacles in the simulation environment: a highly congested environment. The results of the simulation are illustrated in Fig. 18.

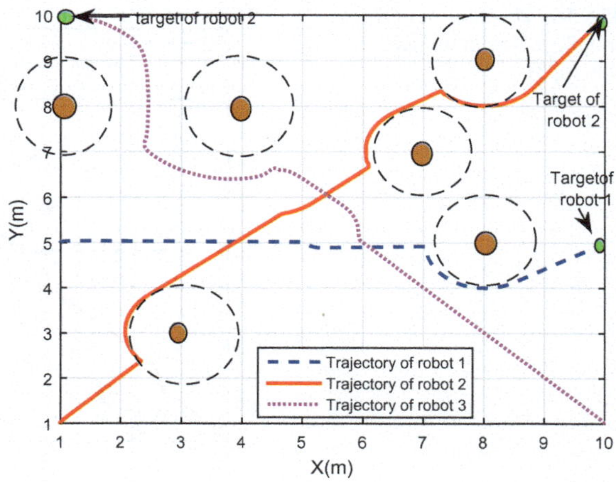


**Fig. 18** Trajectories of three mobile robots in the scenario 3

**Table 3** Simulation parameters in the case of six robots

| | Initial point | Final point |
|---|---|---|
| Robot 1 | (1, 5) | (10, 5) |
| Robot 2 | (1, 1) | (10, 10) |
| Robot 3 | (5, 1) | (5, 10) |
| Robot 4 | (1, 10) | (10, 1) |
| Robot 5 | (10, 8) | (1, 3) |
| Robot 6 | (10, 1) | (2, 7) |

## *7.4 Scenario 4: Cooperation Between Six Mobile Robots*

We increase the number of robots to six mobile robots. The parameters of the robots are grouped in the Table 3.

The simulation results are shown in Fig. 19. Also the distance between the robots during the path planning are shown in Fig. 20.

The application of the decentralized architecture does not require the use of a supervisor to synthesize the trajectories of the robots. But, each robot plans individually its trajectory taking into account the activities of other robots. The use of this architecture for path planning task has yielded impressive results. So, each robot has followed a safe and acceptable trajectory towards its target.

**Fig. 19** Trajectories of six mobiles robots

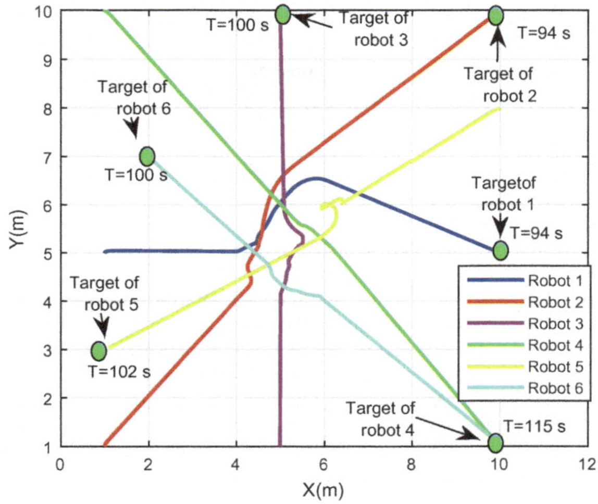


**Fig. 20** Distances between the six mobiles robots during the navigation

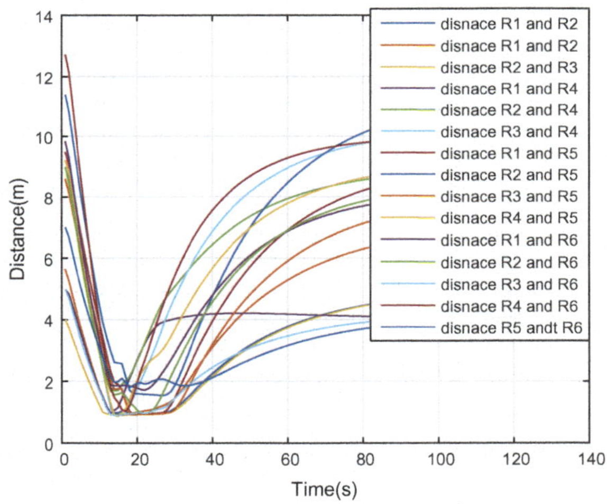


# 8 Conclusion

In this chapter, we studied the control of a wheeled mobile robot group. The question of how to plan robot paths satisfactorily led us to look for a trajectory planning technique that can adapt and manage this problem. This problem is addressed in the case of using multiple robots in a single dynamic workspace. To solve this problem, we used an approach combining the three techniques: the potential field, the neighborhood system and the notion of priority between the robots.

We evaluated the new approach proposed in several possible scenarios. The results of the simulations presented previously are impressive. Thus, the trajectories planned by the robots are admissible and smooth trajectories. In addition, the safety of robots at the time of path planning is improved.

## References


1. Arrichiello, F.: Coordination control of multiple mobile robots. Dipartimento di Automazione, Elettromagnetismo, Ingegneria Dell'informazione e Matematica Industriale (2006)
2. Baxter, J.L., Burke, E., Garibaldi, J.M., Norman, M.: Multi-robot search and rescue: a potential field based approach. Autonomous Robots and Agents, pp. 9–16. Springer, Berlin (2007)
3. Bennewitz, M., Burgard, W., Thrun, S.: Optimizing schedules for prioritized path planning of multi-robot systems. 271–276 (2001)
4. Benzerrouk, A., Adouane, L., Lequievre, L., Martinet, P.: Navigation of multi-robot formation in unstructured environment using dynamical virtual structures. In: 2010 IEEE/RSJ International Conference on Intelligent Robots and Systems (IROS), pp. 5589–5594. IEEE (2010)
5. Dahl, T.S., Matarić, M., Sukhatme, G.S.: Multi-robot task allocation through vacancy chain scheduling. Robot. Auton. Syst. **57**(6–7), 674–687 (2009)
6. Defoort, M., Kokosy, A., Floquet, T., Perruquetti, W., Palos, J.: Motion planning for cooperative unicycle-type mobile robots with limited sensing ranges: a distributed receding horizon approach. Robot. Auton. Syst. **57**(11), 1094–1106 (2009)
7. Dhaouadi, R., Hatab, A.: Dynamic modelling of differential-drive mobile robots using Lagrange and Newton-Euler methodologies: a unified framework. Adv. Robot. Autom. **2**(2) (2013)
8. Fox, D., Burgard, W., Thrun, S.: The dynamic window approach to collision avoidance. IEEE Robot. Autom. Mag. 23–33 (1997)
9. Gaillard, F.: Approche cognitive pour la planification de trajectoire sous contraintes. PhD thesis, Universite des Sciences et Technologie de Lille-Lille I (2012)
10. Glavaski, D., Volf, M., Bonkovic, M.: Robot motion planning using exact cell decomposition and potential field methods. In: Proceedings of the 9th WSEAS International Conference on Simulation, vol. 8, pp. 126–131 (2009)
11. Guys, L.: Aircraft trajectory planning without conflict: biharmonic functions and harmonic navigation function. PhD thesis, Université Toulouse 3 Paul Sabatier (2014)
12. Haj Darwish, A., Joukhadar, A., Kashkash, M.: Using the bees algorithm for wheeled mobile robot path planning in an indoor dynamic environment. Cogent Eng. **5**(1), 1426539 (2018)
13. Hassan, A.M., Elias, C.M., Shehata, O.M., Morgan, E.I.: A global integrated artificial potential field/virtual obstacles path planning algorithm for multi-robot system applications (2017)
14. Kuo, P.-H., Li, T.-H.S., Chen, G.-Y., Ho, Y.-F., Lin, C.-J.: A migrant-inspired path planning algorithm for obstacle run using particle swarm optimization, potential field navigation, and fuzzy logic controller. Knowl. Eng. Rev. **32** (2017)
15. Li, G., Chou, W.: Path planning for mobile robot using self-adaptive learning particle swarm optimization. Sci. China Inf. Sci. **61**(5), 052204 (2018)
16. Ma, Y., Zheng, G., Perruquetti, W., Qiu, Z.: Motion planning for non-holonomic mobile robots using the i-PID controller and potential field. In: IEEE/RSJ International Conference on Intelligent Robots and Systems (IROS 2014), pp. 3618–3623 (2014)
17. Matoui, F., Boussaid, B., Abdelkrim, M.N.: Local minimum solution for the potential field method in multiple robot motion planning task. In: 2015 16th International Conference on Sciences and Techniques of Automatic Control and Computer Engineering (STA), pp. 452–457. IEEE (2015)
18. Matoui, F., Boussaid, B., Abdelkrim, M.N.: Path planning of two robots in dynamic workspace based on potential field. In: 2017 International Conference on Advanced Systems and Electric Technologies (IC_ASET), pp. 267–272. IEEE (2017)

19. Matoui, F., Boussaid, B., Metoui, B., Frej, G., Abdelkrim, M.N.: Path planning of a group of robots with potential field approach: decentralized architecture. IFAC-PapersOnLine **50**(1), 11473–11478 (2017)
20. Matoui, F., Boussaid, B., Abdelkrim, M.N.: Distributed path planning of a multi-robot system based on the neighborhood artificial potential field approach. SIMULATION: Trans. Soc. Model. Simul. Int. SAGE Publications, Sage UK, London, England (2018)
21. Mohanty, P.K., Parhi, D.R.: Controlling the motion of an autonomous mobile robot using various techniques: a review. J. Adv. Mech. Eng. **1**(1), 24–39 (2013)
22. Montiel, O., Orozco-Rosas, U., Sepúlveda, R.: Path planning for mobile robots using bacterial potential field for avoiding static and dynamic obstacles. Expert Syst. Appl. **42**(12), 5177–5191 (2015)
23. Moon, H., Luntz, J.: Prediction of equilibria of lifted logarithmic radial potential fields. Int. J. Robot. Res. **23**(7–8), 747–762 (2004)
24. Neto, A.A., Macharet, D.G., Campos, M.F.: Multi-agent rapidly-exploring pseudo-random tree. J. Intell. Robot. Syst. **89**(1–2), 69–85 (2018)
25. Parasuraman, S., Ganapathy, V., Shirinzadeh, B.: Behaviour based mobile robot navigation technique using AI system: experimental investigation on active media pioneer robot. IIUM Eng. J. **6**(2) (2005)
26. Parker, L.E.: Multiple mobile robot systems. Springer Handbook of Robotics, pp. 921–941. Springer, Berlin (2008)
27. Siegwart, R., Nourbakhsh, I.R., Scaramuzza, D.: Introduction to Autonomous Mobile Robots. MIT Press, Cambridge (2011)
28. Solana, Y., Furci, M., Cortés, J., Franchi, A.: Multi-robot path planning with maintenance of generalized connectivity. In: 2017 IEEE 1st International Symposium on Multi-Robot and Multi-Agent Systems (2017)
29. Song, P., Kumar, V.: A potential field based approach to multi-robot manipulation. In: 2002 Proceedings IEEE International Conference on Robotics and Automation (ICRA'02), vol. 2, pp. 1217–1222. IEEE (2002)
30. Tan, J., Zhao, L., Wang, Y., Zhang, Y., Li, L.: The 3D path planning based on a* algorithm and artificial potential field for the rotary-wing flying robot. In: 2016 8th International Conference on Intelligent Human-Machine Systems and Cybernetics (IHMSC), vol. 2, pp. 551–556. IEEE (2016)
31. Tzafestas, S.G.: Introduction to Mobile Robot Control (2013)
32. Varsos, K., Moon, H., Luntz, J.: Generation of quadratic potential force fields from flow fields for distributed manipulation. IEEE Trans. Robot. **22**(1), 108–118 (2006)
33. Yingchong, M.: Path planning and control of non-holonomic mobile robots. PhD thesis, Ecole Centrale de Lille (2014)

# Which is Better for Mobile Robot Trajectory Optimization: PSO or GA?

**Safa Ziadi, Mohamed Njah and Sana Charfi**

**Abstract** In mobile robot research domain and especially in its path planning sub-domain, the optimization tools the most used are PSO (Particle Swarm Optimization) and GA (Genetic Algorithms). In fact, mobile robot path planning is a multi-objective optimization problem where at least these objectives are considered: the shortest and the safest path to the goal. Sometimes other objectives are added like travel duration. The subject of this paper is a comparison investigation concerning the efficiency of PSO and GA in mobile robot path planning optimization. We start with a comparison of evolutionary performances of PSO and GA. Then, we apply PSO and GA in the optimization of parameters of $DVSF^2$ (Dynamic Variable Speed Force Field) path planning approach. Different environments of different types, statics and dynamics are considered in these simulations to decide which is better for mobile robot path planning optimization: PSO or GA.

**Keywords** Genetic Algorithms · Particle Swarm Optimization · Mobile robot · Path planning

## 1 Introduction

Robots are more and more used in different domains like industry, medicine, spaces applications and also in the agricultural field. In mobile robot research domain, path planning is a problem that pulled much attention in recent years. It can be described as the construction of collision-free paths which manage robots towards

S. Ziadi (✉) · M. Njah
Control and Energy Management Laboratory (CEM-Lab), National Engineering School of Sfax, University of Sfax, Sfax, Tunisia
e-mail: ziadisafa@yahoo.fr

M. Njah
e-mail: mohamed.njah@enis.tn

S. Ziadi · M. Njah · S. Charfi
Digital Research Center of Sfax, Technopole of Sfax, Sfax, Tunisia
e-mail: charfisana@yahoo.fr

 
J. Ghommam et al. (eds.), *New Trends in Robot Control*, Studies in Systems, Decision and Control 270, https://doi.org/10.1007/978-981-15-1819-5_13

their destination. Many objectives are considered at the same time like the path length, distance to obstacles and time to goal.

This research domain is interested also to control the robot during its navigation in different conditions. Many solutions have been proposed in the literature based on many optimization tools mainly Ant Colony Optimization(ACO), PSO and GA. They share a common characteristic that they are evolutionary algorithms, based on natural evolution. The philosophy of PSO is based on the movement of birds and fishes when they are looking for food. And, the philosophy of GA is based on the survival of the genetically fittest individuals. Also, the philosophy of ACO is based on the method of ants when they define a short path to the food.

In [1], the ACO optimization approach was used to search a near optimal path for a mobile robot and avoid obstacles timely both in simple and complex environments. The drawbacks of ACO were principally low convergence rates and the frequent falls into local optima.

Both optimization methods have been also used in the computation of mobile robot path motion [2, 3]. Many research results have been presented in the literature. For example, in [4, 5], GA is used to find optimal mobile robot path. In [5], enhanced Genetic Algorithm (EGA) and Artificial Potential Field (APF) are combined to give a new mobile robot path planning approach [5]. The results of this combination are compared with many other famous approaches to prove the efficiency of this new approach. Also, in [6, 7], PSO is used to find the optimal mobile robot trajectory. These researches proved the capacity of PSO and GA to search the optimal mobile robot trajectory in different types of environments.

In [8], a mobile robot path planning method by utilizing genetic algorithms (GA) was proposed. The main goal considered in this optimization is the shortest secure path between the initial point and the defined goal position in the minimal possible travel time. The result of this paper proved that GA has a good ability to improve the path planning algorithm and generate an optimized path length. Also, in [9], Particle Swarm Optimization (PSO) is used to optimize the path length. The environment in this paper was an unknown dynamic one. The main objective of the problem is to make the robot move from the starting position to the goal position while avoiding all obstacles and follow the shortest possible path. The robot in this paper moved further towards the global best position. These two papers [8, 9] proved the efficiency of these two algorithms (PSO and GA) to search the optimal path length.

The literature contains some comparison investigations between PSO and GA. For example in [10, 11], authors apply on the same so called Pressurized Water Reactor (PWR) core at full power in single-phase flow with forced circulation. In [10], PSO performed better than GA to solve this kind of problem and it is more robust and consistent than GA. Also in [11], PSO produced better and faster results than GA.

In [12], authors used benchmark functions to test the performance of PSO and GA. These comparison tests demonstrate that PSO is more computationally efficient. In fact, it uses a smaller number of function evaluations than the GA.

In [13], for the comparison between PSO and GA, authors use a Simulation Model of a Real Hexapod Robot. The results of this comparison show that PSO is more suitable for this complex problem than GA. In the opposite, some other researches found that GA is better than PSO to solve some problems. For example, in [14], GA is the best approach that can be used for solving a multi-depots multi-vehicles pickup and delivery problems with time windows (m-MPDPTW).

This paper builds upon the PSO-DVSF$^2$ optimized path planning approach proposed in our work [15, 16]. PSO-DVSF$^2$ is a combination between PSO and the famous DVSF$^2$ mobile robot path planning approach. In PSO-DVSF$^2$, PSO is used to define an optimized combination of DVSF$^2$ parameters that produce the shortest path to the goal and the safest trajectory which means the greatest distance to obstacles. For this comparison investigation, we create the GA-DVSF$^2$ optimized mobile robot path planning approach similar to the PSO-DVSF$^2$. Its particularity is that parameters of DVSF$^2$ are optimized with GA and not with PSO as it is the case in PSO-DVSF$^2$.

This document is organized as follows. In Sect. 2, we introduce the characteristics of the PSO and GA optimization methods. The results of a comparison investigation of intrinsic evolutionary performances of PSO and GA are given in Sect. 3. In Sect. 4, we present the comparison simulation results of the DVSF$^2$ mobile robot path planning method combined with PSO and GA and applied in environments of different types. The conclusion and future works are given in Sect. 5.

## 2 PSO and GA Characterization:

### *2.1 PSO*

PSO is a heuristic method of optimization which is invented by Eberhart and Kennedy in 1995 [17]. The algorithm consists of a swarm of particles flying through the search space. Each individual i is characterized by a position $X_i$ (2) and a velocity $V_i$ (1), where $x_i \in R_n$, $v_i \in R_n$ and n is the dimension of the search space. Each position of the particle represents a potential solution of the problem of optimization [17].

$$V_i(t+1) = \vartheta.V_i(t) + c_1.r_1.(X_{pbest;i} - X_i) + c_2.r_2.(X_{gbest} - X_i) \tag{1}$$

$$X_i(t+1) = X_i(t) + V_i(t+1) \tag{2}$$

where $c_1$, $c_2$ are acceleration constants, $r_1$, $r_2$ are random factors in the range [0, 1], $X_{pbest}$ is the personal best position, $X_{gbest}$ is the global best position and $\vartheta$ is called the dynamic inertia (3).

$$\vartheta = \vartheta_{max} - \frac{\vartheta_{max} - \vartheta_{min}}{iter_{max}}.iter \tag{3}$$

where $\vartheta_{max}$ is the initial weight, $\vartheta_{min}$ is the final weight, $iter_{max}$ is the maximum number of iterations and iter is the number of current iterations. The chosen parameters for our simulations are given in Table 1.

**Table 1** Parameters of PSO [18]

| Parameter | Value |
|---|---|
| $\vartheta_{max}$ | 0.9 |
| $\vartheta min$ | 0.4 |
| $c_1$ | 2 |
| $c_2$ | 2 |

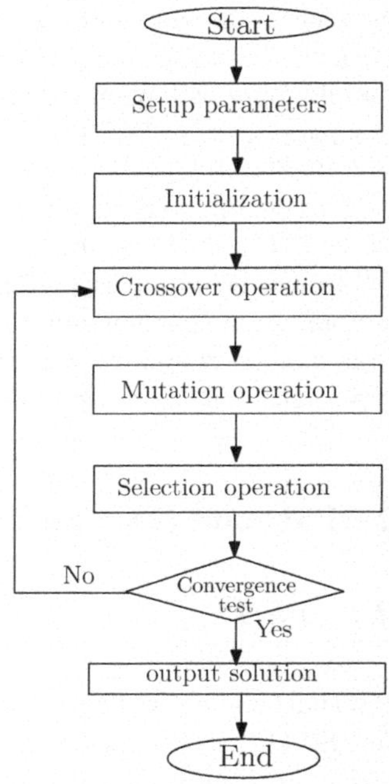


**Fig. 1** The flowchart of GA [2]

## 2.2 *Genetic Algorithms*

Genetic Algorithms is a heuristic approach applied in the resolution of different multi-objective optimisation problems [19]. It is invented by John Holland in the early 1970s and it is based on a population of solutions optimised in parallel. A fitness value is obtained from the objective function for each member of a population. The best solution is that having the best fitness value. During the procedure of GA, three operators are applied, namely selection, crossover and mutation. This procedure is repeated until finding the best solution. The framework of GA is summarized in Fig. 1.

**a-Selection**

The primary objective of the selection operator is to emphasize the good solutions and eliminate the bad solutions in a population while keeping the population size constant. Tournament method is applied in this work as it is considered an efficient method in the literature [20].

**b-Crossover**

With probability pc (the "crossover probability" or "crossover rate"), there is an exchange of chromosome segments between two individuals to form two new offspring. The two chromosomes are broken in a point chosen with a uniform probability. There are also "multi-point crossover" versions of the GA in which the crossover rate for a pair of parents depends on the number of points at which a crossover takes place.

**c-Mutation**

Uniform mutation is applied in this work. The mutation consists on generating a new random value for the selected gene in its domain of variation and mutate the gene with probability pm (the mutation probability or mutation rate), and place the resulting chromosomes in the new population.

## 3 Comparison of Evolutionary Performances of PSO and GA

The objective of this experiment is to compare the intrinsic performances of PSO and GA. In our simulations, performances of PSO and GA are measured with the absolute error and the convergence time after a predefined number of iterations using populations of predefined number of particles. The absolute error is the difference between the real value of the global minimum and the value calculated by the algorithm. It is calculated as follows:

$$Absolute_{error} = |V_{true} - V_{observed}| \tag{4}$$

where $V_{true}$ is the true function value and $V_{observed}$ is the observed value found by PSO or GA. The convergence time is the time consumed by the predefined number of iterations. Some benchmark test functions are used in the literature for such a comparison study.

In this simulation, Benchmark Test Functions are used to compare the efficiency of PSO and GA to find the optimal global minimum. Populations are of 100 individuals for the three benchmark functions. The number of iterations is fixed to 6000 for the two first benchmark functions: the sphere function and the sum squares one. For the last benchmark function, the number of iterations is fixed to 200.

The chosen parameters for the GA in our simulation are given in Table 2.

**Table 2** Parameters of GA

| Parameter | Value |
|---|---|
| Crossover rate | $P_c = 0.9$ |
| Mutation rate | $P_m = 0.01$ |

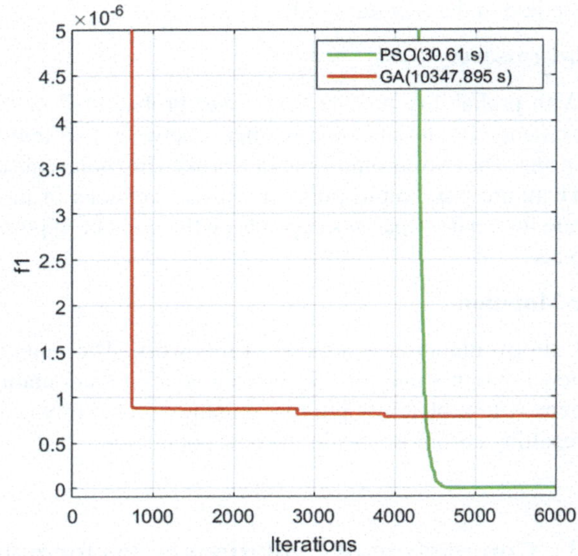


**Fig. 2** The search of the minimum of $f_1$ by GA and PSO

## *3.1 Sphere Function*

The first benchmark test function is the Sphere function formulated as follows [21]:

$$Sphere : f_1(x) = \sum_{i=1}^{n} x_i^2 \tag{5}$$

with n = 30, $x_{max} = 100$, $x_{min} = -50$ and $f_{1,global-min} = 0$.

In this comparison 6000 iterations are allowed with populations of 100 individuals. Figure 2 presents the evolution of the search operation of the global minimum of $f_1(x)$ by PSO and GA with iterations. It is right that GA attains a good result ($\simeq 10^{-6}$) after only 1000 iterations but these 1000 iterations take 18 min. In the other hand, PSO needs about 4500 iterations to give a better result ($\simeq$0). These 4500 iterations take only 19 s. As a conclusion, PSO achieves a better result in a fewer time.

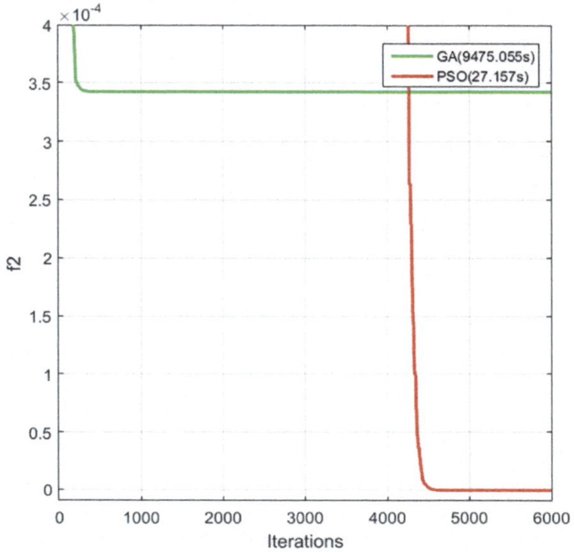


**Fig. 3** The search of the minimum of $f_2$ by GA and PSO

## 3.2 *Sum Squares Function*

The second benchmark test function is the Sum Squares function formulated as follows [22]:

$$f_2(x) = \sum_{i=1}^{n} i x_i^2 \tag{6}$$

with n = 30, $x_{max} = 100$, $x_{min} = -50$ and $f_{2,global-min} = 0$.

Figure 3 presents the evolution of the global minimum of $f_2(x)$ found by PSO and GA with iterations. In this simulation, GA converges to a good result ($\simeq 3.5 \times 10^{-4}$) after only 200 iterations. These 200 iterations take 2 min. In the other hand, PSO converges to a better result ($\simeq 0$) after 4500 iterations. But, these 4500 iterations take only 21s. For this function also, the result achieved by PSO is better than that achieved by GA and in a shorter time.

## 3.3 *Booth Function*

The third benchmark test function is the Booth function formulated as follows [22]:

$$f_3(x) = (x_1 + 2x_2 - 7)^2 + (2x_1 + x_2 - 5)^2 \tag{7}$$

with $n = 2$, $x_{max} = 4$, $x_{min} = -4$ and $f_{3,global-min} = 0$.

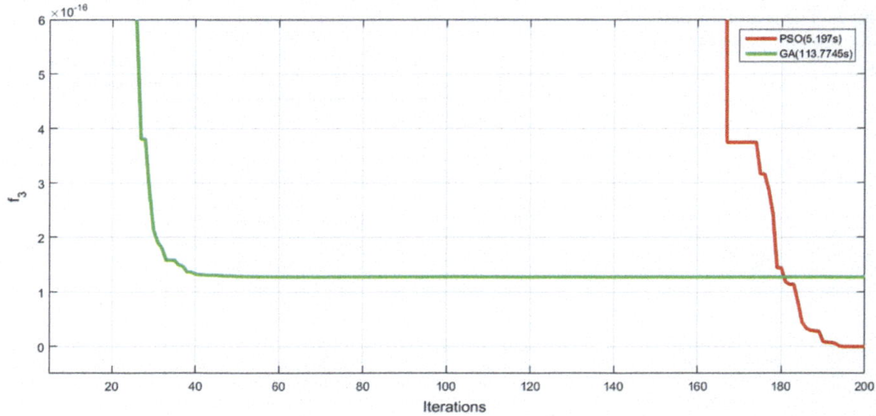


**Fig. 4** The search of the minimum of $f_3$ by GA and PSO

In this comparison 200 iterations are applied with populations of 50 individuals. Figure 4 presents the evolution of the search of the global minimum of $f_3(x)$ by PSO and GA with iterations. For this function also, GA converges to a good result ($\simeq 1.3 \times 10^{-16}$) after only 40 iterations. In the other hand, PSO converges to a better result ($\simeq 0$) but after $\sim 200$ iterations. If we convert iterations to time, we find that the 40 iterations of GA take 1 minute and the 200 iterations of PSO take only 5 s.

As a conclusion, in the 3 simulations, PSO achieves better results in fewer times than GA.

# 4 Comparison of Mobile Robot Path Planning Optimization Efficiencies of PSO and GA

In these simulations, the trajectory of a robot is designed using the famous DVSF$^2$ path planning approach. In a first time, parameters of DVSF$^2$ are optimized using PSO. In this case, we speak about PSO-DVSF$^2$ path planning approach. In a second time, parameters of DVSF$^2$ are optimized using GA. In this case, we speak about GA-DVSF$^2$ path planning approach.

## *4.1 Dynamic Variable Speed Force Field: DVSF$^2$*

DVSF$^2$ [15, 23] is an improved version of the basic Force Field ($F^2$) [24] mobile robot path planning approach.

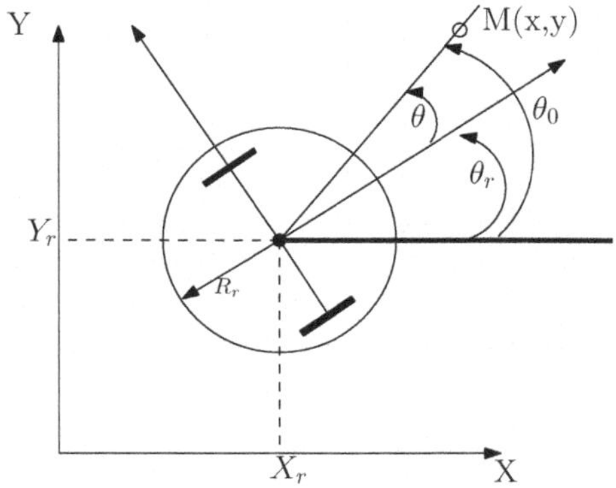


**Fig. 5** Dimension, position and orientation of the robot in 2-D space

For DVSF[2], a robot travels with a variable speed. The concept of the virtual repulsive force is based on some parameters: $R_r$ is the radius of a robot, $V_r$ is the absolute value of a robot's velocity, $V_{max}$ is the maximum absolute value of a robot's velocity and $\theta_r$ is the robot's orientation in global coordinates Fig. 5. The mathematical relationships are defined by the following equations:

$$\theta = \theta_0 - \theta_r \tag{8}$$

$$E_r = \frac{V_r}{V_{max}.C} \quad , \quad C > 1 \ and \ 0 \leq E_r < 1 \tag{9}$$

$$D_{max} = \frac{k.E_r.R_r}{1 - E_r \cos(\theta)}.T_p \quad , \quad k > 1 \tag{10}$$

$$D_{min} = \rho_0.D_{max} \quad , \quad 0 < \rho_0 < 1 \tag{11}$$

where $\theta_0$ is the orientation of the point (x, y) in global coordinates, $\theta_r$ is the orientation of the robot and $\theta$ is the relative angle of the point (x, y) to the robot's orientation. $E_r$ is a positive decimal fraction. C is a positive number which assigns the environmental influence to the force field. k is a positive multiplier which delimits the coverage area of the force field. $D_{max}$ is the maximum active distance of a robot's force field and $D_{min}$ is the distance to the robot before which the repulsive force is maximum. $\rho_0$ is a positive fractional number that heavily influences how close the robot can be separated from obstacles. $T_p$ represents the priority of a task undertaken by the robot. For the case of a single robot, $T_p$ is 1. So, $D_{max}$ and $D_{min}$ depend on the parameters k, C, P, Q and $\rho_0$. Then, in our work, we use PSO or GA in the optimization of these five parameters.

The repulsive force generated by a robot is defined by:

$$|F_{rep-rob}| = \begin{cases} 0 & if \quad D > D_{max} \\ P \cdot \frac{D_{max}-D}{D_{max}-D_{min}} & if \; D_{min} \leq D < D_{max} \\ F_{max} & if \quad D < D_{min} \end{cases} \tag{12}$$

where D is the shortest distance between point (x, y) and the perimeter of the robot. P limits the magnitude of the repulsive force and it's a positive constant scalar. The magnitude of the repulsive force changes from P to 0 when D changes from $D_{min}$ to $D_{max}$. While D is less than $D_{min}$, the magnitude of the repulsive force equals $F_{max}$, the maximum repulsive force. P and $F_{max}$ should be selected based on the robot's characteristics, with $F_{max} \gg P$.

### 4.1.1 The Attractive Force ($F_{att}$)

It is a virtual force applied by the goal on the center of the robot. It attracts the robot to the goal. It's module is constant in the $F^2$ approach.

$$|F_{att}| = Q \tag{13}$$

Furthermore, the attractive force orientation is given by (14)

$$\theta_{F_{att}} = atan\left(\frac{Y_t - Y}{X_t - X}\right). \tag{14}$$

### 4.1.2 The Reaction Force ($F_{rep}$)

It is a virtual force applied by the obstacles or other robots. It repulses the robot away from the obstacles or robots. Its module equals the module of the repulsive force.

$$|F_{rep}| = |F_{rep-rob}|_d \tag{15}$$

where d is the distance between the perimeter of the robot and the obstacle. Furthermore, the reaction force orientation is given by (16)

$$\theta_{F_{rep}} = atan\left(\frac{Y - Y_A}{X - X_A}\right) \tag{16}$$

where A($X_A$, $Y_A$) is a virtual obstacle point at which the largest repulsive force is attained. The position and orientation of the robot state (Cartesian parameterizations) at time s are represented by the state vector:

$$q_s = \begin{bmatrix} x_s \\ y_s \\ \theta_s \end{bmatrix} \tag{17}$$

where $x_s$ and $y_s$ are the robot position coordinates and $\theta_s$ is the orientation of the robot at time s.

The robot motion in the time interval $[s, s+1]$ can be expressed as a function of the current location, orientation and control (18):

$$\begin{bmatrix} x_{s+1} \\ y_{s+1} \\ \theta_{s+1} \end{bmatrix} = \begin{bmatrix} x_s + v_s \cos(\theta_s)\Delta t \\ y_s + v_s \sin(\theta_s)\Delta t \\ \theta_s + \omega_s \Delta t \end{bmatrix} \tag{18}$$

$$u_s = \begin{bmatrix} v_s \\ \omega_s \end{bmatrix} \tag{19}$$

$u_s$ (19) will be computed during the travel of the robot between the start and the goal points, while it searches for a suitable collision-free path.

In this method, the linear velocity and the angular velocity depend on the repulsive force and the attractive force. This variation is formulated with the following equations:

$$F_{Total} = F_{att} + \sum_{i=1}^{n} F_{i_{rep-ob}} \tag{20}$$

$$\omega_s = \frac{\angle F_{Total} - \theta_r}{\Delta t} \tag{21}$$

$$v_s = \frac{v_{max}}{1 + \frac{max(|\overrightarrow{F_{rep-ob}}|)}{P}} \tag{22}$$

The $DVSF^2$ parameters to optimize and their domain of variation are as follows: P ∈ [5, 20], c ∈ [1.5, 3], k ∈ [2, 10], Q ∈ [5, 20] and $\rho_0 \in [0.2, 1]$.

The fitness function is based on two objectives: the first is to minimize the path length to the goal (FF) and the second is to maximize the distance between the nearest obstacle and the robot ($D_{nearest-robot}$). The fitness function is a weighted sum of the first term and the inverse of the second term. It is formulated as follows:

$$fitness = w_1.f_1(x_i) + w_2.f_2(x_i) \tag{23}$$

where

- $f_1$: minimizing the travel length/$f_1(x_i) = FF(x_i)$
- $f_2$: keeping away the robot from obstacles called the danger index/$f_2(x_i) = \sum\sum \frac{1}{D_{nearest-robot}}$
- $w_1 + w_2 = 1$. In our simulations $w_1 = 0.7$ and $w_2 = 3$.

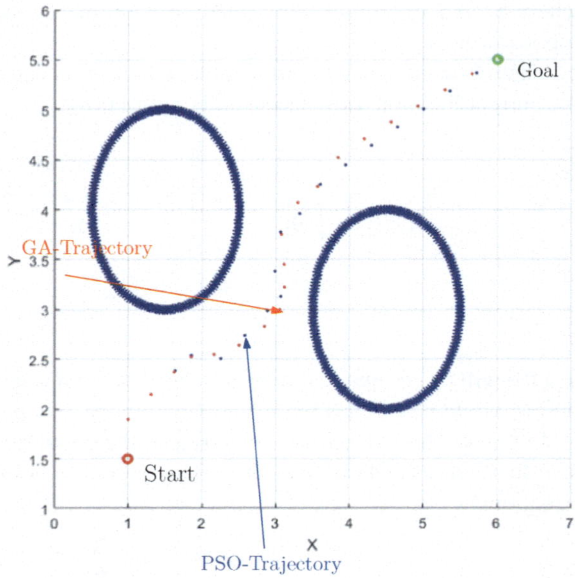


**Fig. 6** Robot trajectories in a two-static-obstacle environment

## 4.2 Simulation Results

In this second part of simulation, $DVSF^2$ mobile robot path planning approach is applied to find the optimal parameters: k, P, Q, c, $\rho_0$. We consider, firstly, a static environment with two obstacles. After that, we will consider a multi-obstacle environment. We will finish by a dynamic environment. All robots are identical and characterized by: vmax = 0.1 m/s, Tp = 1 and Rr = 0.3 m. The maximum number of iterations and the number of particles are respectively 30 and 10.

In this subsection, we are interested to prove the effectiveness of PSO and GA in solving the mobile robot path planning problem.

### 4.2.1 A Static Environment with Two Obstacles

The environment is static and known previously. Consequently, the path planning approach is demanded to propose a full path from the start point to the goal taking into account the positions of the obstacles. The chosen parameters for GA in this case are Pc = 0.5 and Pm = 0.07.

Figure 6 presents two trajectories proposed by PSO-$DVSF^2$ and GA-$DVSF^2$. Figure 7 presents the evolution of the fitness function by PSO and GA. The fitness function converge to 33.39 with GA and to 30.85 with PSO. This necessitates 29.64 s of computation for GA and only 28 s for PSO. The path length and the danger index found by GA are 7 m and 44.7 respectively. The path length and the danger

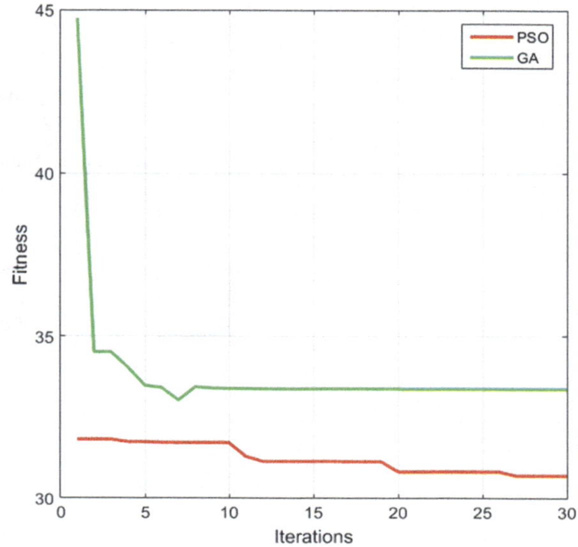


**Fig. 7** Evolution of the fitness function with iterations (GA and PSO)

index found by PSO are 6.66 m and 41.21 respectively. Consequently, PSO-DVSF$^2$ proposed a trajectory which is a little bit shorter than that proposed by GA-DVSF$^2$ with very closed safety distances.

### 4.2.2 A Static Environment with 4 Obstacles

In this simulation, the environment contains 4 static obstacles. The chosen parameters for GA in this case are Pc = 0.9 and Pm = 0.01.

Figure 8 shows the two trajectories proposed by PSO-DVSF$^2$ and GA-DVSF$^2$. The results are not very far but the favor continues on the side of PSO. Figure 9 presents the evolution of the fitness function by PSO and GA. The PSO-DVSF$^2$ fitness function converges to 84 after a computation time of 2 min and the GA-DVSF$^2$ fitness function converges to ∼93 after a computation time of 2 min 11 s. With PSO, the path length and the danger index are 18.6 m and 112.02 respectively. But, with GA, the path length and the danger index are 20 m and 124.28 respectively.

### 4.2.3 A Dynamic Environment

The particularity of this simulation is that the environment contains a dynamic obstacle. It contains also three static obstacles. The chosen parameters for GA in this case are Pc = 0.7 and Pm = 0.02.

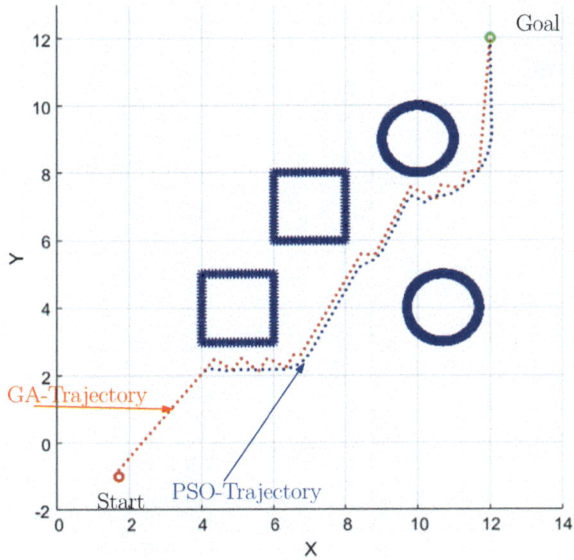


**Fig. 8** Robot trajectories in a multi-static-obstacles environment

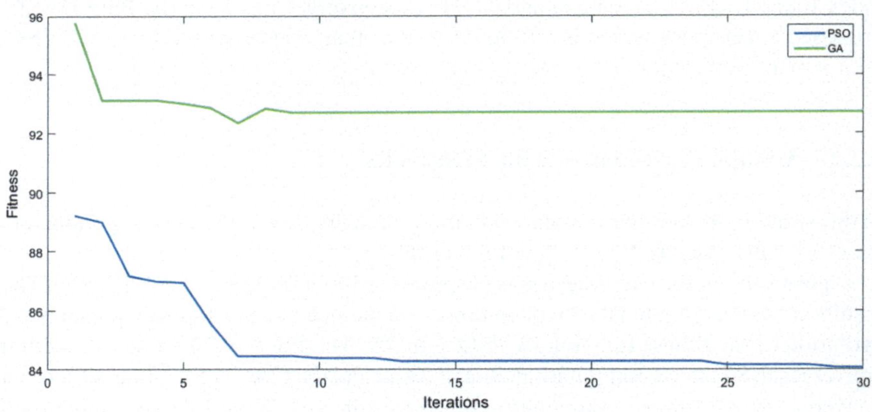


**Fig. 9** Evolution of the fitness function with respect to iterations by GA and PSO

Figure 10 presents the trajectories proposed by PSO-DVSF$^2$ and GA-DVSF$^2$ mobile robot path planning approaches. Figure 11 presents the evolution of the fitness function of PSO-DVSF$^2$ and GA-DVSF$^2$ with iterations. In this simulation, the fitness function related to PSO-DVSF$^2$ converges to 91.83 after 1 min 56 s of computation and the fitness function related to GA-DVSF$^2$ converges to 94.17 after 2 min of computation. This means that the results of PSO-DVSF$^2$ are a little bit better than those given by GA-DVSF$^2$. In fact, with PSO, the path length and the danger index

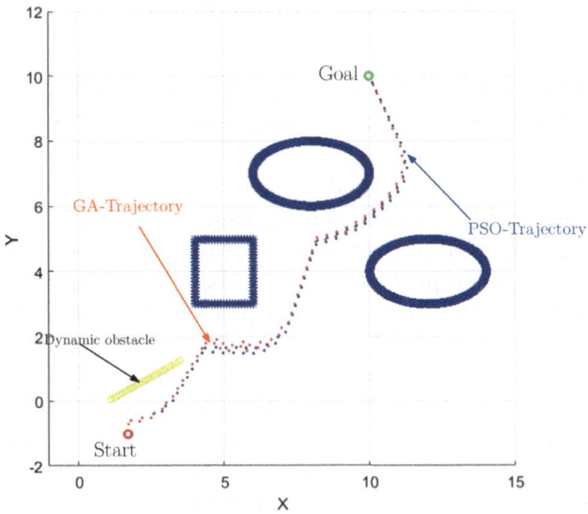


**Fig. 10** Robot trajectories in a dynamic environment

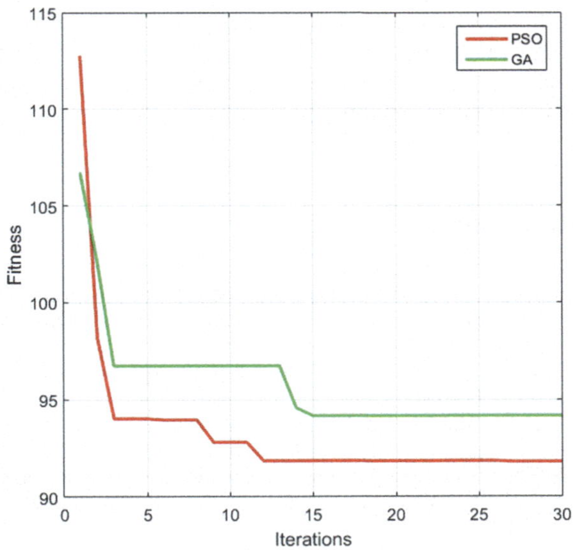


**Fig. 11** Evolution of the fitness function with respect to iterations by GA and PSO

are 18.55 m and 123.23 respectively. But, with GA, the path length and the secure distance are 18.65 m and 126.53 respectively.

In this simulation, the results of the two approaches are very close but the favor continues to be on the PSO-DVSF[2] side.

## 5 Conclusion and Future Work

In this paper, we presented the results of a comparative investigation on the efficiency in optimizing mobile robot path planning between PSO and GA.

We started this investigation with a comparison of evolutionary performances between these two optimization techniques. In the three benchmark tests, PSO achieves better results in fewer times than GA.

In mobile robot path planning, we considered the PSO-DVSF$^2$ and the GA-DVSF$^2$ mobile robot path planning approaches. These two approaches are based on the famous DVSF$^2$ path planning approach. For PSO-DVSF$^2$, parameters of DVSF$^2$ are optimized with PSO and for GA-DVSF$^2$, parameters of DVSF$^2$ are optimized with GA. Three environments have been considered. The first and the second one were static but with different numbers and positions of obstacles and the last one was dynamic. In fact, it contains a dynamic obstacle. Simulation results were not very far and the trajectories proposed by PSO-DVSF$^2$ and GA-DVSF$^2$ were neighbors. That said, the favor is on the side of PSO-DVSF$^2$ which proposes generally the shortest and the safest path taking a shorter time of computation.

In a future work, we will consider real time control of mobile robots and we will compare the efficiency of PSO and GA in the optimization of the proposed paths. In such case, the rapidity of computation is a very interesting parameter especially in dynamic environments.

## References

1. Cong, Y.Z., Ponnambalam, S.G.: Mobile robot path planning using ant colony optimization. In: IEEE/ASME International Conference on Advanced Intelligent Mechatronics, AIM 2009. IEEE (2009)
2. Zhang, H.-y., Lin, W.-m., Chen, A.-x.: Path planning for the mobile robot: a review. Symmetry. **10**, 450–467 (2018)
3. Zafar, Mohd.N., Mohanta, J.C.: Methodology for path planning and optimization of mobile robots: a review. Procedia Comput. Sci. **133**, 141–152 (2018)
4. Zhang, X., Zhao, Y., Deng, N., Guo, K.: Dynamic path planning algorithm for a mobile robot based on visible space and an improved genetic algorithm. Int. J. Adv. Robot. Syst. **13**, 91–108 (2016)
5. Nazarahari, M., Khanmirza, E., Doostie, S.: Multi-objective multi-robot path planning in continuous environment using an enhanced genetic algorithm. Expert Syst. Appl. **115**, 106–120 (2019)
6. Cholodowicz, E., Figurowski, D.: Mobile robot path planning with obstacle avoidance using particle swarm optimization. Pomiary Automatyka Robotyka **21** (2017)
7. Adamu, P.I., Jegede, J.T., Okagbue, H.I., Oguntunde, P.E.: Shortest path planning algorithm-a Particle Swarm Optimization (PSO) approach. In: Proceedings of the World Congress on Engineering, vol. 1 (2018)
8. Samadi, M., Othman, M.F.: Global path planning for autonomous mobile robot using genetic algorithm. In: 2013 International Conference on Signal-Image Technology and Internet-Based Systems (SITIS). IEEE (2013)

9. Rath, M.K., Deepak, B.B.V.L.: PSO based system architecture for path planning of mobile robot in dynamic environment. In: 2015 Global Conference on Communication Technologies (GCCT). IEEE (2015)
10. Souza Lima, C.A., Lapa, C.M.F., do N.A., Pereira, C.M., da Cunha, J.J., Alvim, A.C.M.: Comparison of computational performance of GA and PSO optimization techniques when designing similar systems-Typical PWR core case. Ann. Nucl. Energy **38**, 1339–1346 (2011)
11. Khoshahval, F., Minuchehr, H., Zolfaghari, A.: Performance evaluation of PSO and GA in PWR core loading pattern optimization. Nucl. Eng. Des. **241**, 799–808 (2011)
12. Hassan, R., Cohanim, B., de Weck, O.: A comparison of particle swarm optimization and the genetic algorithm. In: American Institute of Aeronautics and Astronautics (2004)
13. Kecskes, I., Szekacs, L., Fodor, J.C., Odry, P.: PSO and GA optimization methods comparison on simulation model of a real hexapod robot. In: IEEE 9th International Conference on Computational Cybernetics (ICCC), pp. 125–130 (2013)
14. Ben Alaia, E., Harbaoui, I., Borne, P., Bouchriha, H.: A comparative study of the PSO and GA for the m-MDPDPTW. Int. J. Comput. Commun. Control **13**, 8–23 (2018)
15. Ziadi, S., Njah, M., Chtourou, M.: PSO-DVSF2: a new method for the path planning of mobile robots. In: 16th International Conference on Sciences and Techniques of Automatic control and computer engineering - STA'2015, Monastir, Tunisia (2015). Accessed 21–23 Dec 2015
16. Ziadi, S., Njah, M., Chtourou, M.: A*PSO-DVSF2 : an optimized mobile robot path planning approach. Int. J. Electr. Electron. Data Commun. **6**, 63–68 (2018)
17. Kaewkamnerdpong, B., Bentley, P.: Perceptive particle swarm optimization: an investigation. In: Proceedings of the 2005 IEEE Swarm Intelligence Symposium, SIS 2005, pp. 169–176 (2005)
18. Yuhui, S., Eberhart, R.: Empirical study of particle swarm optimization. In: Proceedings of the 1999 Congress on Evolutionary Computation, 1999, CEC 99, vol. 3. IEEE (1999)
19. Katiyar, S.: A comparative study of genetic algorithm and the particle swarm optimization. AKGEC Int. J. Technol. **2**(2), 21–24 (2011)
20. Noura Teixeira, O.: Computaçáo evolucionà: dos aspectos filosóficos à implementaçáo dos algoritmos genèticos na soluçáo do problema do caixeiro viajante simétrico. Trabalho de Conclusãâo de Curso (Bacharelado em Ciência da Computaçáo) - Universidade Federal do Pará, Belém (2003)
21. Aliab, A.F., Tawhid, M.A.: A hybrid particle swarm optimization and genetic algorithm with population partitioning for large scale optimization problems. Ain Shams Eng. J. (2016)
22. Prieto, L., Komínkova-Oplatková, Z., Frías, R., Hernández, J.: A time performance comparison of particle swarm optimization in mobile devices. In: MATEC Web of Conferences, vol. 76. EDP Sciences (2016)
23. Miro, J.V., Taha, T., Wang, D., Dissanayake, G., Liu, D.: An efficient strategy for robot navigation in cluttered environments in the presence of dynamic obstacles. In: The Eighth International Conference on Intelligent Technologies (2007)
24. Wang, D.: A generic force field method for robot real-time motion planning and coordination. A thesis submitted in fulfilment of the degree of doctor of philosophy (2009)

# Part III

# Dynamic Modeling of a Quadrotor UAV Prototype

**W. Mizouri, S. Najar, L. Bouabdallah and M. Aoun**

**Abstract** Quadrotor UAV is a highly coupled and under-actuated system. Estimation of the dynamic model is an important step for the quadrotor control design. In this chapter, two modelling methods are proposed to determine the dynamic model of a quadrotor prototype. Firstly a mathematical model is developed to describe the quadrotor 6 degrees of freedom using Euler–Lagrange formalism. The unknown model parameters are then obtained using calculations and experimental tests. In the second method the roll system is estimated using closed loop identification method and frequency domain analysis. The simulation results are compared with practical responses. Both of the two methods shows good performances.

**Keywords** Quadrotor UAV · Dynamic model · Frequency domain · Close loop identification

## 1 Introduction

A Quadrotor is a vertical take-off and landing aircraft. It is used in many fields such as delivery, surveillance, industrial inspection, rescue. Quadrotor system consists of four rotor fixed on extremities of a cross-like frame. A flight controller board, sensors and power storage are placed in quadrotor body center. In the standard configuration two rotors are spinning in a clockwise direction with the other two spinning in an reverse direction to compensate aerodynamic torques produced by each rotor. Quadrotor

W. Mizouri (✉) · S. Najar · L. Bouabdallah · M. Aoun
National Engineering School of Gabes, University of Gabes, Gabes, Tunisia
e-mail: mizouri.walid@gmail.com

S. Najar
e-mail: slaheddine.najar@gmail.com

L. Bouabdallah
e-mail: drbouabadallah@gmail.com

M. Aoun
e-mail: mohamed.aoun@gmail.com

 
J. Ghommam et al. (eds.), *New Trends in Robot Control*, Studies in Systems, Decision and Control 270, https://doi.org/10.1007/978-981-15-1819-5_14

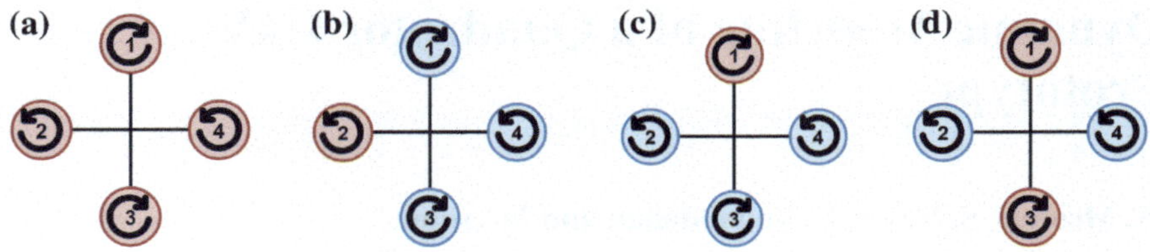


**Fig. 1** Quadrotor UAV motions: **a** Rise and fall, **b** Roll, **c** Pitch, **d** Yaw

motion is controlled using propellers speed variation. In fact, the rise/fall motion is given by increasing or decreasing all motors speed with the same proportion as illustrated in Fig. 1a. To move the helicopter in forward/backward direction, the speed of motor (1) has to be changed over the speed of motor (3), thus the pitch rotation is obtained as shown in Fig. 1b. The rotation around $x_B$ axis (roll) is provided by varying the of motor (2) versus motor (4), this will produce a transversal motion as given in Fig. 1c. The rotation around the vertical axis is given when the second and fourth propellers speeds are changed versus the first and third propellers speeds Fig. 1d.

This work is organized as follows. After presenting the related works in Sect. 2, the Quadrotor prototype used in this study is introduced in Sect. 3. Then mathematical model is detailed and unknown parameters of the obtained model are identified in Sect. 4. In Sect. 5, a second modelling method is proposed to estimate the quadrotor roll system. the obtained results of the both of method are interpreted in Sect. 6. Finally, some concluding remarks are drawn in the last Section.

## 2 Related Works

Quadrotor system modelling becomes an attractive subject for many researchers. [1–3] used Newton–Euler and Euler–Lagrange approach which lead to a non linear motion equations describing the quadrotor 6 degrees of feedom. In [8] a linear parametric model of quadrotor UAV is developed using frequency domain analysis of the open loop responses. Author of [6] proposed a modeling and an attitude stabilization method, the results are provided by free flight tests. [10] developed the system identification of the translation system for a cost open-source quadrotor prototype MikroKopter. In [4] a non linear model was developed using Euler–Lagrange formalism, then the unknown model parameters were identified using mathematical calculations and experimental tests. Authors in [9] have used Black-box approach for a variable-pitch quadrotor identification. The approach was developed and explained. [12] proposed a practical method to develop a linear model for Hummingbird Asctec quadrotor. Pitch loop was estimated as first order plus time-delay system then a fractional order FOPID and Integer Order IOPID controllers for pitch angle control in the work of [7]. Quadrotor transfer function is obtained by linear parameter estimation, then a PID controller is designed and validated with outdoor flight test in [5]. In the Ref. [11] an ARMAX model describing the quadrotor system is obtained basing on closed loop identification method.

## 3 Quadrotor Prototype

In this paper a quadrotor prototype is used as a test platform named Sunbird and shown in Fig. 2. Sunbird is a home-made quadrotor platform designed at Modelling Analysis and Control of System MACS research laboratory.

Sunbird quadrotor is composed as follows:

- An Arduino DUE card used as flight controller (Fig. 3b). It is based on a 32-bit ARM core micro-controller with a 84 MHz clock. It contains 12 analogue inputs, 2 analogue output and 54 digital Input/Output.

- A FlySky 2.4 GHz Radio Control RC with 4 channels is used to remotely control the quadrotor (Fig. 3a).
- The rotor is composed of Hextronik DT750 brushless DC motor, $10 \times 4.5$ propellers and Electronic Speed Controller ESC 30A operating with Pulse Width Modulation (PWM) signal generated by Arduino Due card Fig. 4.
- An ultrasonic sensor HC-SR 05 is aimed to measure the altitude with a 4.5 m of measurement range.

**Fig. 2** Sunbird Quadrotor Prototype

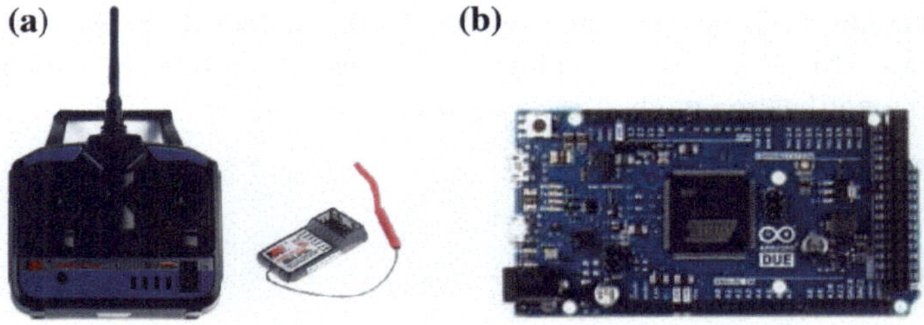


**Fig. 3** Fly Sky 2.4 GHz Rc controller (**a**), Flight controller: Arduino Due board (**b**)

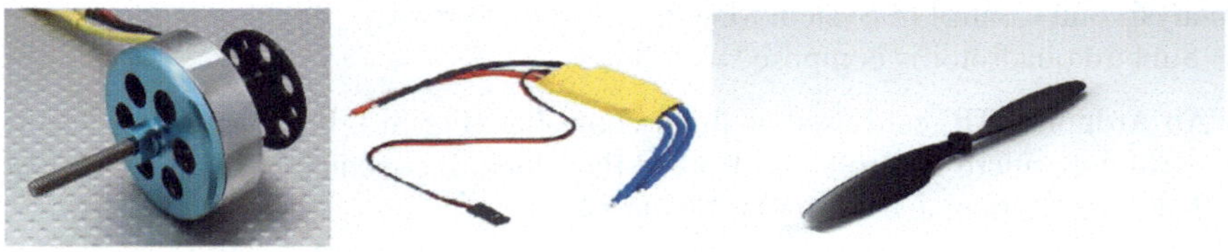

**Fig. 4** Sunbird prototype rotor

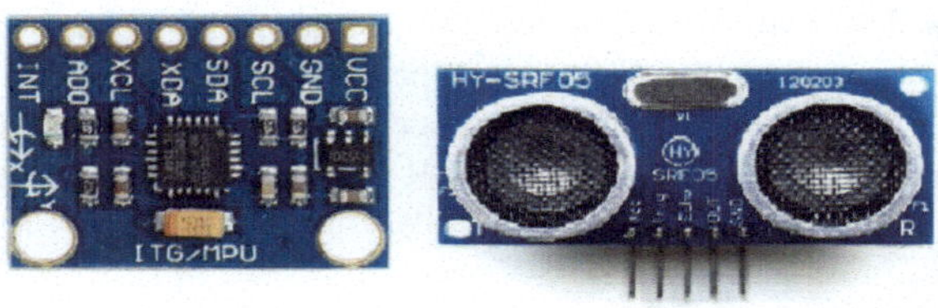


**Fig. 5** Inertial Measurement Unit MPU6050 and SFR05 Ultrasonic sensor

- Measurement Processing Unit MPU6050 is used to estimate attitude (Fig. 5). MPU-6050 is a six axis IMU sensor containing 3 axis gyroscope ITG 3200 giving the angular velocities $(\dot{\phi}, \dot{\theta}, \dot{\psi})$ and 3 axis accelerometer ADXL 345 measuring linear accelerations $(A_x, A_y, A_z)$.

In fact Arduino Due receives data from MPU-6050 via I2C-bus. Then, a complementary filter is used to compensate the drifts affecting gyroscope measurements and attenuate high frequency signals affecting accelerometer like vibration. Denoting $\phi_{acc}, \phi_{Gyr}, \phi_{Comp}$ as the roll angle estimated respectively using accelerometer, gyroscope and complementary filter. Using accelerometer, we can estimate the quadrotor orientation relative to the fixed inertial frame $E$ as follows

$$\phi_{acc} = \arctan\left(\frac{A_x}{\sqrt{A_y^2 + A_z^2}}\right) \tag{1}$$

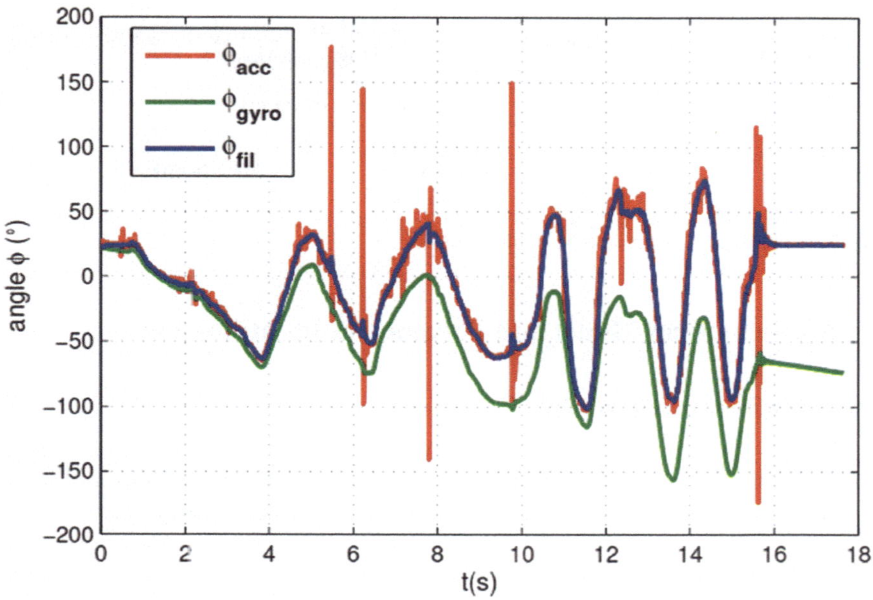


**Fig. 6** Comparison of accelerometer, gyroscope and complementary filter estimated roll angle

where $(A_x, A_y, A_z)$ data raw given by accelerometer. Gyroscope estimated angle is given by simple integration of angular rate $\dot{\phi}$

$$\phi_{Gyr} = \phi_{Gyr} + \dot{\phi} \times dt \tag{2}$$

The complementary filter is a combination between accelerometer and gyroscope measurements

$$\phi_{Comp} = \alpha\, (\phi_{Comp} + \dot{\phi} \times dt) + (1 - \alpha)\, \phi_{acc} \tag{3}$$

where $\alpha \in [0, 1]$ is used to tune the complementary filter and $dt$ is the sample time.

Figure 6 shows a comparison of measured roll angle using respectively accelerometer, gyroscope and complementary filter. We can clearly see the robustness of the used filter. However, the gyroscope estimated angle is drifting whenever the accelerometer estimation is very sensitive to vibrations. The composition of quadrotor prototype is summarized in Fig. 7.

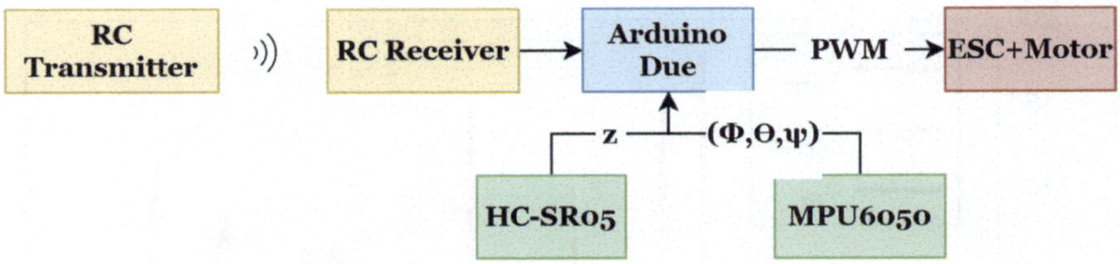


**Fig. 7** Sunbird quadrotor prototype architecture

# 4 Mathematical Model and Parameters Identification

## *4.1 Mathematical Model*

This section describes the quadrotor mathematical model developed using Euler–Lagrange formalism. Denoting $q$ the generalized coordinates of the quadrotor

$$q = (x, y, z, \phi, \theta, \psi) \in \mathbb{R}^6 \tag{4}$$

where $\xi = (x, y, z)$ is the positions of the quadrotor center of mass relative to the fixed inertial frame $\boldsymbol{E}$, the orientation of the quadrotor are expressed by $\vartheta = (\phi, \theta, \psi)$, where $\phi$ is the roll angle around the $x_B$ axis, $\theta$ is the pitch angle around the $y_B$ axis and $\psi$ is the yaw angle around the $z_B$ axis.

The transformation matrix from inertial frame $\boldsymbol{E}$ to the body frame $\boldsymbol{B}$ is given by

$$R = \begin{bmatrix} c\psi\, c\theta & c\psi\, s\theta\, s\phi - s\psi\, c\phi & c\psi\, s\theta\, c\phi + s\psi\, s\phi \\ s\psi\, c\theta & s\psi\, s\theta\, s\phi + c\psi\, c\phi & s\psi\, s\theta\, c\phi - s\phi\, c\psi \\ -s\theta & c\theta\, s\phi & c\theta\, c\phi \end{bmatrix} \tag{5}$$

**Fig. 8** Quadrotor UAV coordinate system

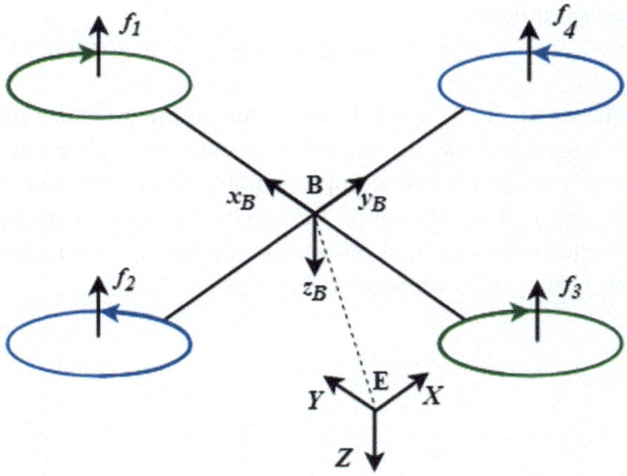

where $s\theta = \sin(\theta)$ and $c\theta = \cos(\theta)$.

Lagrangian equation is defined as follow

$$L(q, \dot{q}) = E_p - E_c \tag{6}$$

where $E_c$ and $E_p$ are respectively the potential and kinetic energy.

$$\begin{cases} E_p = mgz \\ E_c = \frac{m}{2}\dot{\xi}^T\dot{\xi} + \frac{1}{2}\dot{\upsilon}^T I\dot{\upsilon} \end{cases} \tag{7}$$

where $I$ is the inertia matrix, $m$ is the total mass of the quadrotor and $g$ is the gravity acceleration.

Euler–Lagrange formulas is

$$\frac{d}{dt}\left(\frac{\delta L}{\delta \dot{q}}\right) - \frac{\delta L}{\delta q} = \begin{bmatrix} F \\ \tau \end{bmatrix} \tag{8}$$

where $F$ is the force of translation, $\tau$ is the total torque. Let's start with the rotational dynamic of the quadrotor. where $\tau$ is the total torques produced by quadrotor system. We can write

$$\tau = \tau_1 + \tau_2 \tag{9}$$

where $\tau_1$ is the roll, pitch and yaw torques produced by the quadrotor motors

$$\tau_1 = \begin{bmatrix} bl(\Omega_2^2 - \Omega_4^2) \\ bl(\Omega_1^2 - \Omega_3^2) \\ d(\Omega_1^2 - \Omega_2^2 + \Omega_3^2 - \Omega_4^2) \end{bmatrix} \tag{10}$$

$\tau_2$ is the gyroscopic effects given by

$$\tau_2 = \begin{bmatrix} \dot{\theta} I_{rotor}\Omega_g \\ \dot{\phi} I_{rotor}\Omega_g \\ 0 \end{bmatrix} \tag{11}$$

where $\Omega_i$ is the angular speed of the $i$th motor, $b$ is the thrust coefficient, $d$ is the drag coefficient, $l$ is the distance between the motors and the quadrotor center of gravity and $I_{rotor}$ is the rotor inertia moment.

$$\Omega_g = (\Omega_1 - \Omega_2 + \Omega_3 - \Omega_4) \tag{12}$$

The generalized moments are

$$\begin{cases} \frac{d}{dt}\left(\frac{\delta L}{\delta \dot{\phi}}\right) - \frac{\delta L}{\delta \phi} = I_{xx}\ddot{\phi} + (I_{zz} - I_{yy})\dot{\theta}\dot{\psi} \\ \frac{d}{dt}\left(\frac{\delta L}{\delta \dot{\theta}}\right) - \frac{\delta L}{\delta \theta} = I_{yy}\ddot{\theta} + (I_{xx} - I_{zz})\dot{\phi}\dot{\psi} \\ \frac{d}{dt}\left(\frac{\delta L}{\delta \dot{\psi}}\right) - \frac{\delta L}{\delta \psi} = I_{zz}\ddot{\theta} + (I_{yy} - I_{xx})\dot{\phi}\dot{\theta} \end{cases} \tag{13}$$

where $I$ is the inertia matrix:

$$I = \begin{bmatrix} I_{xx} & 0 & 0 \\ 0 & I_{yy} & 0 \\ 0 & 0 & I_{zz} \end{bmatrix}$$

The rotational dynamic equations can be deduced from (10), (11) and (13)

$$\begin{cases} \ddot{\phi} = \dot{\theta}\dot{\psi}\left(\frac{I_{yy} - I_{zz}}{I_{xx}}\right) + \frac{I_{rotor}}{I_{xx}} \times \frac{d(\theta\Omega_g)}{dt} + \frac{1}{I_{xx}}U_2 \\ \ddot{\theta} = \dot{\phi}\dot{\psi}\left(\frac{I_{zz} - I_{xx}}{I_{yy}}\right) + \frac{I_{rotor}}{I_{yy}} \times \frac{d(\theta\Omega_g)}{dt} + \frac{1}{I_{yy}}U_3 \\ \ddot{\psi} = \dot{\phi}\dot{\theta}\left(\frac{I_{xx} - I_{yy}}{I_{zz}}\right) + \frac{1}{I_{zz}}U_4 \end{cases} \tag{14}$$

where $(U_1, U_2, U_3, U_4)$ are the model inputs which are given by the following expressions

$$\begin{cases} U_1 = b\,(\Omega_1^2 + \Omega_2^2 + \Omega_3^2 + \Omega_4^2) \\ U_2 = bl\,(\Omega_2^2 - \Omega_4^2) \\ U_3 = bl\,(\Omega_1^2 - \Omega_3^2) \\ U_4 = d\,(\Omega_1^2 - \Omega_2^2 + \Omega_3^2 - \Omega_4^2) \end{cases} \tag{15}$$

In order to reduce complexity of calculus, gyroscopic effects and moments of inertia terms can be neglected. Thus the motion equations (14) becomes

$$\begin{cases} \ddot{\phi} = \frac{1}{I_{xx}}U_2 \\ \ddot{\theta} = \frac{1}{I_{yy}}U_3 \\ \ddot{\psi} = \frac{1}{I_{zz}}U_4 \end{cases} \tag{16}$$

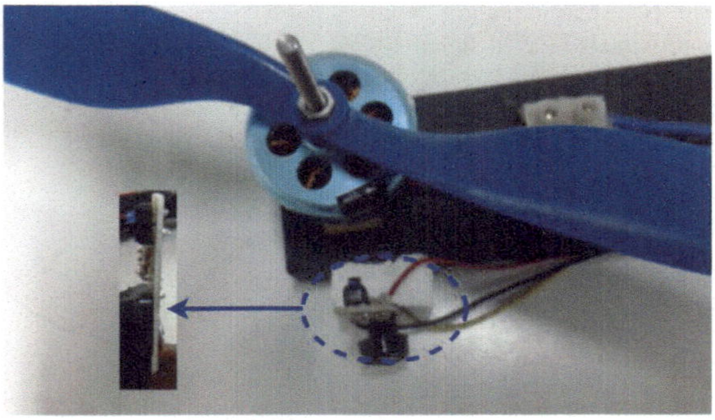

**Fig. 9** Infra-Red sensor used for the motor speed measuring

## 4.2 *Parameters Identification*

In this part the parameters of the quadrotor model are identified using experimental tests or physical formulas.

### 4.2.1 Moment of Inertia

To estimate the moment of inertia, quadrotor system can be decomposed into several parts. Using the formulas (17) we can calculate the inertia moments of each part relative to his center of mass.

$$I = \int x^2 dm \tag{17}$$

The moment of each parts relative to the quadrotor center of gravity is given by Huygens–Steiner theorem (18)

$$I_{cog} = I + m \times d^2 \tag{18}$$

where $d$ is the distance between quadrotor and the part center of gravity. The moment of inertia of the quadrotor system was obtained by summing all inertia moments.

### 4.2.2 Rotor Dynamic

In this subsection characteristics of Hextronik DT750 brushless DC motor was estimated. In fact the motor was operating with a Pulse Width Modulation (PWM) signal. PWM is a periodic signal with variable pulse width between [1000, 2000] microseconds (μs ). Thereby the motor Speed controller ESC was excited by series of PWM step signal generated by Arduino Due. Then the corresponding motor speed was measured using a home made Infra-red sensor shown in Fig. 9.

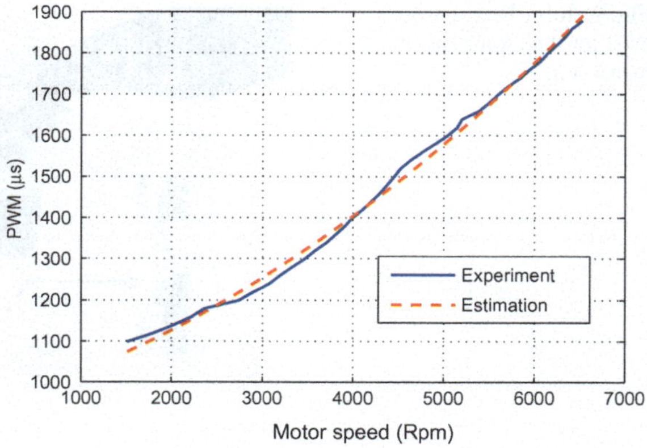


**Fig. 10** Measured and approximated PWM signal and motor speed characteristics

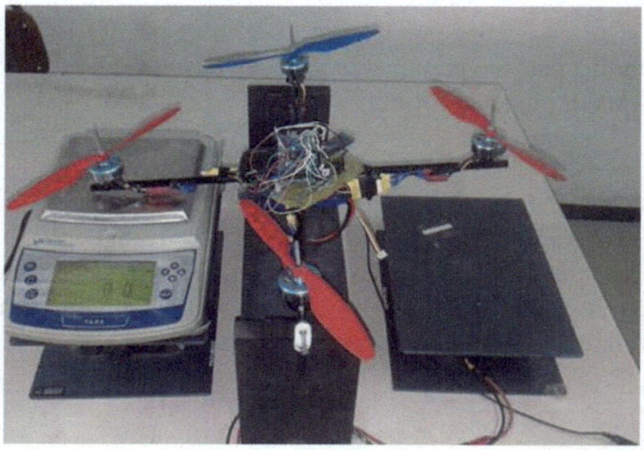

**Fig. 11** Test bench to measure the thrust force of rotor

Using "*polyfit*" Matlab function the motor characteristics can be expressed by the following second order equation:

$$PWM = a_2\Omega^2 + a_1\Omega + a_0 \tag{19}$$

where $a_2 = 1.2093\ 10^{-5}$, $a_1 = 6.5776\ 10^{-2}$ and $a_0 = 948.2$. Figure 10 illustrates the measured and estimated characteristics of PWM signal according to the motor speed.

### 4.2.3 Thrust Coefficient

Basing on the relation between the thrust force the square of the angular rate $F_t = b\Omega^2$, the thrust factor can be estimated. The method consists on varying the motor speed and measuring the produced thrust force $F_t$ using electronic balance with the test bench as shown in Fig. 11.

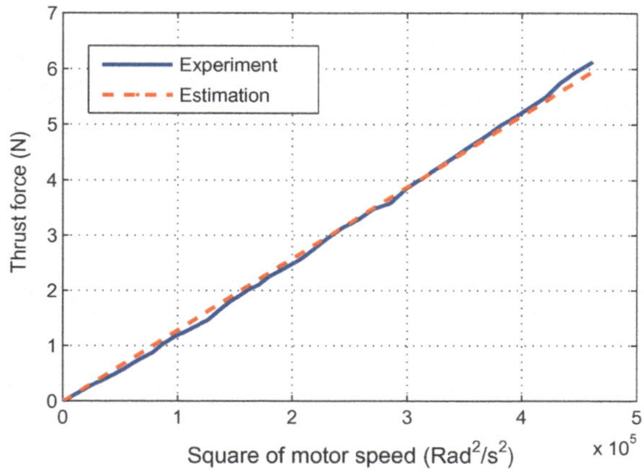


**Fig. 12** Estimation and experimental of the thrust force relative to the square of motor velocity

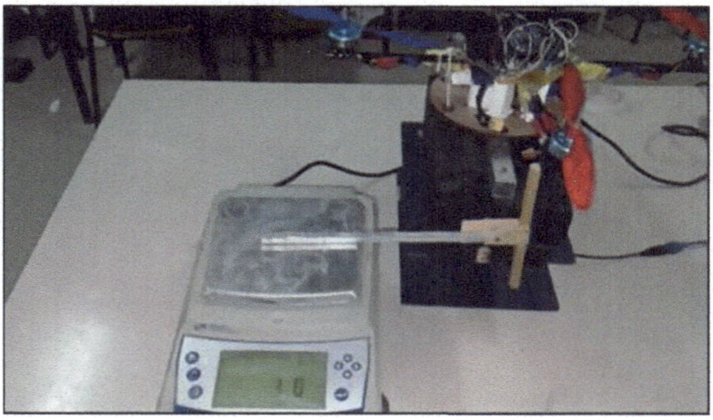

**Fig. 13** Test bench used for drug factor identification

Figure 12 shows the plot of the measured and estimated characteristic of the thrust force over the squared motor speed.

### 4.2.4 Drug Coefficient

The Yaw torque is proportional to the square of the angular rate of the motor $\tau = d\ \Omega^2$. Similarly to the thrust factor using a test bench in Fig. 13 the drug factor can be identified.

The plot of the yaw torque over the squared motor speed is shown in Fig. 14.

The parameter used on the quadrotor dynamic model are listed in the Table 1.

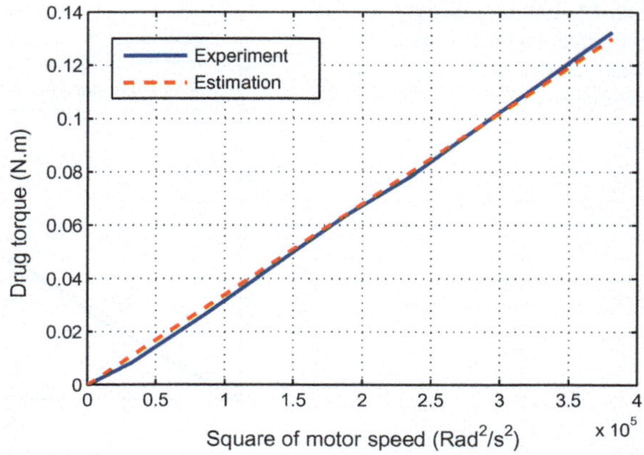


**Fig. 14** Measured and approximated of drug moment relative to the square of the motor speed

**Table 1** Numerical values of quadrotor identified parameters

| Parameter | Value |
|---|---|
| $b$ | $1.1487\ 10^{-5}$ |
| $d$ | $2.69\ 10^{-8}$ |
| $I_{xx}$ | $1.2035\ 10^{-5}$ Kg m$^2$ |
| $I_{yy}$ | $1.1923\ 10^{-5}$ Kg m$^2$ |
| $I_{zz}$ | $2.3149\ 10^{-5}$ Kg m$^2$ |
| $m$ | 0.960 Kg |
| $l$ | 0.24 m |

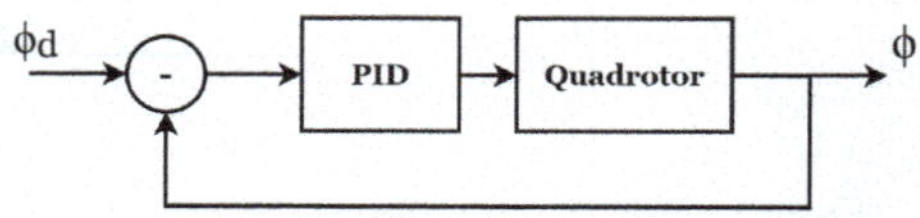


**Fig. 15** Process model in a PID control loop

# 5 Roll System Frequency Identification

In this section, we are interested on roll system modeling using the closed loop identification method. It is based on frequency domain analysis in order to estimate the process model.

The roll system described in (16) is unstable, hence the necessity of a control loop.

## *5.1 Control Loop Identification*

A Proportional Integral Derivative controller was used to ensure the roll loop stability. The PID control loop used in this section was described in Fig. 15.

The PID controller is described by the following expression

$$C(s) = k_p + \frac{k_i}{s} + k_d s \tag{20}$$

where $(k_p, k_i, k_d)$ are respectively the proportional, integral and derivative factors.

The control system is based on reverse model technique. In fact the control signals was used to calculate the desired motors speed $(\Omega_{d1}, \Omega_{d2}, \Omega_{d3}, \Omega_{d4})$ from the Eqs. (15)

$$\begin{cases} \Omega_{d1} = \sqrt{\frac{U_1}{4b} + \frac{U_\theta}{2bl} + \frac{U_\psi}{4d}} \\ \Omega_{d2} = \sqrt{\frac{U_1}{4b} + \frac{U_\phi}{2bl} - \frac{U_\psi}{4d}} \\ \Omega_{d3} = \sqrt{\frac{U_1}{4b} - \frac{U_\theta}{2bl} + \frac{U_\psi}{4d}} \\ \Omega_{d4} = \sqrt{\frac{U_1}{4b} - \frac{U_\phi}{2bl} - \frac{U_\psi}{4d}} \end{cases} \tag{21}$$

Then using the desired motor speed, we can deduce the PWM signal. The PID controller was implemented on the Arduino DUE board and then was experimentally adjusted. The controller parameters are chosen as follow

$$\{k_p = 0.35\ k_i = 0.28,\ k_d = 0.46\} \tag{22}$$

## 5.2 Frequency Identification

The closed loop system in Fig. 15 was excited by a varied sine setpoint with a magnitude of 5° and a frequency between $[10^{-2},\ 10^{2}]$ Hz. The PID control loop is implemented in the Arduino due card operating with a sample time of 0.01 s. The quadrotor was placed on a test bench ensuring rotation around the roll axis with minimum friction effects shown in Fig. 16.

Figures 17, 18 and 19 illustrates the setpoint and the output signals with a frequency of respectively $10^{-2}$, $10^{-1}$ and 1 Hz.

As illustrated by the simulation results, for low frequencies the input and the output signals are quietly non phases and gains are observed. However, for frequency around 1 Hz the output amplitude increases locally then decreases for the highest frequencies, and a lag behaviour is also observed. From several sine responses, the phase and magnitude can be measured for each frequency. Then the Bode diagram can be plotted. Bode plot of the quadrotor control loop is shown in Fig. 20.

Basing on bode plot the closed loop transfer function can be calculated. From the Bode diagram the closed loop system can be estimated as a second order system with resonance. Denoting $F(s)$ the control loop transfer function describes in Fig. 15

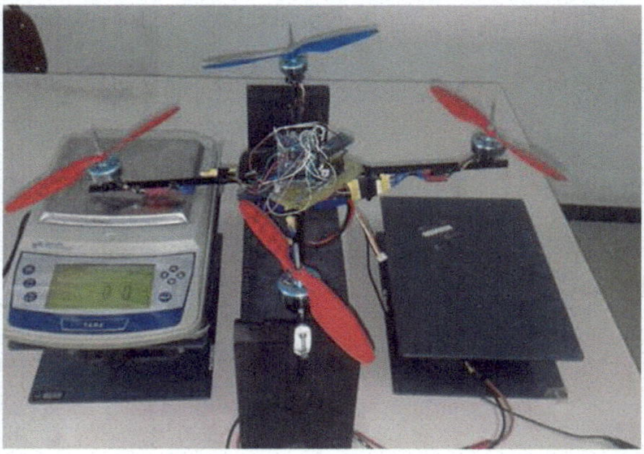

**Fig. 16** Quadrotor prototype placed on a test bench ensuring rotation around roll axis

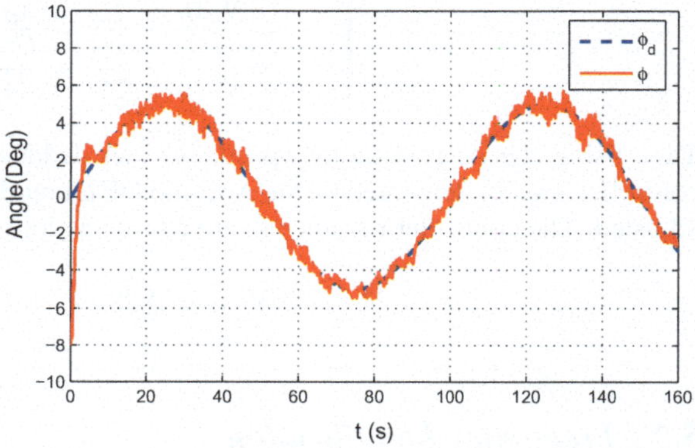


**Fig. 17** Roll angle response for a sine setpoint with a frequency of 0.01 Hz

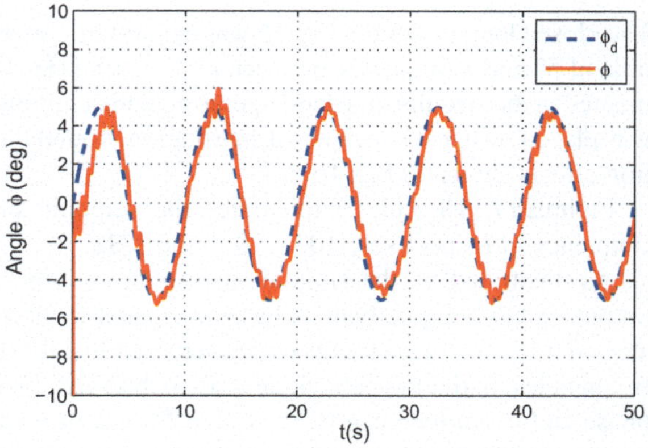


**Fig. 18** Roll angle response for a sine setpoint with a frequency of 0.1 Hz

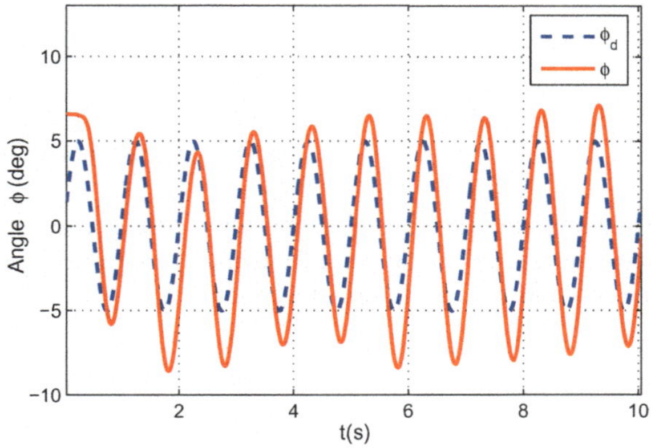


**Fig. 19** Roll angle response for setpoint frequency $f = 1\,\text{Hz}$

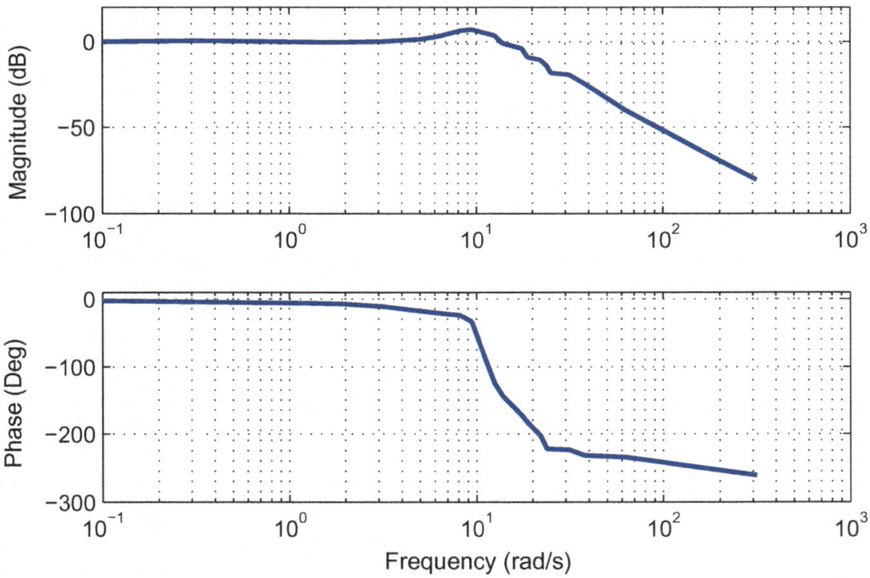


**Fig. 20** Bode plot of the roll system closed loop

$$F(s) = \frac{1}{\frac{1}{\omega_n^2}s^2 + \frac{2\xi}{\omega_n}s + 1} \tag{23}$$

where $\omega_n$ is the normal frequency and $\xi$ is the damping factor.

Denoting $Q$ the resonance factor of the closed loop system which can be described by the following expression

$$Q = \frac{1}{2\xi\sqrt{1-\xi^2}} \tag{24}$$

Thus

$$4Q^2\xi^2(1-\xi^2)=1 \tag{25}$$

We can write also

$$4Q^2\xi^4-4Q^2\xi^2+1=0 \tag{26}$$

The damping factor $\xi$ is given by solving the second order differential equation (26). We denotes $\omega_r$ the resonance frequency. Equation (27) describes the relation between the resonance $\omega_r$ and normal frequency $\omega_n$.

$$\omega_n=\frac{\omega_r}{\sqrt{1-2\xi^2}} \tag{27}$$

We obtained the following normal frequency and damping factor:

$$\xi=0.24\ ,\quad \omega_n=10\,rad/s \tag{28}$$

The transfer function control loop can be written as

$$F(s)=\frac{C(s)G(s)}{1+C(s)G(s)} \tag{29}$$

where $C(s)$ is the PID controller and $G(s)$ is the process model transfer function.

$$G(s)=\frac{F(s)}{C(s)(1-F(s))} \tag{30}$$

Replacing $C(s)$ and $F(s)$ with their expression we can write

$$G(s)=\frac{1}{\left(\frac{k_d s^2+k_p s+k_i}{s}\right)\left(\frac{s^2}{\omega_n^2}+\frac{2\xi s}{\omega_n}\right)} \tag{31}$$

Thus the open loop roll model is following third order transfer function

$$G(s)=\frac{1}{\frac{k_d s^3}{\omega_n^2}+s^2\left(\frac{2\xi k_d}{\omega_n}+\frac{k_p}{\omega_n^2}\right)+s\left(\frac{k_i}{\omega_n^2}+\frac{2\xi k_p}{\omega_n}\right)+\frac{2\xi k_p}{\omega_n}} \tag{32}$$

The deduced process model is a third order transfer function system

$$G(s)=\frac{1}{c_3s^3+c_2s^2+c_1s+c_0} \tag{33}$$

where

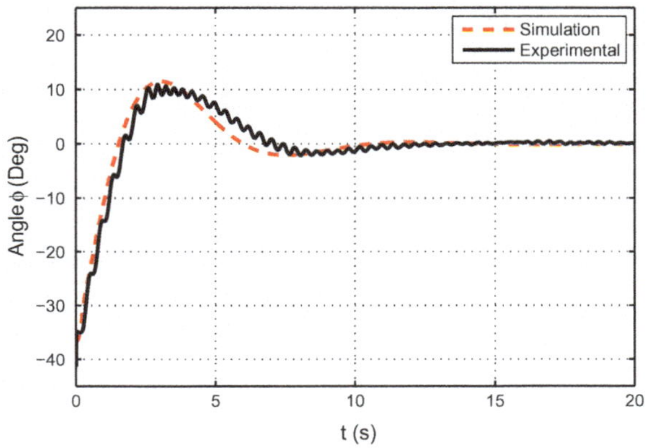


**Fig. 21** Comparison of first model simulation and experimental roll angle step response with ($k_p = 0.35$, $k_i = 0.28$, $k_d = 0.46$)

$$\begin{cases} c_3 = \dfrac{k_d}{\omega_n^2} \\ c_2 = \dfrac{2\xi k_d}{\omega_n} \\ c_1 = \dfrac{k_i}{\omega_n^2} + \frac{2\xi k_p}{\omega_n} \\ c_0 = \dfrac{2\xi k_p}{\omega_n} \end{cases} \tag{34}$$

## 6 Discussion

To ensure the accuracy of the estimated models, the designed control system is implemented with an Arduino DUE board executing with a rate of 100 Hz. The task consists on stabilizing roll, pitch and yaw angle at 0° using test benches used above. Then simulation results has been compared with the experimental results.

Figure 21 illustrates the simulation of first model and measured roll responses with initial angle of −35°. The rising and the settling time of the two responses are nearly the same. Figures 22 and 23 show the results of pitch and yaw loop responses with $\theta_0 = 25°$ and $\psi_0 = 28°$. Simulation with the estimated model and experiment responses are almost close.

Figure 24 illustrates the experimental and simulation step responses of the roll estimated models. Both of these responses are very close, and their rise and settling time are nearly the same.

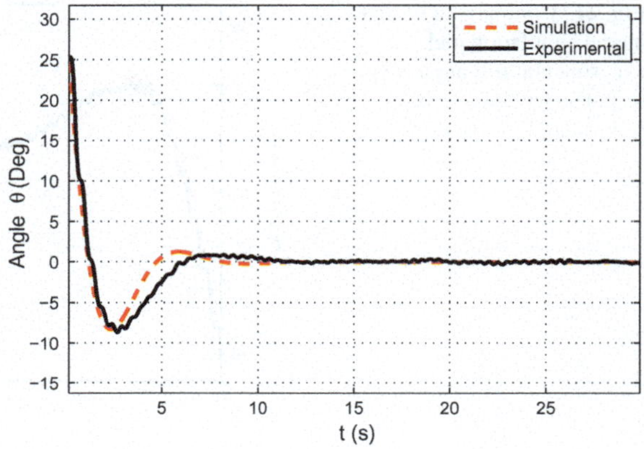


**Fig. 22** Comparison of first model simulation and experimental pitch angle step response with ($k_p = 0.3$, $k_i = 0.29$, $k_d = 0.28$)

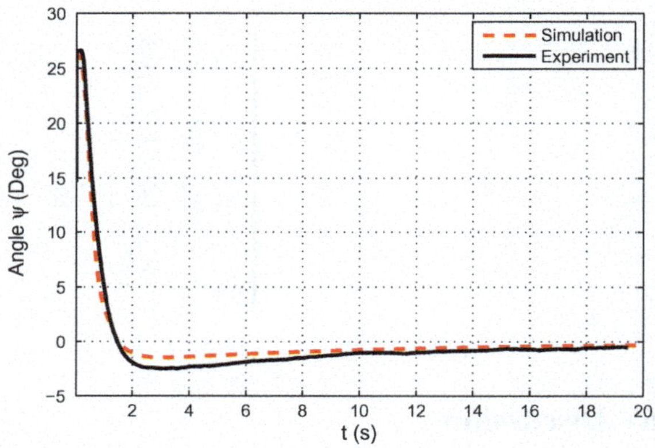


**Fig. 23** Comparison of first model simulation and experimental yaw angle step response with ($k_p = 0.086$, $k_i = 0.006$, $k_d = 0.25$)

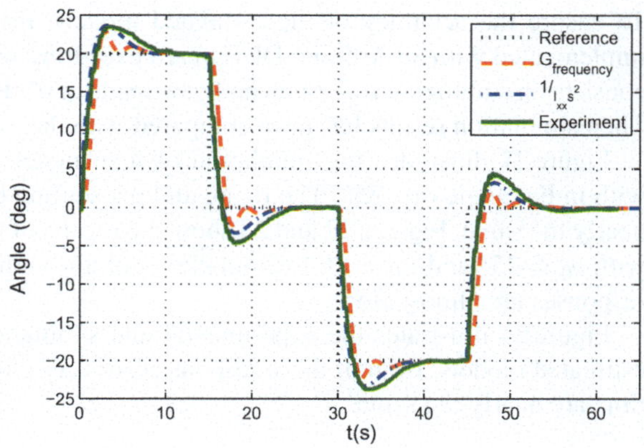


**Fig. 24** Comparison of estimated models and experimental roll responses

## 7 Conclusion

In this work a dynamic model of a quadrotor home made prototype is estimated using two methods. The first method is based on Euler–Lagrange formalism to develop the motion equations of the quadrotor. Then we estimated the unknown parameters of obtained mathematical model using experimental tests and arithmetic calculus. The second method is based on closed loop frequency identification approach. The proposed methods are then evaluated experimentally. Both of the two obtained models proving the performances of the proposed approaches.

**Acknowledgements** Research Laboratory Modelling Analysis and Control of Systems (MACS) LR16S22, Omar Ibn Khattab Street, 6029 Gabes, Tunisia TN.

## References

1. Bouabdallah, S.: Design and control of quadrotors with application to autonomous flying (2007)
2. Carrillo, L.R.G., López, A.E.D., Lozano, R., Pégard, C.: Modeling the quad-rotor mini-rotorcraft. Quad Rotorcraft Control, pp. 23–34. Springer, Berlin (2013)
3. Castillo, P., Lozano, R., Dzul, A.E.: Modelling and Control of Mini-flying Machines. Physica-Verlag, Heidelberg (2006)
4. Chovancová, A., Fico, T., Chovanec, L., Hubinsk, P.: Mathematical modelling and parameter identification of quadrotor (a survey). Procedia Eng. **96**, 172–181 (2014)
5. Dong, W., Gu, G.-Y., Zhu, X., Ding, H.: Modeling and control of a quadrotor UAV with aerodynamic concepts. In: Proceedings of World Academy of Science, Engineering and Technology, no. 77, p. 437 (2013).
6. Goel, R., Shah, S.M., Gupta, N.K., Ananthkrishnan, N.: Modeling, simulation and flight testing of an autonomous quadrotor (2009)
7. Han, J., Di, L., Coopmans, C., Chen, Y.: Pitch loop control of a VTOL UAV using fractional order controller. J. Intell. Robot. Syst. **73**(1–4), 187–195 (2014)
8. Niermeyer, P., Raffler, T., Holzapfel, F.: Open-loop quadrotor flight dynamics identification in frequency domain via closed-loop flight testing. AIAA Guid., Navig., Control Conf. **118**, 21–26 (2015)
9. Panizza, P., Riccardi, F., Lovera, M.: Black-box and grey-box identification of the attitude dynamics for a variable-pitch quadrotor. IFAC-PapersOnLine **48**(9), 61–66 (2015)
10. Sa, I., Corke, P.: System identification, estimation and control for a cost effective open-source quadcopter. In: IEEE International Conference on Robotics and Automation (ICRA), pp. 2202–2209. IEEE (2012)
11. Schreurs, R., Weiland, S., Tao, H., Zhang, Q., Zhu, J., Zhu, Y., Xu, C.: Open loop system identification for a quadrotor helicopter system. In: 2013 10th IEEE International Conference on Control and Automation (ICCA), pp. 1702–1707. IEEE (2013)
12. Yoo, M.G., Hong, S.K.: System identification of the quadrotor flying robot in hover using prediction error method. Electr. Electron. Eng. **118**, 21–26 (2015)

# Model-Based Fault Detection of Permanent Magnet Synchronous Motors of Drones Using Current Sensors

**Guilhem Jouhet, Luis E. González-Jiménez, Marco A. Meza-Aguilar, Walter A. Mayorga-Macías and Luis F. Luque-Vega**

**Abstract** This work proposes a simple, low-cost and effective scheme for the detection of the most common faults in drone actuators which are, usually, permanent magnet synchronous motors (PMSM). The scheme is based on a modelling stage which only requires the current measurements from a faultless motor. From this, a simplified transfer function of the motor is derived. Then, the output of this model and a healthy motor are used as arguments of simple tests to detect the occurrence of a set of characterized faults in the target motor. The setup of the scheme and the development of the tests are straightforward. The faults considered in this work are inter-turn short-circuit, changes in friction constant and flying off propeller and other less common faults. Experimental results show that these faults are accurately detected and characterized by the proposed scheme, opening doors to further work on predictive maintenance and drone adaptive or re-configurable controllers.

**Keywords** PMSM modelling · Fault detection · Drone actuators · Current sensor

G. Jouhet
Ecole Centrale de Lyon, Écully, France
e-mail: guilhem.jouhet@ecl16.ec-lyon.fr

L. E. González-Jiménez (✉) · M. A. Meza-Aguilar · W. A. Mayorga-Macías
Electronics, Systems and Informatics Department, ITESO University, Tlaquepaque, Mexico
e-mail: luisgonzalez@iteso.mx

M. A. Meza-Aguilar
e-mail: mar_ant8@hotmail.com

W. A. Mayorga-Macías
e-mail: ng700810@iteso.mx

L. F. Luque-Vega
Centro de Investigación, Innovación y Desarrollo Tecnológico CIIDETEC-UVM,
Universidad del Valle de México, Mexico City, Mexico
e-mail: luis.luque@uvmnet.edu



J. Ghommam et al. (eds.), *New Trends in Robot Control*, Studies in Systems, Decision and Control 270, https://doi.org/10.1007/978-981-15-1819-5_15

# 1 Introduction

As commercial drones are encountering growing interest on industrial applications such as package delivery, environment monitoring, aerial photography, 3D mapping, among others; their reliability is one of their most important characteristics because it determines whether the drone accomplishes its task or not. This is directly related to the three main blocks of any closed-loop control system: the sensors, the control algorithm and the actuators. Therefore, the research on drones is developed mainly by scientists and engineers in control [3, 4], programming [8], and modelling areas [2]. However, it is difficult to encounter works related to the validation of the reliability of the actuators of a drone. This work proposes a scheme to increase the drones in any application by detecting a set of faults in the drone's actuators with pre-flight tests and real-time monitoring. The most common actuators in drones are small Permanent Magnet Synchronous Motors (PMSM), a type of motor that is overwhelmingly used for drone application due to its huge torque, its efficiency (it is a brushless motor), its compactness, and the increasing industrial feasibility of strong magnets and dense Electronic Speed Controller (ESC). Some researches have been done on very big PMSM used in the automotive industry [1] which presents different dominant dynamics in comparison to the small motors used in drones. The most of works are focusing on one particular fault [10, 11], making harder to set comprehensive fault detection systems. In addition, many of them are using complex and costly methods to detect faults such as Neural Network [6], or detailed simulation that are hardly set up for drone PMSM as their internal parameters (electrical and mechanical) are difficult to encounter and determine, as well as the specs of their ESC [7]. The idea of this work is to combine the results that can simply be implemented to small PMSM [5], the precision and low-computational needs of the others [6, 7] in order to detect and characterize accurately many faults at low costs (time, technical and financial) so that anyone can implement that solution and test his PMSM. Long-run objectives of that study are to use the results of these detections to proceed predictive maintenance of PMSMs and to update drone controllers according to the motors states to increase reliability, and lower damages in case of fatal failure.

The possibility to detect many faults using current data [1, 10, 12] and the simplicity of current of current sensor integration in drone embedded systems turn current sensors into a natural choice for that work. The first part of this work presents the retained fault detection scheme. Then, a commercial drone PMSM, its propeller and its ESC are modelled. Finally, a set of drone PMSM common faults is determined, and each failure is experimentally simulated, characterized and detected.

# 2 Overall Proposed Scheme

The proposed architecture used for PMSM fault detection is depicted in Fig. 1. It can be seen that the studied physical system is driven by a central unit that also collects experimental data. These experimental data are used to identify the parameters of the PMSM dynamic model which are stored once they have been determined, so,

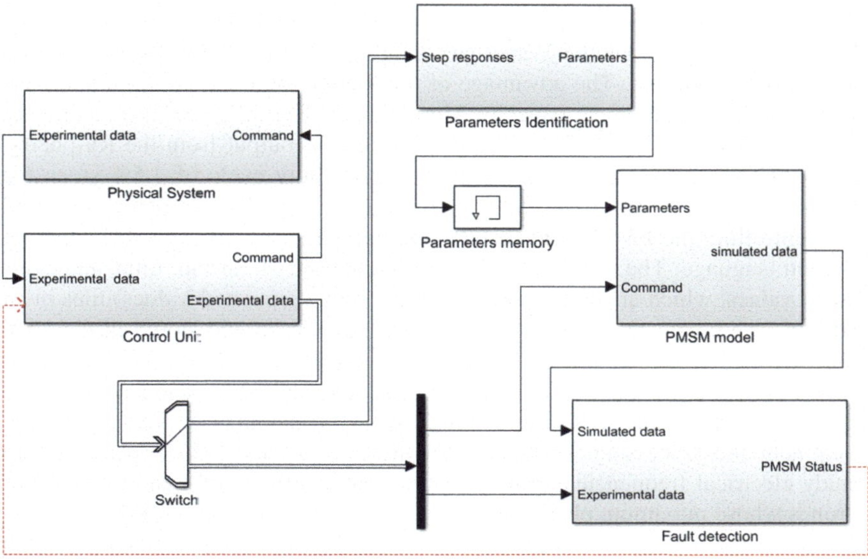


**Fig. 1** Proposed fault detection scheme

it is no necessary to update them automatically. The PMSM model, when filled by saved parameters and by the PMSM command signal, gives simulated data to the fault detection algorithm that detect the PMSM status (e.g. nominal or faulty) and characterize the fault, if relevant, by comparing experimental to simulated data. Future work will focus on using this final information, represented by the dotted line in Fig. 1, to modify the control signals of the drone motors by re-configuring the control gains or changing the control laws.

## 2.1 *Hardware and Software Setup*

The blocks of the proposed scheme (Fig. 1) are related to the hardware and software of the overall system as follows.

- A PMSM coupled with an ESC and a propeller, supplied with a standard 12 V-5000 mAh battery, are fixed on a commercial thrust test-bench from Racerstar (Fig. 2). These elements represent the studied physical system which is used to characterize the motor behaviour in normal and faulty operation. The battery output current and the current of one phase of the PMSM are sensed using two Hall effect current sensors (ref. ACS712-20A). This current sensor range is 20 A, which is sufficient for most of PMSMs of commercial drones, and its impedance is negligible so the PMSM phases are not unbalanced by its normal operation.

- The processing central unit is composed by a Raspberry Pi board operating under Raspbian and coupled with a Waveshare "High-Precision AD/DA Board" for analogical data sampling. The advantage of the Raspberry, which is not a real-time system, is its huge memory capacity. Moreover, the PMSM's ESC can be directly controlled using a PWM (Pulse Width Modulation) output from the Raspberry. The ESC was calibrated so that a PWM with a duty cycle of 4.5% stops the motor and a DC of 9% induces its maximum velocity. The Raspberry program for controlling the PMSM and recollecting experimental data was developed in Python language. The experimental data is composed of current, time, and command values, which are stored using CSV (Comma-Separated Values) files in an USB device connected to the Raspberry. Data is sensed and stored in the CSV files in packs of 200 samples. This operation is repeated as long as necessary, so experimental test length can be adjusted. The sensing and storage frequency are not constant due to the architecture of the system but enable good post-experiment treatment and processing. The sample frequency is around 16 kHz, permitting to study electrical frequencies up to 8 kHz, which is sufficient for most PMSM of drones whose maximum phase fundamental frequency is around 800 Hz.
- The obtained CSV files are post-treated on a desktop computer using Matlab scripts. According to the embedded program that creates coherent experimental data sets of 200 samples, the post-treatment scripts compute one bundle of PMSM state variable values (control signal, phases RMS current, phases current fundamental frequency, battery output current, time) for each of these sets. This way of processing is efficient but imposes precautions for speed calculation, as autocorrelation is used to get the electrical frequencies of the phases in order to reduce computational needs. Indeed, if the 200 samples of each set are not sufficient to cover one electrical period, i.e. sensing very low frequencies, the method is not convenient but can be easily adapted. The data bundles are used as state variables of the PMSM model characterization or monitoring in the following blocks:
  - The model parameters identification block. It was developed as a Matlab script and only uses experimental data from the response of faultless PMSM with a step input of different values. Regressions between PMSM state variables at steady regimen enable to partially fill the PMSM model unknown parameters. The remaining parameters are related to the PMSM transient states and are manually determined by a transfer function identification. The "ident" Matlab built-in tool was helpful for this task.
  - The fault detection block. This algorithm was also developed as a Matlab script and compares the PMSM experimental data to the simulation output to detect and characterize faults using rule-based schemes. This process will be explained in detail in the following sections.

- The PMSM model is implemented as a Simulink model and easily interfaced with the two previous Matlab scripts. Using the identified parameters and the central unit command values, this model supplies the fault detection script with the same PMSM state variables as the experimental CSV files. The PMSM simulated

**Fig. 2** Experimental test bench

behaviour corresponds to the nominal behaviour of a faultless motor assuming that the model's parameter were correctly determined in the previous stage.

The necessary hardware to implement the proposed scheme is low cost, providing an estimated value of \$100 for the thrust test-bench, \$30 for the Raspberry Pi 3, \$10 for the current sensors and \$35 for the ADC shield. Moreover, the software is common in industrial and academic institutions. Finally, these are easily combined to obtain a simple and effective system to tackle the fault detection problem in PMSM.

## 3 PMSM Dynamic Modelling

A classical way of modelling a PMSM is to use its mathematical equations (see [9]) in a Simulink model as in [7]. They are derived from the electrical and mechanical fundamental laws (Newton's laws of motion and Kirchhoff's circuit laws) which are simplified using some assumptions. The final dynamical equations are obtained using the Park transform that enables to reduce the three electrical equations in $a$-$b$-$c$ space to only two in $d$-$q$ space. This transformation is defined as follows

$$\begin{pmatrix} v_d \\ v_q \end{pmatrix} = \frac{2}{3} \begin{pmatrix} \sin\theta & \sin(\theta - \frac{2\pi}{3}) & \sin(\theta + \frac{2\pi}{3}) \\ \cos\theta & \cos(\theta - \frac{2\pi}{3}) & \cos(\theta + \frac{2\pi}{3}) \end{pmatrix} \begin{pmatrix} v_a \\ v_b \\ v_c \end{pmatrix}. \tag{1}$$

Once the system is transformed (see [7]), the electrical equations of the model are obtained as

$$\begin{pmatrix} v_d \\ v_q \end{pmatrix} = \begin{pmatrix} R & -W_e L_q \\ W_e L_d & R \end{pmatrix} \begin{pmatrix} i_d \\ i_q \end{pmatrix} + \begin{pmatrix} L_d & 0 \\ 0 & L_q \end{pmatrix} \frac{d}{dt} \begin{pmatrix} i_d \\ i_q \end{pmatrix} + \begin{pmatrix} 0 \\ P\phi_m W_m \end{pmatrix} \tag{2}$$

where $v_d$ and $v_q$ are the voltages, $i_d$ and $i_q$ are the currents, $R$ is the resistance, $L_d$ and $L_q$ are the inductances, $\phi_m$ the flux linkage, and $W_e$ is the electrical velocity which is related to the mechanical velocity by $W_e = PW_m$ with $P$ as the pairs of poles of the motor. The back-EMFs $e_a$, $e_b$, $e_c$ are related to the system according to the equation

$$\begin{pmatrix} e_a \\ e_b \\ e_c \end{pmatrix} = \begin{pmatrix} W_e\phi_m \cos(W_e t + \frac{2\pi}{3}) \\ W_e\phi_m \cos(W_e t - \frac{2\pi}{3} + \frac{\pi}{2}) \\ W_e\phi_m \cos(W_e t + \frac{2\pi}{3} + \frac{\pi}{2}) \end{pmatrix}. \tag{3}$$

The generated torque is defined as

$$T_e = K_t i_q \tag{4}$$

where $K_t = \frac{3}{2}P\phi_m$ is the torque constant of the motor.

The mechanical equation of the motor is given by

$$J\frac{d}{dt}W_m = T_e - aW_m^2 - bW_m \tag{5}$$

where $J$ is the moment of inertia of the PMSM (including the propeller), and $a$ and $b$ are the viscous and solid friction parameters, respectively. Finally, the thrust generated by the PMSM and the propeller can be obtained approximately as in [2] by

$$F = cW_m^2 \tag{6}$$

where c is a thrust factor determined by the parameters of the propeller.

This dynamical model for the PMSM offers some advantages for control design purposes as the mechanical velocity of its rotor $W_m$ can be controlled by the current $i_q$ which, in turn, can be controlled by the input voltages $v_a$, $v_b$ and $v_c$ by means of (1) and (2). For this, is necessary to define the inverse of the Park's transform as follows

$$\begin{pmatrix} v_a \\ v_b \\ v_c \end{pmatrix} = \begin{pmatrix} \sin\theta & \cos\theta \\ \sin(\theta - \frac{2\pi}{3}) & \cos(\theta - \frac{2\pi}{3}) \\ \sin(\theta + \frac{2\pi}{3}) & \cos(\theta + \frac{2\pi}{3}) \end{pmatrix} \begin{pmatrix} v_d \\ v_q \end{pmatrix}. \tag{7}$$

This modelling procedure is efficient and accurate but only if the parameters of the electrical and mechanical equations, (2) and (5), of the PMSM are precisely known. Despite these parameters are easily known or determined for industrial motors, they remain generally unknown for small PMSM. Moreover, the parameters of the propeller and the ESC are hardly modelled [4] for the same reason. Therefore, an approximated model of the PMSM is obtained in this work by means of two transfer functions, related to the current and the phase frequency, which are obtained using experimental data.

## 4 Model Identification

The proposed model structure is shown in Fig. 3. This model relates the PMSM state variables (the PWM DC, the battery output current $I_{bat}$, the RMS currents $I_{rms}$, the fundamental electrical frequency $F_e$, and the propeller's thrust) with black-box functions that do not need the PMSM parameters.

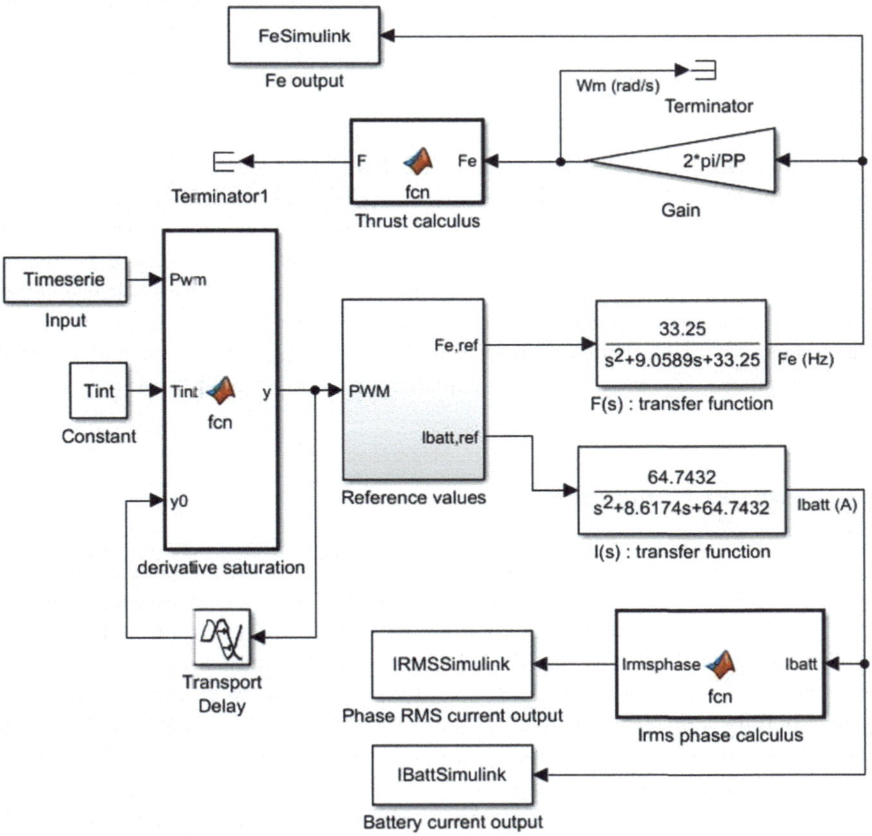


**Fig. 3** Simulink model structure

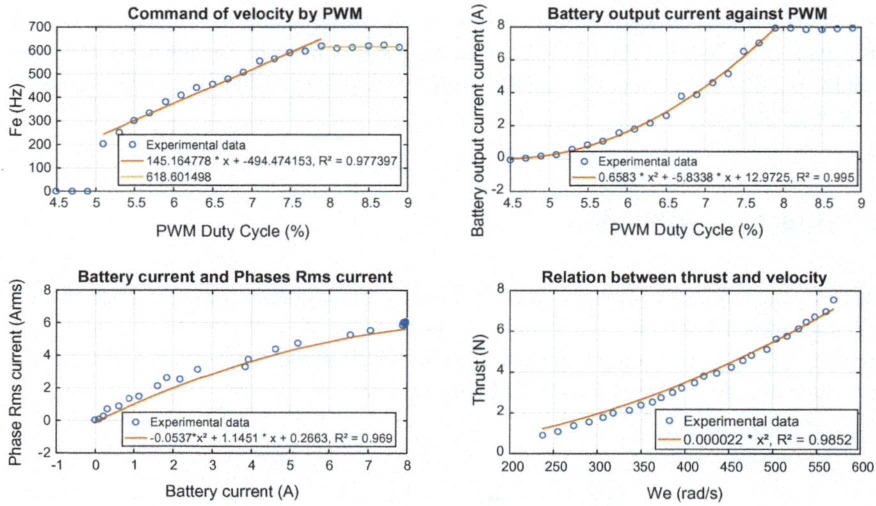


**Fig. 4** Post treatment of studied physical quantities, at steady state

This method firstly proposes to saturate the PWM's DC derivative accordingly to the experimental behaviour of the ESC. The modified DC is then converted to the reference values for $F_e$ and $I_{bat}$, from which their transient responses are computed using two transfer functions. The thrust and the phases RMS currents are respectively calculated from $F_e$ and $I_{bat}$ using two different blocks. As the main idea of this scheme is to avoid to know the PMSM real parameters, the model's transfer function blocks need to be experimentally determined by analyzing the responses of the real systems to known inputs. This process will be explained in detail in the following section.

## 4.1 Transfer Function Blocks

In order to obtain the unknown transfer functions of the model, a sequence of steps inputs of different magnitudes are introduced to the real PMSM by means of the ESC. The range of the magnitude of these DC steps is between 4.6 and 9% with 0.1% increments. Each DC step is held for 5 s so the outputs of the motor can reach steady state, and a 4.5%-DC is held during 5 s between each step so that the motor can fully stop before the next DC step. The obtained experimental data is then post-treated, enabling to obtain the equations of the unknown blocks of the model. These equations are retrieved from the regressions of Fig. 4 and the identified transfer functions depicted in Fig. 5. The regressions are obtained using the steady values of the state variables at different speeds, and the two transfer function are retrieved by using the Systems Identification Toolbox of Matlab.

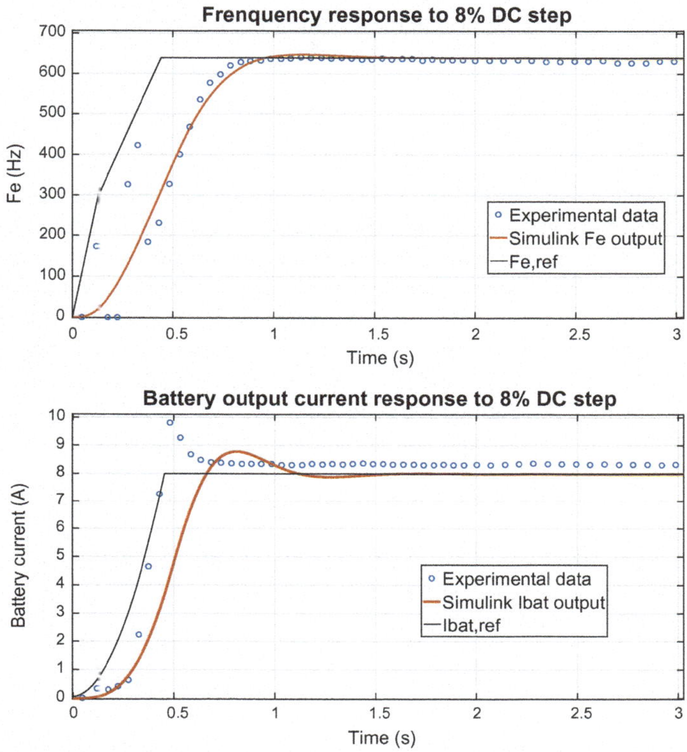


**Fig. 5** $F_e$ and $I_{bat}$ transfer function identification

The variable $F_e$ is saturated at a maximum frequency of 619 Hz and its resulting equations are given by

$$F_e = \begin{cases} 313(DC - 4.5) & \text{if } DC \in [4.5\%, 5.5\%) \\ 145.16(DC - 3.4) & \text{if } DC \in [5.5\%, 9\%] \end{cases} \tag{8}$$

accordingly to the data depicted in the top-left graphic of Fig. 4. Also, its transient behaviour permits to obtain an approximated transfer function, as shown in Fig. 5, which results in

$$F(s) = \frac{33.25}{s^2 + 9.0589\, s + 33.25}. \tag{9}$$

This is obtained as a second order model using the Matlab built-in system identification toolbox. The variable $I_{bat}$ is related to the PWM's DC as

$$I_{bat} = 0.658382DC^2 - 5.833841DC + 12.972593 \tag{10}$$

with a saturation of 8 A. Its transient behaviour is determined as previously resulting in the transfer function (see accuracy in Fig. 5)

$$I_{bat}(s) = \frac{64.7432}{s^2 + 8.6174\, s + 64.7432}. \tag{11}$$

The $I_{bat}$ transient behaviour is difficult to model because its delay and time response depend on the command. Due to this, the model is identified without delay and its response is manually fitted to experimental data using the derivative saturation block of the simulation's model so that the identification error is maintained in a minimum for all commands. A threshold of $9\%/s$ is a good choice to saturate the PWMs DC derivative. However, as it will be shown in Fig. 6, these delay and response time problems rarely exist in real commands, whose changes are lower than the ideal ones used in the model identification process.

As determined in [2], the equation of the thrust $F$ is defined as

$$F = aW^2 \tag{12}$$

where $W$ is the motor's velocity in rad/s, and $a$ is obtained as $2.2 \times 10^{-5}\,\mathrm{Ns}^2$.

Finally, by means of the recollected data, the phases RMS current $I_{rms}$ is then related to the battery output current as

$$I_{rms} = -0.053758 I_{bat}^2 + 1.145492 I_{bat} + 0.3. \tag{13}$$

Hence, all the unknown blocks of the proposed model are fully defined. Now, the overall proposed model must be validated through real time experiments as presented in the next section.

## 4.2 Model Validation

Several experiments to validate the obtained model were carried out. These were based on a comparison between the signals obtained from a simulation of the entire vehicle and the data from real-time experimentation on the PMSM. In order to do this, the motor's input from the simulation was defined as the input of the real PMSM and, then, their outputs compared. All the comparison results were similar, so only a particular experiment is presented as we consider it is representative of the group. In the simulation, a robust controller for the 3D position of the vehicle is designed for the dynamic model of the vehicle considering the actuator's dynamics. The 3D reference of the position depicts a horizontal square with constant height. The comparison results of the closed-loop experiment can be appreciated in Fig. 6.

It can be noted that the approximation error in this experiment remains below the 5% in the steady state of the response, for time >3 s approximately. During this steady state, the vehicle performed the turning maneuvers to follow the square

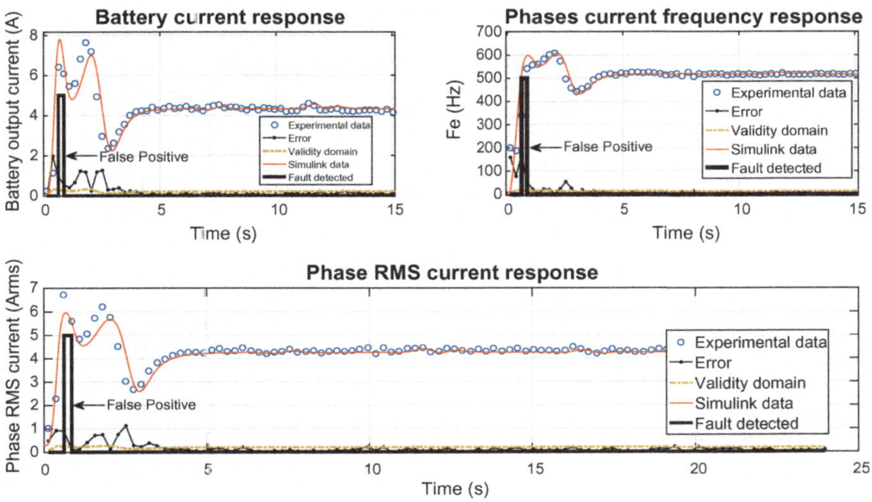


**Fig. 6** Model and experimental outputs (nominal conditions) for a near-real command

reference and, despite this, the error maintained low. However, during the take-off of the vehicle (the transient response) the error was as high as 40% of the reference value. This is expected as a coupled nonlinear system is been approximated by two decoupled linear models. On the other hand, the comparison results were evaluated as satisfactory enough for the considered application.

## 5 Faults Characterization and Detection

According to the literature, the most common faults concerning drone PMSMs are: rapid change of friction coefficients, motor turning off, propellers detaching during flight, partial short-circuits and propeller damage. Each of these faults is presented, experimentally reproduced, its consequences on $F_e$, $I_{bat}$, and $I_{rms}$ determined, and, finally, a strategy is set up in order to detect them. The strategy is based on the characteristic response of the PMSM variables during a particular fault.

### *5.1 High Friction Fault*

This is one of the most common motor fault in a UAV and its generated by a mechanical damage or obstruction in the components of the actuator. As a result, the solid friction between the rotor and the armature of the motor is increased, see [6]. This fault is experimentally generated pressing a rubber object against the external central

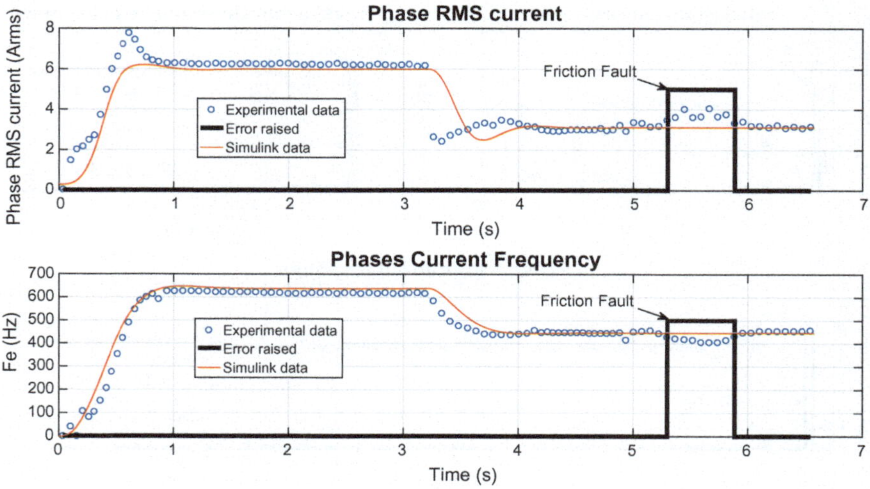


**Fig. 7** Detection of motor friction fault

point of the propeller, soft enough to generate a decreasing in its velocity but without stopping it. The characteristics of this fault are: lower speed and increasing in battery and phase RMS currents. In terms of the variables of the proposed model, if $F_e$ is low, $I_{bat}$ is high, and that the product of their errors between the simulation and the experimental values exceed 25%, the fault is detected. In addition, the frequency $F_e$ must be higher than a threshold (between 200 and 300 Hz is a good choice) while evaluating this fault in order to not take in consideration the error of modelling in low regimens and transient states. Figure 7 shows the detection results for this fault using the proposed characteristics and method. In this figure, the fault is experimentally simulated between time 5.2 and 5.9 s, It can be noted, that the fault is correctly detected and identified.

## 5.2 *Motor-Off Fault*

It is also possible to detect if the motor is accidentally shut down by bad electrical contact, disconnected wire, unbalanced phases, or dead ESC. This error can be identified if the $F_e$ reference is upper than 300 Hz and that the experimental speed is dramatically decreased (lower than 50% of the previous reference). To reproduce this fault, the ESC command wire was disconnected during the experimental test, but it can also be generated by disconnecting a phase wire, In the latter, it is recommended to perform the disconnection before running the test to avoid damaging the motor and its ESC by over tensions. The Fig. 8 shows the accuracy of the proposed features and method, with a phase disconnected. This is more difficult to detect than if the

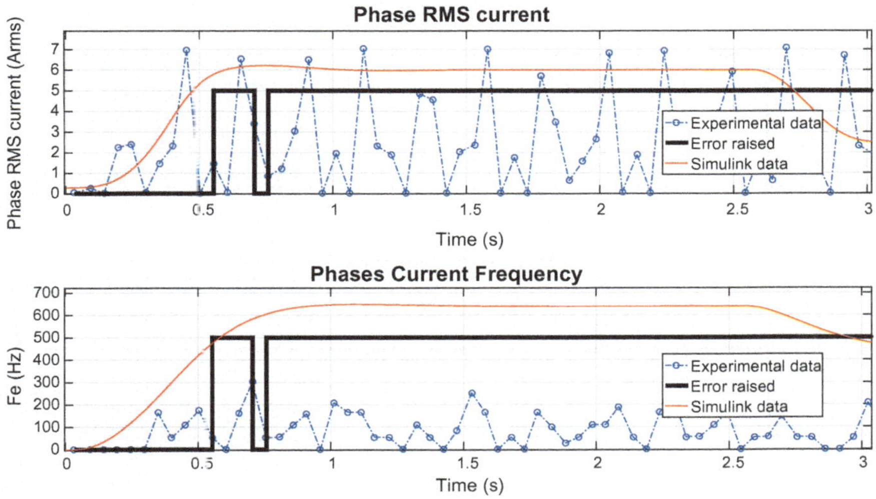


**Fig. 8** Detection of motor off fault

ESC is disconnected because the motor is rotating (in the same way as if the phases are unbalanced) instead of being at rest. The fault is simulated during the whole experimental test and it can be noted that the fault is not detected during the starting acceleration of the motor but it is correctly detected during its deceleration. This is expected as the variables of the proposed model presented the lowest error during the steady state response.

## 5.3 *Detached Propeller Fault*

This fault is the most common for UAVs actuators. It is usually generated as a result of a badly fasten propeller. It can be easily detected using the same approach as for the motor-off fault detection. However, it is isolated only if the $F_e$ frequency is upper than 150% of its reference. This is a straightforward conclusion as the inertia and the load torque for the motor is greatly reduced if this fault is presented. Detecting of this fault is only necessary during flight as a pre-flight visual inspection can easily discard it. The obtained results for the detection of this fault are shown in Fig. 9. The fault is simulated during the whole experimental test, however, the accuracy of the detection is higher during the steady state and presents some variations during its transient response. This is congruent with the analysis of the previous faults results.

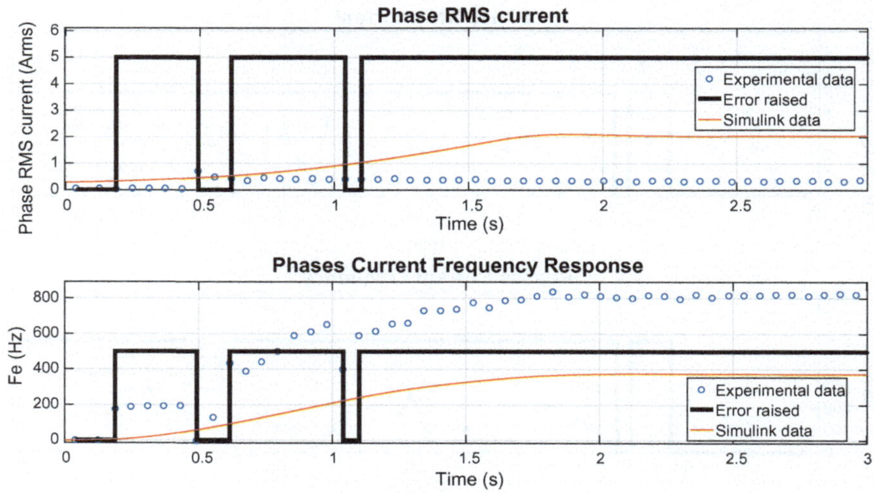


**Fig. 9** Propeller flying off fault

## 5.4 Inter-turn Short-Circuit Fault

PMSM overheating can damage its stator turns and create partial short-circuits during its normal operation. As one of the most common and studied failure, it has been established in [1, 7] that it affects harmonics and RMS values of phases current. An inter-turn is current. An inter-turn is created in a working PMSM by removing the varnish of one coil top turns. According to the amount of tin that is added to these turns, the gravity of short-circuit can be adjusted. Experimentally the consequences of this fault depend on its gravity. This fault lowers the velocity and rises the currents as friction fault does. However, they can be distinguished because the inter-turn short-circuit fault does not change the relation between $I_{bat}$ and $I_{rms}$ on the two phases that are not under short-circuit conditions, while the damaged phase relation between $I_{bat}$ and $I_{rms}$ is affected as in friction fault case: recalculated phase of $I_{rms}$ from experimental $I_{bat}$ is lower than experimental one in these cases. Consequently, at least two phases must be equipped for accurate detection of this fault. The results of this detection are shown in Fig. 10 on the case of a full time, high gravity fault. Again, the accuracy of the detection is higher during steady state.

## 5.5 Damaged Propeller Fault

This type of fault occurs when the propellers of the drone hits a solid object (trees, buildings, poles, etc.) which were not detected by the pilot nor by an embedded anti-collision system [6]. This results in the deformation or breakage of the blades of

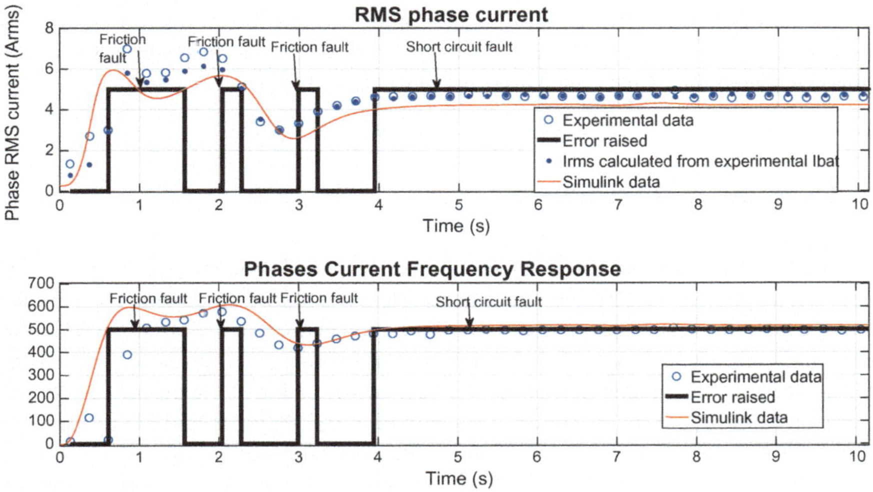


**Fig. 10** Detection of inter-turn short-circuit

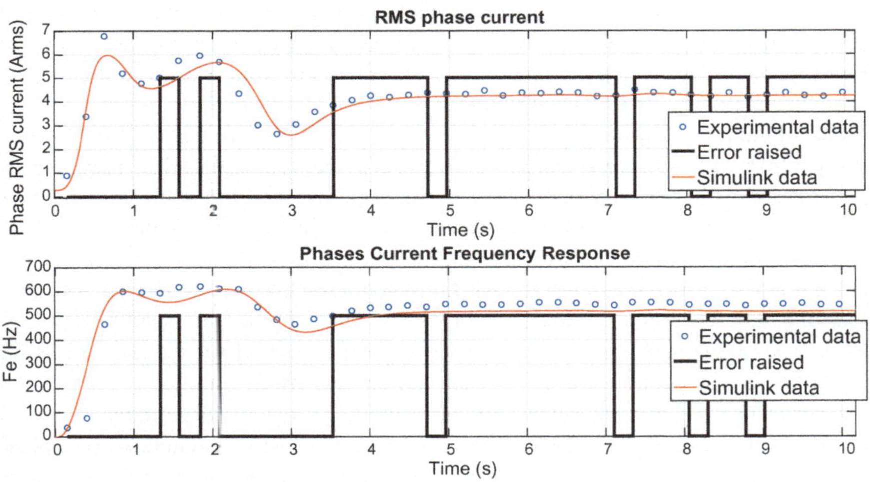


**Fig. 11** Damaged propeller fault detection

the affected propeller and a poor performance of the vehicle's flight. The fault was generated by trimming 2.5 cm of the end of a propeller's blade so it recreates the drone motorization state after the fault occurrence. Experimentally, the consequences of this fault are similar to the detached propeller except that the motor speed is weakly increased and the currents dimly lowered. Accordingly, both faults are detected in the same way and the magnitude of the error values for velocity and currents permit to distinguish between them. These errors can also be used in order to characterize the gravity of the propeller damage, so that other works will be able to redefine the controllers in accordance to the new equations that drive the motorization. Figure 11 shows the detection of the damaged propeller fault only using current sensors.

## 6 Experimental Results Analysis

In order to validate the proposed detection scheme, a black-box model was generated based on a commercial PMSM whose parameters are totally unknown. The simulation results for the model are accurate enough to detect the five faults described in the previous section which are: high friction, motor-off, detached propeller, inter-turn short-circuit, and damaged propeller. The best performance of the proposed scheme is obtained during the steady state as shown in Figs. 6, 7, 8, 9, 10, and 11. In contrast, a poor detection performance was obtained during the transient of the response. This is expected as a linear, decoupled approximation of the nonlinear, coupled actuator was used as a proposed model. Also, it can be noted in Fig. 6 where the difference between the model and the real actuator is bigger during the transient response. On the other hand, the proposed scheme (model and detection criteria) is simple and easy to develop for offline and online applications which are desired features for real time implementations. hence, we consider that the compromise done in this regard evaluates positively the proposed scheme.

The figures of experimental results show that the faults are detected with some false positive occurrences. Nonetheless, these are only featured on low regimens or transient states that are not part of the normal operation of the vehicle's actuators during flight. Hence, they can easily be ignored and the detection scheme activated only during the state of interest (see Fig. 6, where a false positive friction fault is detected at $t = 0.6$ s). Due to practical implementation constrains, the friction fault is simulated between $t = 5.2$ and 5.9 s in Fig. 7 while the other faults are held during their whole experiment. As one detection function is independently set for each fault, the nature and the occurring time for each fault are easily determined.

Figure 10 shows that the inter-turn short-circuit fault is hardly characterized in transient states, and Fig. 11 shows only a partial detection of a damaged propeller. This detection criterion uses the data from a healthy phase in order to detect an inter-turn short-circuit fault. However, it is possible to refine the obtained detection rate by applying post-treatments of the data, using rule-based schemes or inertial sensors data which are available in almost all commercial drones.

It is worth to note that during all the experiments, the five designed detection criteria were activated. Hence, the results also demonstrate the absence of false positives for the proposed global detection scheme and model. This is an important feature for a real time implementation of our proposal.

## 7 Conclusions and Future Work

The presented work proposes and validates a fault detection scheme for drone actuators which are usually PMSM. The scheme considers only the most commons faults: high friction, motor-off, detached propeller, inter-turn short-circuit, and damaged propeller. In addition, a linear, decoupled model based on transfer functions was obtained for the PMSM using experimental data for known inputs and considering current sensors for the phases of the motor. For these five faults, detection criteria were defined based on extensive comparisons between the response of the obtained model and the real actuator with the same input signals.

The designed criteria were tested by artificially generating the faults of interest in a healthy motor. The experimental results showed that the proposed detection scheme is able to detect all the proposed faults, independently. The detection accuracy is high during the steady state response but can be improved for the transient state of the experiment. This can be done by considering other common sensors of the drone as inertial sensors or GPS or increasing the number of the phase current sensors. This, definitively would improve the inter-turn short-circuit and damaged propeller faults detection.

The proposal is composed by two stages: the modelling phase developed offline and the detection phase developed online. The modelling phase is simple as it is based on transfer functions identification. The detection phase is simple and fast as the criteria are constructed basically using thresholds. This eases its embedded implementation as it is not time or computation demanding, hence, permitting its application in most of the commercial drones with open access to the main microcontroller. Moreover, the necessary current sensors are low cost and easy to integrate in the global architecture of the vehicle.

Further work will be oriented on treating experimental and simulated data on real time during flight. It would permit to implement fault-reactive controllers for the vehicle. In addition, transient states may be better characterized using non-linear techniques for pattern identification. As mentioned before, this always imposes a compromise between the complexity of the proposed scheme and its real time embedded implementation.

Finally, the fusion of all the available sensors of the vehicle would improve the detection of the considered set of faults in steady and transient state. Among these sensors, the inertial measurement unit, barometric sensor, GPS sensor and vision-based sensors are considered.

**Acknowledgements** This research is funded by National Council of Science and Technology (CONACyT) of México under grants 261774 and 227601.

## References

1. Alvarez-Gonzalez, F., Griffo, A., Sen, B., Wang, J.: Real-time hardware-in-the-loop simulation of permanent-magnet synchronous motor drives under stator faults. IEEE Trans. Ind. Electron. **64**(9), 6960–6969 (2017)
2. Bouabdallah, S., Siegwart, R.: Backstepping and sliding-mode techniques applied to an indoor micro quadrotor. In: 2005 IEEE International Conference on Robotics and Automation ICRA, pp. 2247–2252. IEEE (2005)
3. Carrillo, L.R.G., Lopez, A.E.D., Lozano, R., Pegard, C.: Quad-rotor control (2011)
4. Kayalvizhi, M., Akilandeswari, M.: Design and implementation of speed regulator for a PMSM using genetic algorithm. Int. J. Innov. Res. Sci. Eng. Technol. **3**(1), 866–872 (2014)
5. Lakshminarayan, I.: Model-based fault detection for low-cost UAV actuators. Ph.D. thesis, University of Minnesota (2016)
6. Li, C., Zhang, Y., Li, P.: Extreme learning machine based actuator fault detection of a quadrotor helicopter. Adv. Mech. Eng. **9**(6), 1–10 (2017)
7. Liu, B.: Fault detection of brushless permanent magnet machine drives. Ph.D. thesis, University of Sheffield (2014)
8. Meier, L., Tanskanen, P., Fraundorfer, F., Pollefeys, M.: The PIXHAWK open-source computer vision framework for MAVs. Int. Arch. Photogramm. Remote Sens. Spat. Inf. Sci. **38**(1), 13–18
9. Pillay, P., Krishnan, R.: Modeling of permanent magnet motor drives. IEEE Trans. Ind. Electron. **35**(4), 537–541 (1988)
10. Ruiz, J.R.R., Rosero, J.A., Espinosa, A.G., Romeral, L.: Detection of demagnetization faults in permanent-magnet synchronous motors under nonstationary conditions. IEEE Trans. Magn. **45**(7), 2961–2969 (2009)
11. Singh, S., Kumar, A., Kumar, N.: Detection of bearing faults in mechanical systems using motor current signature and acoustic signatures. In: 21st International Congress on Sound Vibration, pp. 1–8 (2014)
12. Medeiros, R.L.V., Ramos, J.G.G.S., Nascimento, T.P., Lima Filho, A.C., Brito, A.V.: A novel approach for brushless DC motors characterization in drones based on chaos. Drones **2**(2), 14 (2018)

# ENMPC Versus PID Control Strategies Applied to a Quadcopter

**Nadia Miladi, Taoufik Ladhari and Salim Hadj Said**

**Abstract** Trajectory tracking is a necessary function for an unmanned aerial vehicle (UAV). In this context, we investigate in this chapter the path following problems for the quadcopter dynamic system, which is coupled, underactuated and highly nonlinear. We adopt the Newton–Euler approach to describe the quadcopter model. Two continuous time strategies of control are presented to solve this issue: the technique based on cascade proportional-integral-derivative (PID) controllers and the explicit nonlinear model predictive control (ENMPC) technique. A comparison through numerical simulations, between the performances resulting from these two control strategies in terms of helical trajectory tracking, shows that the ENMPC technique is more effectiveness than the technique based on the PID controllers.

**Keywords** Quadcopter system · Cascade PID · Explicit NMPC · Trajectory tracking

## 1 Introduction

Nowadays, with the calculators evolution and the appearance of a variety of software tools, the implementation of complex control algorithms becomes feasible for hard and coupled nonlinear plant.

Among these complex systems we find the quadcopter, which is small helicopter equipped with four rotors. This kind of vehicles is an underactuated mechanical system, because it has six degrees of freedom and only four actions (four rotors), it has also a strongly coupled and a highly nonlinear dynamic model. Despite that,

N. Miladi (✉) · T. Ladhari · S. H. Said
Departement of Electrical Engineering, National Engineering School of Monastir, LA2SE, 5019 Monastir, Tunisia
e-mail: nadia_miladi@hotmail.fr

T. Ladhari
e-mail: taoufik_ladhari@hotmail.com

S. H. Said
e-mail: salim.hadjsaid@issig.rnu.tn

 
J. Ghommam et al. (eds.), *New Trends in Robot Control*, Studies in Systems, Decision and Control 270, https://doi.org/10.1007/978-981-15-1819-5_16

the quadcopter has many advantages compared by the helicopter, where the most important of which is the simplicity of its construction and maintenance. Thanks to these benefits, the quadcopter was occupying different sectors of activity such as: agriculture, mapping, inspections, search and rescue, surveillance and trade like the delivery quadcopter. This kind of drone requires a trajectory tracking algorithm based on techniques of control to deliver the goods to the client.

Many works were focused on the tracking control of the quadcopter are available in the literature of automatic control. In Hoffmann et al. [3], the authors used the proportional-integral-derivative (PID) controller to fix the trajectory tracking of the quadcopter. In Ganga and Dharmana [2], a linear model predictive control (MPC) was proposed to achieve the path following. In Islam et al. [4], a linear Quadratic regulator (LQR) was designed to control the quadcopter. In Raffo et al. [8], the author combined two strategies of control: a predictive technique which was used to control the quadcopter position dynamic in an outer-loop, with a nonlinear $H_\infty$ technique for the angular dynamic which was proposed in an inner-loop to achieve the quadcopter stabilization. The trajectory tracking control problems for the quadcopter were solved using a feedback linearization controller and an adaptive sliding mode controller in Lee et al. [5]. In Yu et al. [10], the author suggested a robust backstepping controller for uncertain multi-input multi-output (MIMO) nonlinear system and they applied the proposed strategy on a quadrotor UAV platform. In Liu et al. [7], a robust decentralized and linear time-invariant controller was designed to achieve the trajectory tracking, which consists of three loops: an attitude loop to obtain the desire angle tracking, a position loop to achieve the position tracking and a robust compensating loop to limit the effects of uncertainties in the rotational and translational dynamics. In Rios et al. [9], a continuous sliding mode control was proposed to solve the path following problem.

The majority of these methods require a large number of design parameters which should be well adjusted. In this paper, we propose two techniques of control, whose purpose of each technique is to solve the problems of the trajectory tracking for the quadcopter. The first strategy is cascade PID where the position and the altitude controllers are used in the outer loop, whereas the attitude controllers are adopted in the inner loop. The second controller design is the explicit nonlinear model predictive control (ENMPC), which is a form of nonlinear MPC (NMPC). The ENMPC technique was explored in continuous time in order to solve the problem of the NMPC technique, generally need a high computational power. This high computational power needs is due to the on-line optimization that is normally required for implementing a NMPC strategy. While the ENMPC technique don't need to the on-line optimization for its implementation. In this way the problem of NMPC strategy is solved in the ENMPC approach. The purpose of this paper is to compare the performance of these two techniques to solve the trajectory tracking problem for the quadcopter.

The paper is organized as follows: In Sect. 2, we introduce the quadcopter model. In Sect. 3, we describe the method based on the cascade PID controllers . In Sect. 4, we present the algorithm of ENMPC and its application on the quadcopter. A comparison study is given in Sect. 5 that illustrates the tracking performances arising from the both controllers. Finally we conclude with some remarks in Sect. 6.

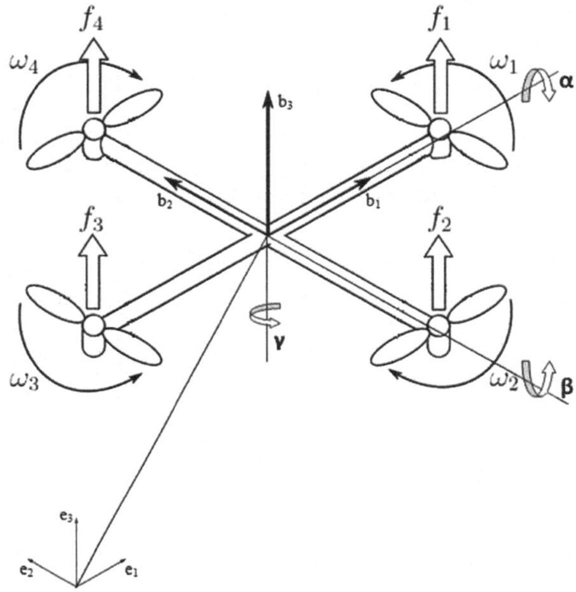


**Fig. 1** Quadcopter configuration scheme

## 2 System Modelling

In this section, the Kinematics and dynamics models of a quad-copter will be derived based on a Newton–Euler formalism [1].

### 2.1 Kinematics Model

The quadcopter's dynamics are defined using two frames of reference: the body fixed frame $B = \{b_1, b_2, b_3\}$ and the earth fixed frame "the inertial frame" $E = \{e_1, e_2, e_3\}$. As shown in Fig. 1

The vector $\xi = [x\ y\ z]'$ is the position of the quadcopter mass center expressed in the inertial frame.

The vector $\eta = [\alpha\ \beta\ \gamma]'$ represents the quadcopter's Euler angles (the orientation of the quadcopter) in the inertial frame.

The rotation matrix from the body frame to the inertial frame is

$$R = \begin{bmatrix} C\gamma C\beta & C\gamma S\beta S\alpha - S\gamma C\alpha & C\gamma S\beta C\alpha + S\gamma S\alpha \\ S\gamma C\beta & S\gamma S\beta S\alpha + C\gamma C\alpha & S\gamma S\beta C\alpha - C\gamma S\alpha \\ -S\beta & C\beta S\alpha & C\beta C\alpha \end{bmatrix} \tag{1}$$

where $C. = cos(.)$ and $S. = sin(.)$

The translational kinematic can be written as:

$$v_E = R.v_B \tag{2}$$

where $v_E$ and $v_B$ are the linear velocities of the quadcopter mass center expressed in the inertial frame and the body fixed frame, respectively.

The rotational kinematic can be written as:

$$\dot{\eta} = W_\eta^{-1}.\omega$$
$$\begin{bmatrix} \dot{\alpha} \\ \dot{\beta} \\ \dot{\gamma} \end{bmatrix} = \begin{bmatrix} 1 & S\alpha \tan\beta & C\alpha \tan\beta \\ 0 & C\alpha & -S\alpha \\ 0 & S\alpha \sec\beta & C\alpha \sec\beta \end{bmatrix} \begin{bmatrix} p \\ q \\ r \end{bmatrix} \tag{3}$$

where $\omega = [p \quad q \quad r]^T$ is the angular velocity vector expressed in the body fixed frame.

## 2.2 Dynamics Model

The quadcopter is supposed to be rigid body, so we can use Newton–Euler equations to describe its dynamics.

### 2.2.1 Rotational Equations of Motion

The rotational equations of motion are expressed in the body frame, based on the Newton–Euler approach, with the following expression:

$$I\dot{\omega} + \omega \wedge (I\omega) + \Gamma = \tau \tag{4}$$

where

I quadcopter diagonal inertia matrix
$\Gamma$ the gyroscopic forces
$\tau$ the external torque

$$\begin{bmatrix} \dot{p} \\ \dot{q} \\ \dot{r} \end{bmatrix} = \begin{bmatrix} (I_{yy} - I_{zz})qr/I_{xx} \\ (I_{zz} - I_{xx})pr/I_{yy} \\ (I_{xx} - I_{yy})pq/I_{zz} \end{bmatrix} - I_r \begin{bmatrix} q/I_{xx} \\ -p/I_{yy} \\ 0 \end{bmatrix} \omega_\Gamma + \begin{bmatrix} U_2/I_{xx} \\ U_3/I_{yy} \\ U_4/I_{zz} \end{bmatrix}. \tag{5}$$

The angular accelerations in the inertial frame are:

$$\ddot{\eta} = \frac{d}{dt}(W_\eta^{-1}\omega) = \frac{d}{dt}(W_\eta^{-1})\omega - W_\eta^{-1}\dot{\omega} \tag{6}$$

where

$$\frac{d}{dt}(W_\eta^{-1}) = \begin{bmatrix} 0 & \dot{\alpha}C\alpha T\beta + \dot{\beta}S\alpha/C^2\beta & -\dot{\alpha}S\alpha C\beta + \dot{\beta}C\alpha/C^2\beta \\ 0 & -\dot{\alpha}S\alpha & -\dot{\alpha}C\alpha \\ 0 & \dot{\alpha}C\alpha/C\beta + \dot{\alpha}S\alpha T\beta/C\beta & -\dot{\alpha}S\alpha/C\beta + \dot{\beta}C\alpha T\beta/C\beta \end{bmatrix}.$$

### 2.2.2 Translational Equations of Motion

The translation equations of motion for the quadrotor are based on Newton's second law and they are derived in the Earth inertial frame

$$m\ddot{\xi} = \begin{bmatrix} 0 \\ 0 \\ -mg \end{bmatrix} + R \begin{bmatrix} 0 \\ 0 \\ U_1 \end{bmatrix}$$

$$\begin{bmatrix} \ddot{x} \\ \ddot{y} \\ \ddot{z} \end{bmatrix} = \begin{bmatrix} 0 \\ 0 \\ -g \end{bmatrix} + \frac{U_1}{m} \begin{bmatrix} C\gamma S\beta C\alpha + S\gamma S\alpha \\ S\gamma S\beta C\alpha - C\gamma S\alpha \\ C\beta C\alpha \end{bmatrix} \tag{7}$$

where $g$ is the acceleration due to the gravity, $m$ is the mass of the quadcopter and $U_1$ is the thrust force.

The quadcopter model can be expressed in a general affine form:

$$\begin{cases} \dot{X} = f(X) + g(X)u \\ Y = h(X) \end{cases} \tag{8}$$

where $X = [\alpha \;\; \dot{\alpha} \;\; \beta \;\; \dot{\beta} \;\; \gamma \;\; \dot{\gamma} \;\; x \;\; \dot{x} \;\; y \;\; \dot{y} \;\; z \;\; \dot{z}]^T$ is the quadcopter state and $u = [U_1 \;\; U_2 \;\; U_3 \;\; U_4]^T$ is the control input. In the trajectory tracking control of a quadcopter, the outputs interest are the position and heading angle. Thus let $Y = [x \;\; y \;\; z \;\; \gamma]^T$.

## 3 The Technique Based on the PID Controllers

In this section we will use the PID controller to achieve the trajectory tracking. For that first of all it is necessary to linearize the quadcopter model around the working point $(X_0, u_0)$.

We choose $X_0 = [0_{12\times1}]$ and $u_0 = [U_{10}, U_{20}, U_{30}, U_{40}]^T$ where the values of $U_{i0}$ for $i = 1, \ldots, 4$ verify this condition $\dot{X} = f(X_0) + g(X_0)u_0 = [0_{12\times1}]$.

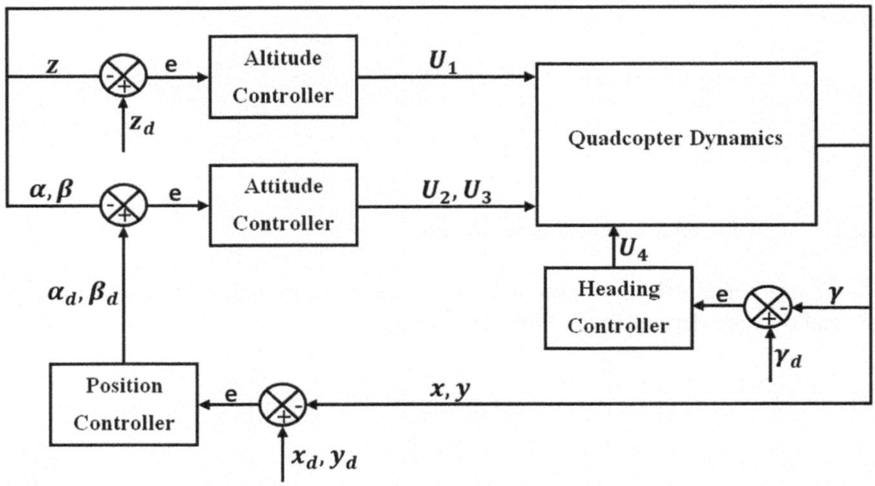


**Fig. 2** The control structure of the quadcopter based on the PID controller

Then the linearized model of the quadcopter is given by:

$$\begin{cases} \ddot{\alpha} = \dfrac{U_2}{I_{xx}} \quad ; \quad \ddot{\beta} = \dfrac{U_3}{I_{yy}} \quad ; \quad \ddot{\gamma} = \dfrac{U_4}{I_{zz}} \\ \ddot{x} = 0 \ ; \quad \ddot{y} = 0 \ ; \quad \ddot{z} = -g + \dfrac{U_1}{m} \end{cases} . \tag{9}$$

The control structure of the quadcopter based on the PID controllers is illustrated in Fig. 2, where each controller block is described in the following.

- **Altitude Controller**

A PID controller, whose expression is below, is developed to control $z$:

$$\ddot{z}_d = K_p(z_d - z) + K_d(\dot{z}_d - \dot{z}) + K_i \int (z_d - z)dt. \tag{10}$$

From the system (9) and the Eq. (10), we can deduce the necessary value of the thrust force $U_1$, which is given by:

$$U_1 = m(\ddot{z}_d + g). \tag{11}$$

- **Heading Controller**

To control the yaw angle $\gamma$, we use another PID controller whose expression is as follows:

$$\ddot{\gamma}_d = K_p(\gamma_d - \gamma) + K_d(\dot{\gamma}_d - \dot{\gamma}) + K_i \int (\gamma_d - \gamma)dt. \tag{12}$$

By invoking the system (9) and the Eq. (12), we can deduce the necessary value of $U_4$, which is expressed by:

$$U_4 = I_{zz}\ddot{\gamma}_d. \tag{13}$$

- **Attitude Controller**

Two others PID controllers, are used to control the roll and pitch angles $\alpha$ and $\beta$, have the expressions as follows:

$$\ddot{\alpha}_d = K_p(\alpha_d - \alpha) + K_d(\dot{\alpha}_d - \dot{\alpha}) + K_i \int (\alpha_d - \alpha)dt. \tag{14}$$

$$\ddot{\beta}_d = K_p(\beta_d - \beta) + K_d(\dot{\beta}_d - \dot{\beta}) + K_i \int (\beta_d - \beta)dt. \tag{15}$$

From the system (9) and the Eqs. (14) and (15), we can deduce the necessary values of $U_2$ and $U_3$, which are expressed by:

$$U_2 = I_{xx}\ddot{\alpha}_d. \tag{16}$$

$$U_3 = I_{yy}\ddot{\beta}_d. \tag{17}$$

- **Position Controller**

The desired roll and pitch angles $\alpha_d$, $\beta_d$ are not available, from the equations below we can calculate them as follows:

$$\ddot{x}_d = \frac{U_1}{m}(\cos\gamma \sin\beta_d \cos\alpha_d + \sin\gamma \sin\alpha_d). \tag{18}$$

$$\ddot{y}_d = \frac{U_1}{m}(\sin\gamma \sin\beta_d \cos\alpha_d - \cos\gamma \sin\alpha_d). \tag{19}$$

As the quadcopter is hovering around the working point, we can assume that: $sin\alpha_d = \alpha_d$ ; $sin\beta_d = \beta_d$ ; $cos\alpha_d = cos\beta_d = 1$ (small angle assumption).

Then, we can rewrite the Eqs. (18) and (19) as:

$$\ddot{x}_d = \frac{U_1}{m}(\beta_d \cos\gamma + \alpha_d \sin\gamma). \tag{20}$$

$$\ddot{y}_d = \frac{U_1}{m}(\beta_d \sin\gamma - \alpha_d \cos\gamma). \tag{21}$$

The matrix form of the system, which is composed of Eqs. (20) and (21) is given by:

$$\frac{m}{U_1}\begin{bmatrix} \ddot{x}_d \\ \ddot{y}_d \end{bmatrix} = \begin{bmatrix} \sin\gamma & \cos\gamma \\ -\cos\gamma & \sin\gamma \end{bmatrix} \begin{bmatrix} \alpha_d \\ \beta_d \end{bmatrix}. \tag{22}$$

Then:

$$\begin{bmatrix}\alpha_d \\ \beta_d\end{bmatrix} = \frac{m}{U_1}\begin{bmatrix}\sin\gamma & -\cos\gamma \\ \cos\gamma & \sin\gamma\end{bmatrix}\begin{bmatrix}\ddot{x}_d \\ \ddot{y}_d\end{bmatrix} = \frac{m}{U_1}\begin{bmatrix}\ddot{x}_d sin\gamma - \ddot{y}_d cos\gamma \\ \ddot{x}_d cos\gamma + \ddot{y}_d sin\gamma\end{bmatrix}. \tag{23}$$

To calculate $\ddot{x}_d$ and $\ddot{y}_d$, two PID controllers are used, so that:

$$\ddot{x}_d = K_p(x_d - x) + K_d(\dot{x}_d - \dot{x}) + K_i \int (x_d - x)dt. \tag{24}$$

$$\ddot{y}_d = K_p(y_d - y) + K_d(\dot{y}_d - \dot{y}) + K_i \int (y_d - y)dt. \tag{25}$$

# 4 Explicit Nonlinear Model Predictive Control

In this section, the algorithm of ENMPC and its implementation on the quadcopter are discussed in detail.

## *4.1 The Algorithm of ENMPC*

Trajectory tracking is among the necessary functions of quadcopter. To realize this function, many control techniques exist. In this study ENMPC strategy [6] is proposed where the tracking control can be achieved by minimising a receding horizon performance index.

$$J = \frac{1}{2}\int_0^T (\widehat{y}(t+\tau) - \widehat{w}(t+\tau))^T(\widehat{y}(t+\tau) - \widehat{w}(t+\tau))d\tau. \tag{26}$$

For a nonlinear MIMO system, the $i$th output in the receding horizon can be approximated by its Taylor series expansion up to order $\rho_i$ as follows:

$$\begin{aligned}\widehat{y}_i(t+\tau) &\approx y_i(t) + \tau\dot{y}_i(t) + \cdots + \frac{\tau^{\rho_i}}{\rho_i!}y_i^{(\rho_i)}(t) \\ &= \begin{bmatrix}1 & \tau & \ldots & \frac{\tau^{\rho_i}}{\rho_i!}\end{bmatrix}[y_i(t) \quad \dot{y}_i(t) \quad \ldots \quad y_i^{(\rho_i)}]^T\end{aligned} \tag{27}$$

where $0 \le \tau \le T$ and $\rho_i$ is the relative degree of the $i$th output. Thus we can write:

$$\widehat{y}(t+\tau) = \begin{bmatrix} \widehat{y}_1(t+\tau) \\ \widehat{y}_2(t+\tau) \\ \vdots \\ \widehat{y}_m(t+\tau) \end{bmatrix} = [T_f \quad T_s] \begin{bmatrix} \bar{Y}_1(t) \\ \vdots \\ \bar{Y}_m(t) \\ y_1^{(\rho_1)}(t) \\ \vdots \\ y_m^{(\rho_m)}(t) \end{bmatrix} \tag{28}$$

where

$\bar{Y}_i = [y_i(t) \quad \dots \quad y_i^{(\rho_i-1)}], i = 1, \dots, m$

$T_s = \text{diag}\begin{bmatrix} \frac{\tau^{\rho_1}}{\rho_1!} & \dots & \frac{\tau^{\rho_m}}{\rho_m!} \end{bmatrix}$

and $T_f = \begin{bmatrix} \bar{\tau}_1 & \dots & 0_{1\times\rho_m} \\ \dots & \dots & \dots \\ 0_{1\times\rho_1} & \dots & \bar{\tau}_m \end{bmatrix}$ with $\bar{\tau}_i = \begin{bmatrix} 1 & \tau & \dots & \frac{\tau^{(\rho_i-1)}}{(\rho_i-1)!} \end{bmatrix}$.

In the same way, the reference trajectories in the receding horizon $\omega_i(t+\tau)$ where $0 \le \tau \le T$ can also be approximated by:

$$\widehat{w}(t+\tau) = \begin{bmatrix} \widehat{w}_1(t+\tau) \\ \widehat{w}_2(t+\tau) \\ \vdots \\ \widehat{w}_m(t+\tau) \end{bmatrix} = [T_f \quad T_s] \begin{bmatrix} \bar{W}_1(t) \\ \vdots \\ \bar{W}_m(t) \\ w_1^{(\rho_1)}(t) \\ \vdots \\ w_m^{(\rho_m)}(t) \end{bmatrix}. \tag{29}$$

By using the lie notation, we obtain:

$$\begin{bmatrix} y_1^{(\rho_1)} \\ y_2^{(\rho_2)} \\ \vdots \\ y_m^{(\rho_m)} \end{bmatrix} = \begin{bmatrix} L_f^{\rho_1} h_1(X) \\ L_f^{\rho_2} h_2(X) \\ \vdots \\ L_f^{\rho_m} h_m(X) \end{bmatrix} + A(X)\bar{u}. \tag{30}$$

According to Eqs. (26), (28) and (29), we can rewrite the cost function as:

$$J = \frac{1}{2}(\bar{Y}(t) - \bar{W}(t))^T \begin{bmatrix} T_1 & T_2 \\ T_2^T & T_3 \end{bmatrix} (\bar{Y}(t) - \bar{W}(t)) \tag{31}$$

where

$$\bar{Y}(t) = \begin{bmatrix} \bar{Y}_1 \dots \bar{Y}_m \; y_1^{(\rho_1)} \dots y_m^{(\rho_m)} \end{bmatrix}^T$$
$$\bar{W}(t) = \begin{bmatrix} \bar{W}_1 \dots \bar{W}_m \; w_1^{(\rho_1)} \dots w_m^{(\rho_m)} \end{bmatrix}^T$$

$T_1 = \int_0^T T_f^T T_f d\tau; \quad T_2 = \int_0^T T_f^T T_s d\tau; \quad T_3 = \int_0^T T_s^T T_s d\tau.$

Thus, the control law is elaborated from the minimization with respect to $u$ of the approximated index Eq. (31)

$$\frac{\partial J}{\partial \bar{u}} = 0. \tag{32}$$

After solving the equality Eq. (32), the control law is explicitly obtained as follows:

$$\bar{u} = -A(X)^{-1}(KM_\rho + M_1) \tag{33}$$

where $K = T_3^{-1} T_2^T \in \mathbb{R}^{m \times \sum_{i=1}^m \rho_i}$ with the $i$th block of $T_2$ is $\rho_i \times m$ matrix, and all its elements are zeros except the $i$th column is given by:

$$\left[ \frac{T^{(\rho_i+1)}}{\rho_i!(\rho_i+1)} \cdots \frac{T^{(2\rho_i)}}{\rho_i!(\rho_i-1)!(2\rho_i)} \right]^T$$

for $i = 1, \ldots, m$ and $T_3$ is given by

$$\text{diag}\left\{ \frac{T^{2\rho_1+1}}{\rho_1!\rho_1!(2\rho_1+1)} \cdots \frac{T^{2\rho_m+1}}{\rho_m!\rho_m!(2\rho_m+1)} \right\}.$$

The matrix $M_\rho \in \mathbb{R}^{\sum_{i=1}^m \rho_i}$ and matrix $M_1 \in \mathbb{R}^m$ are defined as:

$$M_\rho = \begin{bmatrix} \bar{Y}_1(t) \\ \vdots \\ \bar{Y}_m(t) \end{bmatrix} - \begin{bmatrix} \bar{W}_1(t) \\ \vdots \\ \bar{W}_m(t) \end{bmatrix}$$

$$M_1 = \begin{bmatrix} L_f^{\rho_1} h_1(t) \\ L_f^{\rho_2} h_2(t) \\ \vdots \\ L_f^{\rho_m} h_m(t) \end{bmatrix} - \begin{bmatrix} w_1^{(\rho_1)}(t) \\ w_2^{(\rho_2)}(t) \\ \vdots \\ w_m^{(\rho_m)}(t) \end{bmatrix}.$$

## 4.2 Application of the ENMPC on the Quadcopter Model

The ENMPC strategy is based on the derivation of each output until the control inputs appears.

The 1st and the 2nd derivatives with respect to the time allow to have:

$$\begin{bmatrix} \dot{x} \\ \dot{y} \\ \dot{z} \end{bmatrix} = v_E \tag{34}$$

$$\dot{\gamma} = q\frac{\sin\alpha}{\cos\beta} + r\frac{\cos\alpha}{\cos\beta} \tag{35}$$

$$\begin{bmatrix} \ddot{x} \\ \ddot{y} \\ \ddot{z} \end{bmatrix} = \begin{bmatrix} 0 \\ 0 \\ -g \end{bmatrix} + R\begin{bmatrix} 0 \\ 0 \\ \frac{U_1}{m} \end{bmatrix} \tag{36}$$

$$\begin{aligned} \ddot{\gamma} = {} & \frac{\sin\alpha}{I_{yy}\cos\beta}[(I_{zz} - I_{xx})pr + I_r p\omega_\Gamma] + q\dot{\alpha}\frac{\cos\alpha}{\cos\beta} + q\dot{\beta}\frac{\sin\alpha\sin\beta}{\cos^2\beta} \\ & + \frac{\cos\alpha}{I_{zz}\cos\beta}[(I_{xx} - I_{yy})pq] - r\dot{\alpha}\frac{\sin\alpha}{\cos\beta} + r\dot{\beta}\frac{\cos\alpha\sin\beta}{\cos^2\beta} \\ & + \frac{\sin\alpha}{I_{yy}\cos\beta}U_3 + \frac{\cos\alpha}{I_{zz}\cos\beta}U_4. \end{aligned} \tag{37}$$

We remark that the control input $U_1$ appears in Eq. (36), whereas the other control inputs do not. So we have to continue differentiating the position outputs. To simplify the time derivation, we use the relationship $\dot{R} = RS$ with $S \in \mathbb{R}^{3\times 3}$ is an anti-symmetric matrix given as follows:

$$S = \begin{bmatrix} 0 & -r & q \\ r & 0 & -p \\ -q & p & 0 \end{bmatrix}.$$

Thus, the 3rd and 4th derivatives of the position outputs are given by:

$$\begin{bmatrix} x^{(3)} \\ y^{(3)} \\ z^{(3)} \end{bmatrix} = RS\begin{bmatrix} 0 \\ 0 \\ \frac{U_1}{m} \end{bmatrix} + R\begin{bmatrix} 0 \\ 0 \\ \frac{\dot{U}_1}{m} \end{bmatrix}. \tag{38}$$

$$\begin{aligned} \begin{bmatrix} x^{(4)} \\ y^{(4)} \\ z^{(4)} \end{bmatrix} = {} & RS^2\begin{bmatrix} 0 \\ 0 \\ \frac{U_1}{m} \end{bmatrix} + 2RS\begin{bmatrix} 0 \\ 0 \\ \frac{\dot{U}_1}{m} \end{bmatrix} + \frac{U_1}{m}R\begin{bmatrix} \frac{(I_{zz}-I_{xx})pr}{I_{yy}} + \frac{I_r p}{I_{yy}}\omega_\Gamma \\ \frac{(I_{zz}-I_{yy})pr}{I_{yy}} + \frac{I_r q}{I_{xx}}\omega_\Gamma \\ 0 \end{bmatrix} \\ & + R\begin{bmatrix} 0 \\ 0 \\ \frac{\ddot{U}_1}{m} \end{bmatrix} + \frac{U_1}{m}R\begin{bmatrix} \frac{U_3}{I_{yy}} \\ -\frac{U_2}{I_{xx}} \\ 0 \end{bmatrix}. \end{aligned} \tag{39}$$

At this stage, the control inputs explicitly appear in Eq. (39). Then, the vector relative degree for the quadcopter model is $\rho = [4 \quad 4 \quad 4 \quad 2]^T$.

Therefore we have:

$$A(X) = \begin{bmatrix} \frac{C\gamma S\beta C\alpha + S\gamma S\alpha}{m} & -\frac{U_1(S\alpha C\gamma S\beta - C\alpha S\gamma)}{mI_{xx}} & \frac{U_1 C\gamma C\beta}{mI_{yy}} & 0 \\ \frac{S\gamma S\beta C\alpha - C\gamma S\alpha}{m} & -\frac{U_1(S\alpha S\gamma S\beta + C\alpha C\gamma)}{mI_{xx}} & \frac{U_1 S\gamma C\beta}{mI_{yy}} & 0 \\ \frac{C\beta C\alpha}{m} & -\frac{U_1 S\alpha C\beta}{mI_{xx}} & -\frac{U_1 S\beta}{mI_{yy}} & 0 \\ 0 & 0 & \frac{S\alpha}{I_{yy} C\beta} & \frac{C\alpha}{I_{zz} C\beta} \end{bmatrix}$$

where

$$\bar{u} = [\ddot{U}_1 \quad U_2 \quad U_3 \quad U_4]^T$$

$$T_2 = \begin{bmatrix} \frac{T^5}{120} & 0 & 0 & 0 \\ \frac{T^6}{144} & 0 & 0 & 0 \\ \frac{T^7}{336} & 0 & 0 & 0 \\ \frac{T^8}{1152} & 0 & 0 & 0 \\ 0 & \frac{T^5}{120} & 0 & 0 \\ 0 & \frac{T^6}{144} & 0 & 0 \\ 0 & \frac{T^7}{336} & 0 & 0 \\ 0 & \frac{T^8}{1152} & 0 & 0 \\ 0 & 0 & \frac{T^5}{120} & 0 \\ 0 & 0 & \frac{T^6}{144} & 0 \\ 0 & 0 & \frac{T^7}{336} & 0 \\ 0 & 0 & \frac{T^8}{1152} & 0 \\ 0 & 0 & 0 & \frac{T^3}{6} \\ 0 & 0 & 0 & \frac{T^4}{8} \end{bmatrix} \quad T_3 = \begin{bmatrix} \frac{T^9}{5184} & 0 & 0 & 0 \\ 0 & \frac{T^9}{5184} & 0 & 0 \\ 0 & 0 & \frac{T^9}{5184} & 0 \\ 0 & 0 & 0 & \frac{T^4}{2} \end{bmatrix}.$$

## 5 Simulation Results

Simulations are performed in order to compare between the two control strategies: ENMPC technique and the technique based on the PID controllers.

The parameters of the quadcopter model used for the simulations are taken as in Table 1.

The proposed reference trajectory used is a vertical helix defined by:

$$x_d = \frac{1}{2}\cos\left(\frac{t}{2}\right)m; \quad y_d = \frac{1}{2}\sin\left(\frac{t}{2}\right)m; \quad z_d = \frac{t}{10}m; \quad \gamma_d = \frac{\pi}{3}rad.$$

The initial condition of the quadcopter state is chosen as:

$$X(0) = [0, 0, 0, 0, 0.5, 0, 0, 0, 0, 0, 0.5, 0].$$

**Table 1** Quadcopter parameters

| Parameter | Value | Unit |
|---|---|---|
| g | 9.81 | $m/s^2$ |
| m | 0.468 | kg |
| l | 0.225 | m |
| k | $2.98 \cdot 10^{-6}$ | |
| b | $1.14 \cdot 10^{-7}$ | |
| $I_r$ | $3.357 \cdot 10^{-5}$ | $kg\,m^2$ |
| $I_{xx}$ | $4.856 \cdot 10^{-3}$ | $kg\,m^2$ |
| $I_{yy}$ | $4.856 \cdot 10^{-3}$ | $kg\,m^2$ |
| $I_{zz}$ | $8.801 \cdot 10^{-3}$ | $kg\,m^2$ |

**Table 2** Quadcopter parameters

| States | $K_P$ | $K_I$ | $K_D$ |
|---|---|---|---|
| $\alpha$ | 1.75 | 0.005 | 7 |
| $\beta$ | 1.75 | 0.005 | 6 |
| $\gamma$ | 0.25 | 0.008693 | 0.9739 |
| $x$ | 1.5 | 0.009 | 5 |
| $y$ | 1.5 | 0.015 | 5 |
| $z$ | 0.5 | 0.008693 | 1 |

The parameters of the PID controllers are shown in Table 2, which were tuned through many simulations on simulink.

The sole ENMPC parameter (which is the prediction horizon) is adjusted as follows: $T = 0.5$ s.

The performances for both the techniques are illustrated in Figs. 3 and 4. Figure 3 represents the way in which the quadcopter tracks the reference trajectory (in the 3D space) in both cases (by using the ENMPC strategy and by using that based on the PID controllers). The position tracking with respect to the time for the both techniques is illustrated in Fig. 4. The position of the tracking errors of the technique based on the PID controllers and that of the ENMPC are plotted in Fig. 5.

From these figures cited above we remark that the ENMPC technique performs better than the technique based on the PID controllers. Indeed contrary to the ENMPC strategy, that based on the PID controllers do not converge to zero and this seems clear in Fig. 5.

Figure 6 shows the control signals of the both techniques. In the both cases, the control signals are well conditioned, they are more dynamic in the case of ENMPC which allows to have the best performances in terms of tracking objective. Such feature shows the feasibility of the proposed control structures, as well as the superiority and effectiveness of the ENMPC.

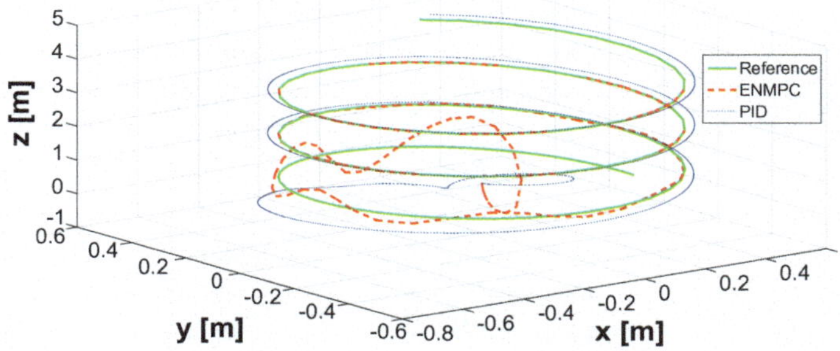


**Fig. 3** Path following

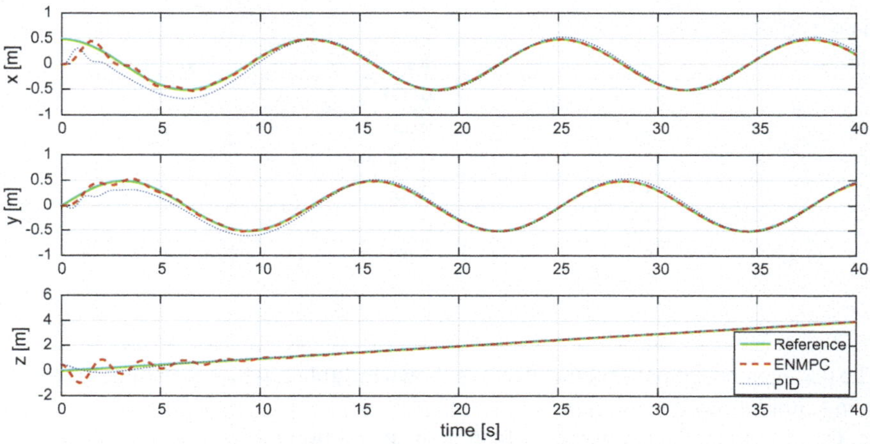


**Fig. 4** Position $(x, y, z)$

# 6 Conclusions

In this paper, we have solved the problem of the trajectory tracking for the quadcopter by using two strategies. The one based on the PID controllers, which is consisting of two loops: in the outer-loop the position and the altitude controllers are used, while that the attitude controllers are adopted in the inner loop. The other strategy is of the ENMPC, which its principal advantage is to achieve optimal performance without the need of on-line optimization. From the comparison between these strategies, we conclude that the ENMPC strategy is better than that based on the PID controllers, thanks to its ability to make the quadcopter follow the reference path and the simplic-

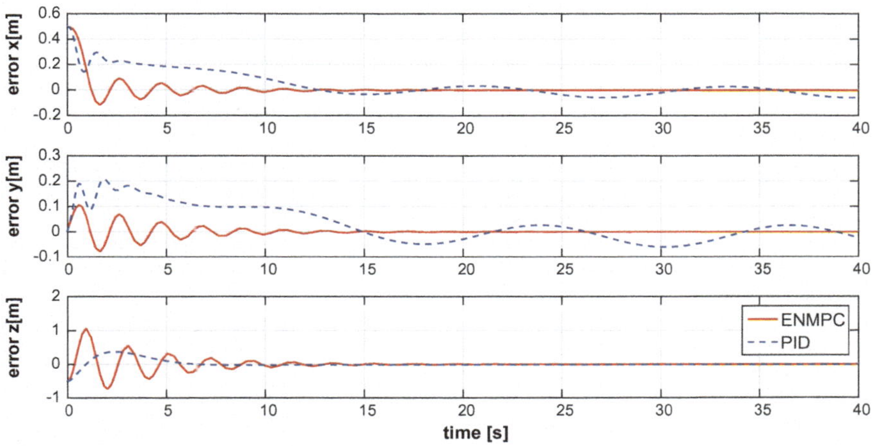


**Fig. 5** Position error $(x, y, z)$

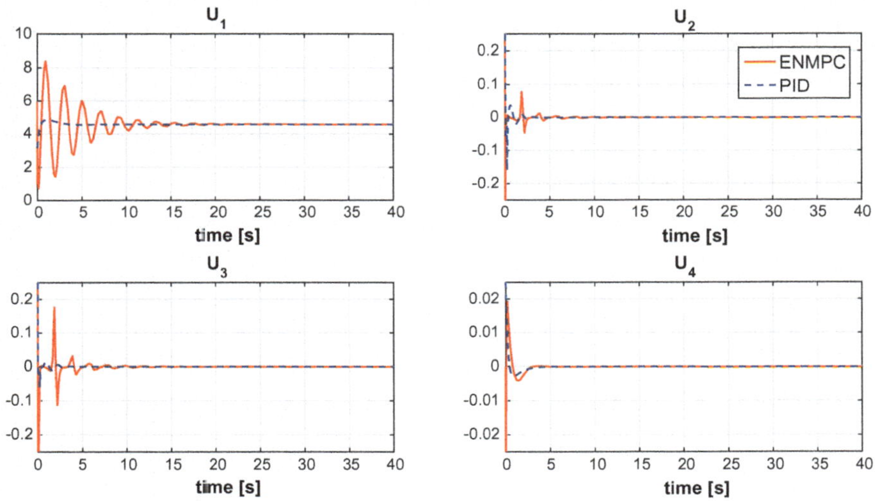


**Fig. 6** Control inputs $(U_1, U_2, U_3, U_4)$

ity of its implementation where only one design parameter should be manipulated while the technique based on the PID controllers have eighteen design parameters.

In our future works, we should investigated the aptitude to compensate the disturbance which is an effective issue for the quadcopter system.

## References

1. Bouabdallah, S., Murrieri, P., Siegwart, R.: Design and control of an indoor micro quadrotor. In: IEEE International Conference on Robotics and Automation, Zurich, Switzerland, pp. 4393–4398 (2004)
2. Ganga, G., Dharmana, M.M.: MPC controller for trajectory tracking control of quadcopter. In: 2017 IEEE International Conference on Circuit, Power and Computing Technologies (ICCPCT), Sasthamcotta, Kollam, Kerala, India, pp. 1–6 (2017)
3. Hoffmann, G., Huang, H., Waslander, S., Tomlin, C.: Quadrotor helicopter flight dynamics and control: theory and experiment. In: AIAA Guidance, Navigation and Control Conference and Exhibit, Hilton Head, South Carolina, p. 6461 (2007)
4. Islam, M., Okasha, M., Idres, M.M.: Trajectory tracking in quadrotor platform by using PD controller and LQR control approach. In: IOP Conference Series: Materials Science and Engineering, Kuala Lumpur, Malaysia, pp. 1–7 (2017)
5. Lee, D., Kim, H.J., Sastry, S.: Feedback linearization vs. adaptive sliding mode control for a quadrotor helicopter. Int. J. Control Autom. Syst. **7**, 419–428 (2009)
6. Liu, C., Chen, W.H., Andrews, J.: Tracking control of small-scale helicopters using explicit nonlinear MPC augmented with disturbance observers. Control Eng. Pract. **20**, 258–268 (2012)
7. Liu, H., Li, D., Zuo, Z., Zhong, Y.: Robust three-loop trajectory tracking control for quadrotors with multiple uncertainties. IEEE Trans. Ind. Electron. **63**, 2263–2274 (2016)
8. Raffo, G.V., Ortega, M.G., Rubio, F.R.: An integral predictive/nonlinear $H_\infty$ control structure for a quadrotor helicopter. Automatica **46**, 29–39 (2010)
9. Rios, H., Falcon, R., Gonzalez, O.A., Dzul, A.E.: Continuous sliding-modes control strategies for quad-rotor robust tracking: real-time application. IEEE Trans. Ind. Electron. (2018)
10. Yu, Y., Guo, Y., Pan, X., Sun, C.: Robust backstepping tracking control of uncertain MIMO nonlinear systems with application to quadrotor UAVs. In: International Conference on Information and Automation, Lijiang, China, pp. 2868–2873 (2015)

# Robust and Adaptive State Estimation of UAV Quadrotors with a High Gain Approach

**Habib Dimassi, Nadia Miladi, Salim Hadj Said and Faouzi M'Sahli**

**Abstract** In this chapter, we study the problem of joint states and external aerodynamic disturbances as well as the problem of adaptive estimation of quadrotor systems. We investigate two realistic configurations of the quadrotor system: in the first configuration, the dynamics of longitudinal and angular velocities are corrupted by unknown time varying disturbances and in the second one, the quadrotor system is subject to constant unknown parameters. For the two latter cases of study, only positions and angles are available for measurements. In particular, we focus on the problem observer matching condition which is necessary to solve both robust and adaptive estimation problems. The latter restrictive assumption is not verified for the quadrotor system. To overcome this problem, we adopt the approach of generation and estimation of auxiliary outputs based on high gain approximate differentiators. For the case of presence of time varying disturbances, we present a first order sliding mode observer in cascade with a high gain observer to reconstruct both the states and the unknown disturbances of the quadrotor system. For the case of presence of constant unknown parameters, we introduce an adaptive observer in cascade with the same high gain observer used in the first configuration to reconstruct both the states and the unknown parameters. Numerical simulations are depicted to illustrate the effectiveness and the good performances of the proposed robust and adaptive estimation approaches.

---

H. Dimassi (✉)
Institut supérieur des sciences appliquées et de technologie de Sousse, University of Sousse, Sousse, Tunisia
e-mail: dimassihabib2013@gmail.com

H. Dimassi · N. Miladi · S. H. Said · F. M'Sahli
ESIER, Ecole Nationale d'Ingénieurs de Monastir, University of Monastir, 5019 Monastir, Tunisia
e-mail: nadia_miladi@hotmail.fr

S. H. Said
e-mail: salim.hadjsaid06@yahoo.fr

F. M'Sahli
e-mail: faouzi.msahli@enim.rnu.tn



J. Ghommam et al. (eds.), *New Trends in Robot Control*, Studies in Systems, Decision and Control 270, https://doi.org/10.1007/978-981-15-1819-5_17

**Keywords** Quadrotor · Robust estimation · Sliding mode observer · High gain observer · Unknown inputs reconstruction

## 1 Introduction

The observation problem of quadrotors has received increasing attention in the last decade. In many existing observation methods designed for quadrotor models, it is assumed that the external aerodynamics forces and the torque disturbances are negligible, however this is unrealistic in real-world applications. Many recent works were focused on the problem of joint states and unknown time varying disturbances estimation as well as the problem of joint states and unknown parameters estimation for quadrotors by exploiting different unknown inputs observer and adaptive design methods which are available in the literature of observation theory and automatic control.

In [1], a high order sliding mode observer was designed for robust estimation of quadrotor model but the method was applied for the linearized model. In [14], a nonlinear disturbance observer was designed to reconstruct the effect of the external unknown force/torque: the unknown inputs error system was shown to be exponentially stable but the proposed method was developed under the strong assumption that all the states of the quadrotor model are measured. In [3], an adaptive second-order sliding mode was proposed to deal with faults and noise corrupting the position measurement, but the method was only applied to the quadrotor attitude model and moreover the external disturbances which may exist in the dynamics of the quadrotors were not considered in this paper. More recently, a nonlinear disturbance observer was employed in [10] to compensate the effect of the wind perturbations, but the considered disturbances does not affect the dynamics of longitudinal and angular velocities which is unrealistic in practice.

On the other hand, adaptive observer design have attracted many researchers in the literature of adaptive control. In [20], an adaptive observer has been proposed by Zhang for a class of linear time-varying systems under the persistency of excitation condition. In [8, 12, 15, 16, 18], adaptive high gain observers have been proposed for a class of nonlinear systems with linear or/and nonlinear parameterization. More recently, adaptive high gain observers have been extended for the case of non uniformly observable nonlinear systems based on the augmented model approach in [9] and in the case of uniformly observable systems subject to sampled outputs in [7]. Furthermore, a class of nonlinear systems satisfying the Lipschitz and the persistency of excitation assumptions has been considered in [2, 4] where authors have designed adaptive observers under the so-called observer matching condition. The latter condition restricts the applicability of these class of adaptive observers. Indeed, in many physical systems and particularly in quadrotor systems, the observer matching condition is not satisfied.

In this chapter, we consider the problem of joint states and time varying unknown disturbances present in the dynamics of the longitudinal and angular velocities as well as the problem of joint states and constant unknown parameters estimation for quadrotors, based on the only positions and angular measurements. Considering the latter configurations, classical unknown inputs observers and adaptive observers couldn't solve the problem of robust and adaptive estimation problems because the observer matching condition also known as "the relative degree" assumption is not satisfied. In [11], the authors have shown that the observer matching condition may be weakened if a classical sliding mode observer is combined with a high order sliding mode exact differentiator which generates additional auxiliary outputs. However, the latter observer is very sensitive to measurements noise and has relatively complicated implementation architecture. More recently, the idea of generating auxiliary outputs was adopted in [13] where high gain exact differentiators with simpler architecture have been designed to generate the estimates of the auxiliary outputs exciting a classical sliding mode observer [19] to achieve both states and unknown inputs estimation.

Motivated by the good performances of the approach presented in [13], we propose in this paper, a high gain observer for auxiliary outputs estimation purpose combined with a first order sliding mode observer for the quadrotor model which is subject to external aerodynamic disturbances in both longitudinal and angular velocities dynamics. Thanks to the high gain differentiator, the relative degree hypothesis is relaxed and the robust estimation of both the states and unknown perturbations is established thanks to the well known insensitivity of the sliding mode observer to the matched unknown inputs.

In the same spirit and exploiting the same idea, the problem of adaptive estimation is solved for the quadrotor system subject to constant unknown parameters such that the observer matching assumption satisfied by combining the adaptive observer of [4] and the high gain observer of [13]. The paper is organized as follows. In the second section, we introduce the quadrotor model. In the third section, we present the two main configurations to be considered for the robust estimation and adaptive estimation problems. In the fourth section, we introduce the technique of generating auxiliary outputs and the high gain approximate differentiator design for the quadrotor system. In the Sect. 5, we present the robust estimation method for the quadrotor model subject to disturbances and we present some simulation results to show effectiveness of the proposed observer. Correspondingly, in Sect. 6, we introduce the adaptive estimation method for the quadrotor model subject to constant unknown parameters with numerical simulation. Finally, we conclude with some remarks.

*Notation*: Throughout this paper, $\|.\|$ denotes the Euclidean norm for vectors and the induced norm for matrices. $I_n$ denotes the identity matrix of size $n$.

## 2 Quadrotor Modeling

Based on the Newton–Euler modeling approach, the dynamic model of the quadrotor could be expressed by the following system

$$\begin{cases} \ddot{\phi} = \frac{l}{I_{xx}} U_2 - \frac{J_r}{I_{xx}} \dot{\theta} \Omega + \frac{(I_{yy} - I_{zz})}{I_{xx}} \dot{\theta} \dot{\psi} + \frac{A_\phi}{I_{xx}} \\ \ddot{\theta} = \frac{l}{I_{yy}} U_3 + \frac{J_r}{I_{yy}} \dot{\phi} \Omega + \frac{(I_{zz} - I_{xx})}{I_{yy}} \dot{\phi} \dot{\psi} + \frac{A_\theta}{I_{yy}} \\ \ddot{\psi} = \frac{1}{I_{zz}} U_4 + \frac{(I_{xx} - I_{yy})}{I_{zz}} \dot{\phi} \dot{\theta} + \frac{A_\psi}{I_{zz}} \\ \ddot{z} = -g + \frac{U_1}{m} \cos\theta \cos\phi + \frac{A_z}{I_m} \\ \ddot{x} = \frac{U_1}{m} (\cos\psi \sin\theta \cos\phi + \sin\phi \sin\psi) + \frac{A_x}{I_m} \\ \ddot{y} = \frac{U_1}{m} (\sin\psi \sin\theta \cos\phi - \sin\phi \cos\psi) + \frac{A_y}{I_m} \end{cases} \quad (1)$$

where $(x, y, z)$ and $(\phi, \theta, \psi)$ are the linear and the angular position of the quadrotor expressed in the inertial frame, respectively. $g$ is the acceleration of gravity, $l$ is the length between the center of gravity of the quadrotor and the center of the propeller, $J_r$ is the rotor inertia, $m$ is the total mass of the quadrotor, $I_{xx}$, $I_{yy}$ and $I_{zz}$ are the moments of inertia with respect to the axes, $\Omega = w_1 - w_2 + w_3 - w_4$ is the relative speed of the rotors. $A_\phi$, $A_\theta$, $A_\psi$, $A_z$, $A_x$ and $A_y$ represent the external aerodynamic disturbances. Finally, $U_1$, $U_2$, $U_3$ and $U_4$ represent the virtual control inputs defined as follow

$$\begin{cases} U_1 = k(\omega_1^2 + \omega_2^2 + \omega_3^2 + \omega_4^2) \\ U_2 = kl(\omega_4^2 - \omega_2^2) \\ U_3 = kl(\omega_3^2 - \omega_1^2) \\ U_4 = b(-\omega_1^2 + \omega_2^2 - \omega_3^2 + \omega_4^2) \end{cases} \quad (2)$$

where $\omega_1$, $\omega_2$, $\omega_3$ and $\omega_4$ are the angular velocities of the four rotors, $k$ and $b$ are the lift and the drag constants.

From the Euler–Newten Model, we deduce that the state space representation of the quadrotor can be expressed by the following system

$$\begin{cases} \dot{x_1} = x_2 \\ \dot{x_2} = a_1 x_4 x_6 - a_2 x_4 \Omega + b_1 U_2 + \frac{A_\phi}{I_{xx}} \\ \dot{x_3} = x_4 \\ \dot{x_4} = a_3 x_2 x_6 + a_4 x_2 \Omega + b_2 U_3 + \frac{A_\theta}{I_{yy}} \\ \dot{x_5} = x_6 \\ \dot{x_6} = a_5 x_2 x_4 + b_3 U_4 + \frac{A_\psi}{I_{zz}} \\ \dot{x_7} = x_8 \\ \dot{x_8} = -g + \frac{U_1}{m} \cos x_3 \cos x_1 + \frac{A_z}{m} \\ \dot{x_9} = x_{10} \\ \dot{x}_{10} = \frac{U_1}{m} (\cos x_5 \sin x_3 \cos x_1 + \sin x_1 \sin x_5) + \frac{A_x}{m} \\ \dot{x}_{11} = x_{12} \\ \dot{x}_{12} = \frac{U_1}{m} (\sin x_5 \sin x_3 \cos x_1 - \cos x_5 \sin x_1) + \frac{A_y}{m} \end{cases} \quad (3)$$

$\mathbf{x} = [\phi \;\; \dot{\phi} \;\; \theta \;\; \dot{\theta} \;\; \psi \;\; \dot{\psi} \;\; z \;\; \dot{z} \;\; x \;\; \dot{x} \;\; y \;\; \dot{y}]^T$ is the state vector of the quadrotor; $a_1 = \frac{(I_{yy}-I_{zz})}{I_{xx}}$, $a_2 = \frac{J_r}{Ixx}$, $a_3 = \frac{(I_{zz}-I_{xx})}{I_{yy}}$, $a_4 = \frac{J_r}{I_{yy}}$, $a_5 = \frac{(I_{xx}-I_{yy})}{I_{zz}}$, $b_1 = \frac{l}{I_{xx}}$, $b_2 = \frac{l}{I_{yy}}$ and $b_1 = \frac{1}{I_{zz}}$.

## 3 Study Cases: Quadrotor Subject to Disturbances and Unknown Parameters

We consider the problem of joint states and external disturbances/unknown parameters based on the only positions and angular measurements of the quadrotor. Let $\mathbf{y}$ the measurement vector such that $\mathbf{y} = [\phi \;\; \theta \;\; \psi \;\; z \;\; x \;\; y]^T = [x_1 \;\; x_3 \;\; x_5 \;\; x_7 \;\; x_9 \;\; x_{11}]^T$.

The quadcopter state model (3) may be written as a linear time-invariant system with unknown inputs:

$$\begin{cases} \dot{\mathbf{x}} = A\mathbf{x} + Bu + D\xi \\ \mathbf{y} = C\mathbf{x} \end{cases} \tag{4}$$

where

$$A = \begin{bmatrix} 0\,1\,0\,0\,0\,0\,0\,0\,0\,0\,0\,0\,0 \\ 0\,0\,0\,0\,0\,0\,0\,0\,0\,0\,0\,0\,0 \\ 0\,0\,0\,1\,0\,0\,0\,0\,0\,0\,0\,0\,0 \\ 0\,0\,0\,0\,0\,0\,0\,0\,0\,0\,0\,0\,0 \\ 0\,0\,0\,0\,0\,1\,0\,0\,0\,0\,0\,0\,0 \\ 0\,0\,0\,0\,0\,0\,0\,0\,0\,0\,0\,0\,0 \\ 0\,0\,0\,0\,0\,0\,0\,0\,1\,0\,0\,0\,0 \\ 0\,0\,0\,0\,0\,0\,0\,0\,0\,0\,0\,0\,0 \\ 0\,0\,0\,0\,0\,0\,0\,0\,0\,0\,1\,0\,0 \\ 0\,0\,0\,0\,0\,0\,0\,0\,0\,0\,0\,0\,0 \\ 0\,0\,0\,0\,0\,0\,0\,0\,0\,0\,0\,0\,1 \\ 0\,0\,0\,0\,0\,0\,0\,0\,0\,0\,0\,0\,0 \end{bmatrix} \tag{5}$$

$$B = \begin{bmatrix} 0 \; b_1 \; 0 \; 0 \; 0 \; 0 \; 0\,0\,0\,0\,0\,0 \\ 0 \; 0 \; 0 \; b_2 \; 0 \; 0 \; 0\,0\,0\,0\,0\,0 \\ 0 \; 0 \; 0 \; 0 \; 0 \; b_3 \; 0\,0\,0\,0\,0\,0 \end{bmatrix}^T \tag{6}$$

$$D = \begin{bmatrix} 0\,1\,0\,0\,0\,0\,0\,0\,0\,0\,0\,0 \\ 0\,0\,0\,1\,0\,0\,0\,0\,0\,0\,0\,0 \\ 0\,0\,0\,0\,0\,1\,0\,0\,0\,0\,0\,0 \\ 0\,0\,0\,0\,0\,0\,0\,1\,0\,0\,0\,0 \\ 0\,0\,0\,0\,0\,0\,0\,0\,0\,1\,0\,0 \\ 0\,0\,0\,0\,0\,0\,0\,0\,0\,0\,0\,1 \end{bmatrix}^T \tag{7}$$

$$u = \left[ U_2, U_3, U_4 \right]^T \tag{8}$$

$$C = \begin{bmatrix} 1\,0\,0\,0\,0\,0\,0\,0\,0\,0\,0\,0 \\ 0\,0\,1\,0\,0\,0\,0\,0\,0\,0\,0\,0 \\ 0\,0\,0\,0\,1\,0\,0\,0\,0\,0\,0\,0 \\ 0\,0\,0\,0\,0\,0\,1\,0\,0\,0\,0\,0 \\ 0\,0\,0\,0\,0\,0\,0\,0\,1\,0\,0\,0 \\ 0\,0\,0\,0\,0\,0\,0\,0\,0\,0\,1\,0 \end{bmatrix} \tag{9}$$

The unknown input vector $\xi$ includes all nonlinear terms and unknown aerodynamic disturbances as

$$\xi = \begin{bmatrix} a_1 x_4 x_6 - a_2 x_4 \Omega + \frac{A_\phi}{I_{xx}} \\ a_3 x_2 x_6 + a_4 x_2 \Omega + \frac{A_\theta}{I_{yy}} \\ a_5 x_2 x_4 + \frac{A_\psi}{I_{zz}} \\ -g + \frac{U_1}{m} \cos x_3 \cos x_1 + \frac{A_z}{m} \\ \frac{U_1}{m} (\cos x_5 \sin x_3 \cos x_1 + \sin x_1 \sin x_5) + \frac{A_x}{m} \\ \frac{U_1}{m} (\sin x_5 \sin x_3 \cos x_1 - \cos x_5 \sin x_1) + \frac{A_y}{m} \end{bmatrix} \tag{10}$$

In this chapter of book, we focus on two cases of study: in the first case of study, we assume that the quadrotor system is subject to additive external time varying aerodynamic disturbances and in the second one we assume that the quadrotor system is subject to unknown parameters:

***Study Case 1: Quadrotor System Subject to Additive External Disturbances***

In this study case, we assume that the sate model of the quadrotor is exactly known and subject to additive external time varying aerodynamic disturbances which appear in the dynamics of the linear and angular velocities $d(t) = [\frac{A_\phi}{I_{xx}}, \frac{A_\theta}{I_{yy}}, \frac{A_\psi}{I_{zz}}, \frac{A_z}{m}, \frac{A_x}{m}, \frac{A_y}{m}]^T \neq 0$.

In this case, the quadrotor state model may be written in the following form:

$$\begin{cases} \dot{\mathbf{x}} = A\mathbf{x} + Bu + Df_1(x) + Dd(t) \\ \mathbf{y} = C\mathbf{x} \end{cases} \tag{11}$$

and the unknown input $\xi(t) = f_1(x) + d(t)$ where

$$f_1(x) = \begin{bmatrix} a_1 x_4 x_6 - a_2 x_4 \Omega \\ a_3 x_2 x_6 + a_4 x_2 \Omega \\ a_5 x_2 x_4 \\ -g + \frac{U_1}{m} \cos x_3 \cos x_1 \\ \frac{U_1}{m}(\cos x_5 \sin x_3 \cos x_1 + \sin x_1 \sin x_5) \\ \frac{U_1}{m}(\sin x_5 \sin x_3 \cos x_1 - \cos x_5 \sin x_1) \end{bmatrix} \quad (12)$$

***Study Case 2: Quadrotor System Subject to Unknown Parameters***

In this case, the sate mode of the quadrotor is uncertain and subject to constant unknown parameters. We assume that the parameter $\theta = a_1$ is unknown and that the dynamics of the quadrotor is noise free, that is $d(t) = 0$. As a consequence, the quadrotor state model may be written in the following form:

$$\begin{cases} \dot{\mathbf{x}} = A\mathbf{x} + Bu + Df_2(x) + Dg_2(x)\theta \\ \mathbf{y} = C\mathbf{x} \end{cases} \quad (13)$$

and the unknown input $\xi(t) = f_2(x) + g_2(x)\theta$ where

$$f_2(x) = \begin{bmatrix} -a_2 x_4 \Omega \\ a_3 x_2 x_6 + a_4 x_2 \Omega \\ a_5 x_2 x_4 \\ -g + \frac{U_1}{m} \cos x_3 \cos x_1 \\ \frac{U_1}{m}(\cos x_5 \sin x_3 \cos x_1 + \sin x_1 \sin x_5) \\ \frac{U_1}{m}(\sin x_5 \sin x_3 \cos x_1 - \cos x_5 \sin x_1) \end{bmatrix} \quad (14)$$

and

$$g_2(x) = \begin{bmatrix} x_4 x_6, & 0, & 0, & 0, & 0, & 0 \end{bmatrix}^T \quad (15)$$

# 4 Generation and Estimation of the Auxiliary Outputs

## *4.1 The Observer Matching Condition Problem*

The problem of unknown input observer (sliding mode observer) design for the class of systems in the form (11) as well as the problem of adaptive observer design for the class of systems (13) have been investigated in the literature of automatic control, however the proposed observers have been designed under the following well-known restrictive structural assumptions

*Assumption 1*: The invariant zeros of $(A, D, C)$ are in the open left-hand complex plane. That is the Rosenbrock matrix given by

$$R = \begin{bmatrix} sI - A & D \\ C & 0 \end{bmatrix} \quad (16)$$

is of full rank.

*Assumption 2*: The observer matching condition is satisfied. That is $\text{Rank}(CD) = \text{Rank}(D)$.

For the two quadrotor state models (11) and (13) in both study cases (1 and 2), it is easy to check that the Rosenbrock matrix introduced in *Assumption 1* is of full rank. However, the observer matching condition introduced in *Assumption 2* is clearly not satisfied and should be relaxed.

For the general class of linear time-invariant system with unknown inputs of the form (4), the idea of generating auxiliary outputs has been firstly proposed by the authors of the paper [11] to relax the restrictive *Assumption 2* and a high order exact sliding mode differentiator have been developed. More recently, this approach was improved by the authors of [13] and a new high gain exact differentiator have been proposed for the auxiliary outputs estimation purpose.

Motivated by the approach of [13], we adopt in this research work the method of generating auxiliary outputs to overcome the observer matching condition restriction. The new generated outputs need to be estimated because they are not available, so a high gain exact differentiator is to be designed for this purpose.

Then, in the case of study 1, the objective is to solve the problem of joint states and unknown inputs (the external disturbances) reconstruction of the quadrotor state model (11) based on a sliding mode observer which exploits the estimates of the auxiliary outputs given by the high gain observer which will be introduced in Sect. 4. In the case of study 2, the objective consists in solving the problem of joint states and unknown parameters estimation of the quadrotor state model (13) based on an adaptive observer which uses also the estimated auxiliary outputs by the high gain observer.

## 4.2 Relaxation of the Observer Matching Condition

In this section, we consider again the general class of linear systems with unknown inputs of the form

$$\begin{cases} \dot{\mathbf{x}} = A\mathbf{x} + Bu + D\xi \\ \mathbf{y} = C\mathbf{x} \end{cases} \quad (17)$$

Let $C = [C_1 \cdots C_p]^T$

*Definition* the relative degree of the $i$th output $y_i = C_i\mathbf{x}$ with respect to the unknown input $\xi$ is the smallest integer $r_i$ such that:
$C_i A^n D = 0$, for $n = 0, \ldots, r_i - 2$
$C_i A^{r_i - 1} D \neq 0$.

We assume that the observer matching condition is not verified for System (17). That is, there exists at least an integer $j \in \{1, \ldots, p\}$ such that the relative degree $r_j$ of the $j$th output with respect to the unknown output signal $\xi$ is such that $r_j \geq 2$. To relax this restriction, the idea is to construct an auxiliary output vector $\mathbf{z} = H\mathbf{x}$ such that the matrix

$$H = \left(C_1 \ldots C_1 A^{q_1 - 1} \ldots, C_p \ldots C_p A^{q_p - 1}\right)^T$$

is such that $\text{Rank}(D) = \text{Rank}(HD)$ where $q_i$ $(1 \leq q_i \leq r_i)$ are appropriate integers to be chosen, for $i = 1, \ldots, p$.

In the rest of this chapter, we will rather focus on the new system with the auxiliary output $\mathbf{z} = H\mathbf{x}$

$$\begin{cases} \dot{\mathbf{x}} = A\mathbf{x} + Bu + D\xi \\ \mathbf{z} = H\mathbf{x} \end{cases} \tag{18}$$

## 4.3 High Gain Approximate Differentiator Design

At this stage, the observer matching condition is relaxed, however the auxiliary output $\mathbf{z} = H\mathbf{x}$ could not be used yet because it is not available. To overcome this problem, high gain approximate differentiators have been employed in [13] to generate an estimate $z_h$ of the auxiliary output $\mathbf{z}$.

Let $\mathbf{z} = [z_1^T, \ldots, z_p^T]^T$ with $z_i = [z_{i1}, \ldots, z_{iq_i}]^T$ such that $z_{ij} = C_i A^{j-1}\mathbf{x}$ for $i = 1, \ldots, p$ and $j = 1, \ldots, q_i$. If $q_i > 1$, the dynamics of $z_i$ are given by

$$\begin{cases} \dot{z}_i = \mathbf{a}_i z_i + b_{i1} f_i(\mathbf{x}, \xi) + b_{i2} u \\ z_{i1} = c_i z_i \end{cases} \tag{19}$$

where $(\mathbf{a}_i, b_{i1})$ is in the canonical form, $f_i(\mathbf{x}, \xi) = C_i A^{q_i}\mathbf{x} + C_i A^{q_i - 1} D\xi$, $b_{i2} = [C_i B \cdots C_i A^{q_i - 1} B]^T$ and $c_i = [1, 0, \ldots, 0]$. A high gain observer for the latter system is given by

$$\begin{cases} \dot{z}_{hi} = \mathbf{a}_i z_{hi} + \Gamma_i c_i (z_i - z_{hi}) + b_{i2} u \\ z_{hi1} = c_i z_{hi} \end{cases} \tag{20}$$

where $z_h = [z_{h1}^T, \ldots, z_{hp}^T]^T$ with $z_{hi} = [z_{hi1}, \ldots, z_{hiq_i}]^T$, $\Gamma_i = [\frac{\gamma_{i1}}{\delta} \cdots \frac{\gamma_{iqi}}{\delta^{qi}}]^T$ such that $\delta$ and $\gamma_{ij}$ are appropriate design parameters. The convergence of the high gain observer (20) to the system (19) has been established in Lemma 1, in the Ref. [13].

That is the auxiliary outputs are estimated with an appropriate choice of the design parameters.

## 4.4 Generation and Estimation of Auxiliary Outputs for the Quadrotor Model

We propose a high-gain approximate differentiator observer for the quadrotor state model (3). The output vector $\mathbf{y} = [y_1, \ldots, y_6]^T = C\mathbf{x} = [C_1\mathbf{x}, \ldots, C_6\mathbf{x}]^T$ where $C = [C_1 \cdots C_6]$. It is easy to check that all the outputs $y_1$, ..., $y_2$ are of a relative degree equal to $r = r_1 = \cdots = r_6 = 2$ with respect to the unknown input vector $\xi$. In this case, a classical unknown input observer could not be designed for our quadrotor since the observer matching condition is not verified, so to deal with this problem, we generate an auxiliary output vector $\mathbf{z} = H\mathbf{x}$ such that the augmented matrix

$$H = \begin{bmatrix} C_1 \\ C_1 A \\ C_2 \\ C_2 A \\ C_3 \\ C_3 A \\ C_4 \\ C_4 A \\ C_5 \\ C_5 A \\ C_6 \\ C_6 A \end{bmatrix} = I_{12} \tag{21}$$

is of full rank and the condition $\text{rank}(HD) = \text{rank}D$ is satisfied. Let $\mathbf{z} = [z_1, \ldots, z_6]^T = [H_1\mathbf{x}, \ldots, H_6\mathbf{x}]^T$ where $z_i = [z_{i1}, z_{i2}]^T$ for $i = 1, \ldots, 6$. Differentiating each $z_i$ and using the definition of the relative degree (See *Definition 1* in the Ref. [13]), we obtain the dynamics of the auxiliary outputs given by

$$\dot{z}_{i1} = z_{i2} + C_i B u \tag{22}$$

$$\dot{z}_{i2} = C_i A^2 \mathbf{x} + C_i A D \xi + C_i B u \tag{23}$$

$$z_{i1} = c_i z_i \tag{24}$$

which is of the form of System (19) with $\mathbf{a}_i = \begin{bmatrix} 0 & 1 \\ 0 & 0 \end{bmatrix}$, $b_{i1} = \begin{bmatrix} 0 \\ 1 \end{bmatrix}$

$b_{i2} = \begin{bmatrix} C_i B \\ C_i A B \end{bmatrix}$, $c_i = \begin{bmatrix} 1 & 0 \end{bmatrix}$

$f_i(\mathbf{x}, \xi) = C_i A^2 \mathbf{x} + C_i A D \xi$, for $i = 1, \ldots, 6$.

Now, for $i = 1, \ldots, 6$, we propose the following high gain observers for Systems (24)

$$\dot{z}_{hi1} = z_{hi2} + \tfrac{\gamma_{i1}}{\delta}(z_{i1} - z_{hi1}) + C_i Bu \tag{25}$$

$$\dot{z}_{hi2} = \tfrac{\gamma_{i2}}{\delta^2}(z_{i1} - z_{hi1}) + C_i ABu \tag{26}$$

$$z_{hi1} = c_i z_{hi} \tag{27}$$

where $z_{hi} = [z_{hi1}, z_{hi2}]^T$ for $i = 1, \ldots, 6$ represent the estimated auxiliary outputs.

Using *Lemma 1* in [13], we deduce that for the system (24) and the high gain approximate differentiator (27), there exists a finite time $T_i(\delta)$ such that the convergence of the high gain observation error is ensured for all $t \geq T_i(\delta)$ and that the time convergence $T_i(\delta)$ may be diminished at will by choosing the design parameter $\delta$ sufficiently small.

## 5 Robust State Estimation with High Gain Approach for the Quadrotor System in the Study Case 1

### *5.1 Sliding Mode Observer Synthesis for the Quadrotor Model Subject to Disturbances*

In this section, we propose a robust observer for the quadrotor system subject to time varying disturbances in the study case 1 [see System (11)] which exploits the estimates of the auxiliary outputs generated from the high gain observer (20). The robust observer that we propose is a sliding mode observer whose the objective is to ensure both states and unknown inputs estimation despite that the observer matching condition is not satisfied for the couple of matrices $(D, C)$. The structure of the latter observer is given by

$$\begin{cases} \dot{\hat{x}} = A\hat{x} + Bu + L(z_h - \hat{z}) + D\nu \\ \hat{z} = H\hat{x} \end{cases} \tag{28}$$

$$\nu = \begin{cases} \eta \frac{M(z_h - \hat{z})}{\|M(z_h - \hat{z})\|} & \text{if } M(z_h - \hat{z}) \neq 0 \\ 0 & \text{if } M(z_h - \hat{z}) = 0. \end{cases} \tag{29}$$

where $L$ and $M$ are such that for some definite positive matrices $P$ and $Q$

$$\begin{cases} (A - LH)^T P + P(A - LH) = -2Q \\ MH = D^T P \end{cases} \tag{30}$$

The convergence of the state estimation error was proved to be ensured in finite time for sufficiently small $\epsilon \in (0, 1)$ and sufficiently large design parameter $\eta$ ([13],

*Theorem 1*). The reconstruction of the unknown input vector $\xi(t) = f_1(x) + d(t)$ may also be achieved in finite time once the sliding mode is reached. Consequently, the proposed high-gain approximate differentiator based sliding-mode observer allows to solve the problem of joint states and unknown inputs estimation despite that the observer matching condition is not verified for System (17).

We use the estimates of the auxiliary outputs to construct a sliding-mode observer taking the form (28). To solve the system of linear matrix equations (30), we consider the following convex optimization problem [5]:

Minimize $\lambda$ subject to

$$P > 0 \tag{31}$$

$$PA - KH + (PA - KH)^T < 0 \tag{32}$$

$$\begin{bmatrix} \lambda I_6 & D^T P - MH \\ (D^T P - MH)^T & \lambda I_{12} \end{bmatrix} > 0 \tag{33}$$

When the above optimization problem has a minimum $\lambda = 0$, System (30) is guaranteed with the corresponding values of $P$, $M$, and $L = P^{-1}K$. This problem may be solved using convex optimization algorithms well developed in LMI toolbox of MATLAB. Another alternative consists in using the package *cvx* specific to LMIs problems and compatible with MATLAB. To compute the design matrices $M$ and $L$ we have proceeded as the latter alternative and we have obtained the following numerical design matrices

$$M = \begin{bmatrix} 0\,1\,0\,0\,0\,0\,0\,0\,0\,0\,0\,0 \\ 0\,0\,0\,1\,0\,0\,0\,0\,0\,0\,0\,0 \\ 0\,0\,0\,0\,0\,1\,0\,0\,0\,0\,0\,0 \\ 0\,0\,0\,0\,0\,0\,0\,1\,0\,0\,0\,0 \\ 0\,0\,0\,0\,0\,0\,0\,0\,0\,1\,0\,0 \\ 0\,0\,0\,0\,0\,0\,0\,0\,0\,0\,0\,1 \end{bmatrix} \tag{34}$$

The matrix $L$ is of size $(12 \times 12)$ such that $L(i, i) = 25$ for all $i = 1, \ldots, 12$, $L(2j, 2j-1) = 1$, for all $j = 1, \ldots, 6$ and the rest of its elements are equal to zero. Using the computed design matrices $L$ and $M$, the sliding mode observer designed for the quadrotor state model subject to aerodynamic external disturbances may be expressed in the following extended form:

$$\begin{cases} \dot{\hat{x}}_1 = \hat{x}_2 + 25(z_{h11} - \hat{x}_1) \\ \dot{\hat{x}}_2 = b_1 U_2 + (z_{h11} - \hat{x}_1) + 25(z_{h12} - \hat{x}_2) + \nu_1 \\ \dot{\hat{x}}_3 = \hat{x}_4 + 25(z_{h21} - \hat{x}_3) \\ \dot{\hat{x}}_4 = b_2 U_3 + (z_{h21} - \hat{x}_3) + 25(z_{h22} - \hat{x}_4) + \nu_2 \\ \dot{\hat{x}}_5 = \hat{x}_6 + 25(z_{h31} - \hat{x}_5) \\ \dot{\hat{x}}_6 = b_3 U_4 + (z_{h31} - \hat{x}_5) + 25(z_{h32} - \hat{x}_6) + \nu_3 \\ \dot{\hat{x}}_7 = \hat{x}_8 + 25(z_{h41} - \hat{x}_7) \\ \dot{\hat{x}}_8 = (z_{h41} - \hat{x}_7) + 25(z_{h42} - \hat{x}_8) + \nu_4 \\ \dot{\hat{x}}_9 = \hat{x}_{10} + 25(z_{h51} - \hat{x}_9) \\ \dot{\hat{x}}_{10} = (z_{h51} - \hat{x}_9) + 25(z_{h52} - \hat{x}_{10}) + \nu_5 \\ \dot{\hat{x}}_{11} = \hat{x}_{12} + 25(z_{h61} - \hat{x}_{11}) \\ \dot{\hat{x}}_{12} = (z_{h61} - \hat{x}_{11}) + 25(z_{h62} - \hat{x}_{12}) + \nu_6 \end{cases} \tag{35}$$

such that for each $i = 1, \ldots, 6$ and $j = 2i$

$$\nu_i = \eta \frac{(z_{hi2} - \hat{x}_j)}{\sqrt{\sum_{i=1}^{6} (z_{hi2} - \hat{x}_j)^2}} \tag{36}$$

For the quadrotor model (3) and the associated sliding mode observer (35) combined with the high-gain observer (27) and based on the arguments of *Theorem 1* in [13], we deduce that there exists a finite time $T_2(\delta)$ such that for all $t \geq T_2(\delta)$, the observation error $e = x - \hat{x}$ is uniformly ultimately bounded with an upper bound $\beta(\delta)$. Moreover, the convergence time $T_2(\delta)$ and the upper bound $\beta(\delta)$ may be reduced at will by choosing small values of the design parameter $\delta \in (0, 1)$. Furthermore, the unknown inputs vector $\xi$ reconstruction may be deduced once the sliding mode is reached and for any positive constant $\delta$, there exists a finite time $T_3$ and a positive constant $\epsilon$ such that for all $t > T_3$, $\|\nu(y_h, \hat{z}, \eta) - \xi\| < \epsilon$ which means that the unknown inputs vector $\xi$ may be estimated by the discontinuous input $\nu(z_h, \hat{z}, \eta)$. Finally, using the estimates of the quadrotor states and those of the unknown inputs, we can reconstruct the aerodynamic disturbances using the equation $d(t) = f_1(x) - \xi(t)$ as follows:

$$\begin{cases} \hat{A}_\phi = I_{xx}(\nu_1 - a_1 \hat{x}_4 \hat{x}_6 + a_2 \hat{x}_4 \Omega) \\ \hat{A}_\theta = I_{yy}(\nu_2 - a_3 \hat{x}_2 \hat{x}_6 - a_4 \hat{x}_2 \Omega) \\ \hat{A}_\psi = I_{zz}(\nu_3 - a_5 \hat{x}_2 \hat{x}_4) \\ \hat{A}_z = m\nu_4 + mg - U_1 \cos \hat{x}_3 \cos \hat{x}_1 \\ \hat{A}_x = m\nu_5 - U_1(\cos \hat{x}_5 \sin \hat{x}_3 \cos \hat{x}_1 + \sin \hat{x}_1 \sin \hat{x}_5) \\ \hat{A}_y = m\nu_6 - U_1(\sin \hat{x}_5 \sin \hat{x}_3 \cos \hat{x}_1 - \cos \hat{x}_5 \sin \hat{x}_1) \end{cases} \tag{37}$$

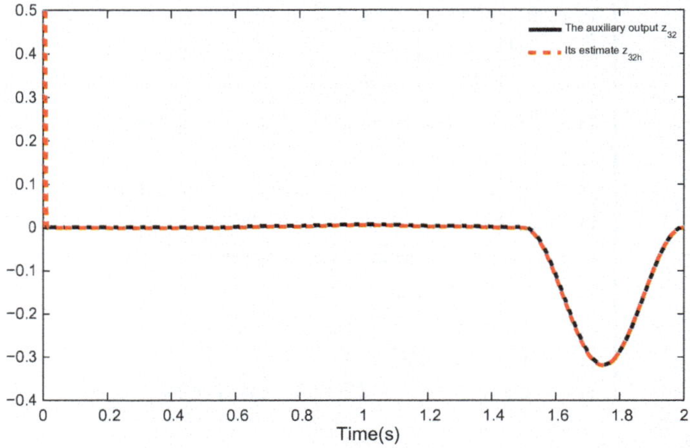


**Fig. 1** The auxiliary output $z_{32}$ and its estimate $z_{32h}$ by the high gain observer (20)

## *5.2 Numerical Simulations for the Case of Quadrotor System Subject to Time Varying Disturbances (Study Case 1)*

Numerical simulations are performed in order to illustrate the effectiveness of the proposed high-gain approximate differentiator based sliding-mode observer. The numerical values of the physical parameters of the quadrotor system are chosen as follows: $I_{xx} = 4.856e^{-3}$, $I_{yy} = 4.856e^{-3}$, $I_{zz} = 8.80e^{-3}$, $l = 0.225$, $g = 9.81$, $m = 0.468$ and $Jr = 3.357e^{-5}$. The control inputs are designed based on the angular velocities of the four rotors $w_1$, $w_2$, $w_3$ and $w_4$ [see the system of Eq. (2)]. The latter signals are selected based on some appropriate sinusoidal signals. The initial condition of the quadrotor state is chosen as $x(0) = [0, 0, 0, 0, 0.5, 0, 0, 0, 0.5, 0, 0, 0]^T$. For illustration, we assume that the aerodynamic disturbance $A_x$ is not negligible and it is equal to a delayed step signal $u(t-1)$. The design parameters of the high gain observer (27) are chosen to be $\gamma_{i1} = 2$, $\gamma_{i2} = 1$ for $i = 1, \ldots, 6$ and $\delta = 0.005$. The design parameter of the sliding mode observer $\eta$ is equal to 100. The initial states of both high-gain and sliding-mode observers are set to zero. In practice, in order to avoid the peaking phenomenon, we saturate the signals $z_{hi}$ for $i = 1, \ldots, 6$ as follows:

$$z_{hi}^{s} = \left[ S_{i1}\textbf{sat}\left(\frac{z_{hi1}}{S_{i1}}\right), S_{i2}\textbf{sat}\left(\frac{z_{hi2}}{S_{i2}}\right) \right]^T \tag{38}$$

We choose $S_{11} = S_{12} = S_{21} = S_{22} = S_{31} = S_{32} = 1$ and $S_{41} = S_{42} = S_{51} = S_{52} = S_{61} = S_{62} = 90$.

In Fig. 1, we illustrate the estimation of the auxiliary outputs $z_{32}$. The quadrotor state estimation is illustrated in Figs. 2 and 3. The reconstruction of the unknown input

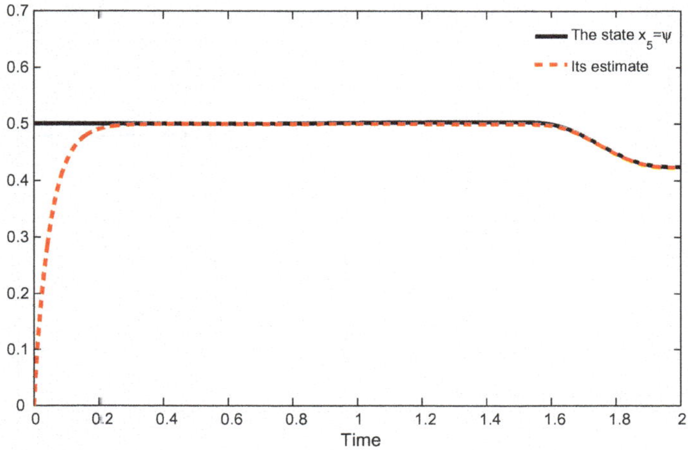


**Fig. 2** The state $x_5 = \psi$ and its estimate $\hat{x}_5$ by the sliding mode observer (35) with the discontinuous inputs (36) and the associated high gain observer (20)

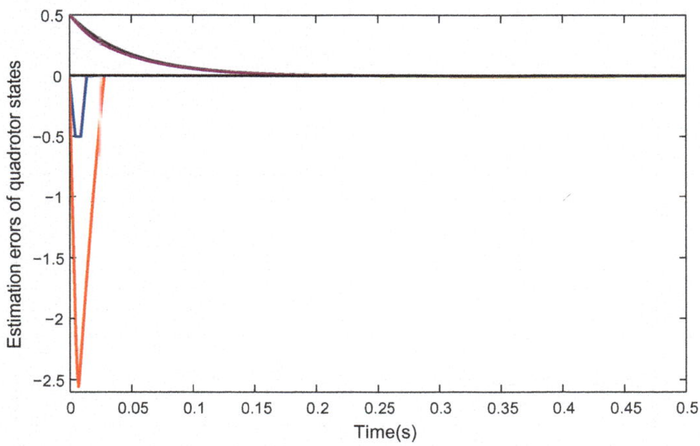


**Fig. 3** The estimation errors: $e_1 = x_1 - \hat{x}_1, \ldots, e_{12} = x_{12} - \hat{x}_{12}$ of by the sliding mode observer (35)

$\xi_5$ is shown in Fig. 4. Finally The recovery of the aerodynamic external disturbance $A_x$ is depicted in Fig. 5.

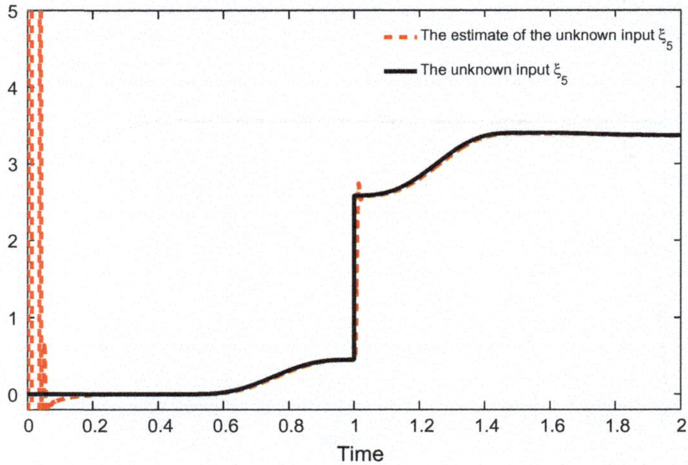


**Fig. 4** The unknown input $\xi_5$ and its estimate by the sliding mode observer (35) with the discontinuous inputs (36) and the associated high gain observer (20)

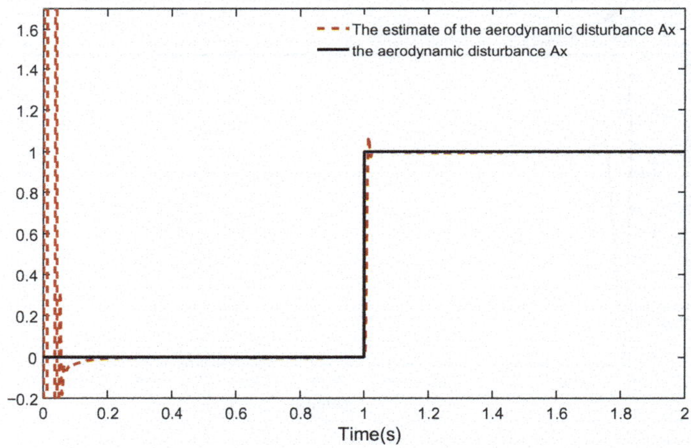


**Fig. 5** The aerodynamic disturbance $A_x$ and its estimate by the sliding mode observer (35) with the discontinuous inputs (36) and the associated high gain observer (20)

# 6 Adaptive State Estimation with High Gain Approach for the Quadrotor System in the Study Case 2

## *6.1 Adaptive Observer Design for the Quadrotor Model Subject to Unknown Parameters*

In this section, we propose an adaptive observer for the quadrotor system subject to constant unknown parameters in the study case 2 [see System (13)] which exploits the estimates of the auxiliary outputs generated from the high gain observer (20). The objective adaptive observer that we propose is to ensure both states and unknown parameters estimation despite that the observer matching condition is not satisfied for the couple of matrices $(D, C)$ and is to be relaxed following the procedure of generation and estimation of auxiliary outputs detailed above in Sect. 4. Using the estimated auxiliary output vector $z_h$ generated by the high gain observer (20) introduced in the Sect. 4, the adaptive observer that we propose is given by the following dynamics:

$$\dot{\hat{x}} = A\hat{x} + Bf_2(\hat{x}) + Bg_2(\hat{x})\hat{\theta} + Eu + L(z_h - H\hat{x}) \tag{39}$$

where $\hat{x}$ denotes the estimated state and $\hat{\theta}$ is an adaptive parameter updated online following the adaptation law:

$$\dot{\hat{\theta}} = \rho g_2(\hat{x})^T M(z_h - H\hat{x}) - \sigma\hat{\theta} \tag{40}$$

The constants $\rho$ and $\sigma$ are positive real numbers imposed by the designer. The design matrices $L$ and $M$ are regular matrices of appropriate dimensions such that for some symmetric positive definite matrices $P$ and $Q$, we have

$$(A - LH)^T P + P(A - LH) = -2\bar{Q} \tag{41}$$

$$B^T P = MH \tag{42}$$

$$|B|\,(K_f + K_g K_\theta) < \frac{\lambda_{min}(Q)}{\lambda_{max}(P)} \tag{43}$$

The design matrices $L$ and $M$ are computed in the same way as for the sliding mode observer in Sect. 5.

It is to be noticed that for the design of the above adaptive observe, the nonlinear functions $f_2(x)$ and $g_2(x)$ much satisfy the Lipshitz condition. In the case of the quadrotor system, the nonlinear functions $f_2(x)$ and $g_2(x)$ described by the Eqs. (14) and (15) are just continuously differentiable. However, assuming that the trajectories of the quadrotor system are uniformly bounded which is easily feasible in a complete observer-based output-feedback control system by appropriately designing the control inputs $U_1$, $U_2$, $U_3$ and $U_4$, one can apply the so-called Lipschitz extension technique using saturation functions on the states of the quadrotor system to make these nonlinear functions globally Lipschitz without changing the dynamics of the system (for more details about the Lipschitz extension technique, see for instances the Refs. [6, 8, 17]).

Note also that we have used the so-called $\sigma$-modification technique by adding the term $-\sigma\hat{\theta}$ in the adaptation law (40) which prevents the adaptive parameter $\theta$ to increase unboundedly and ensures the convergence of the adaptive parameter independent whether the persistency of excitation assumption is satisfied or not. Indeed, the latter assumption is a classical condition usually used in the literature of adaptive control to ensure the asymptotic parametric convergence. Of course, with the use of the $\sigma$-modification technique, we don't need the latter restrictive assumption, however we just obtain the convergence in finite time of the states and the adaptive parameters to a compact set which may be arbitrarily reduced by appropriately selecting the design parameters $\rho$, $\sigma$ and $\epsilon$.

Assuming that parameter $\theta = a_1 = -0.81$ is unknown and to be estimated, the adaptive observer that we design for the quadrotor state model subject to the latter unknown parameter may be expressed in the following extended form:

$$\begin{cases} \dot{\hat{x}}_1 = \hat{x}_2 + 25(z_{h11} - \hat{x}_1) \\ \dot{\hat{x}}_2 = \hat{\theta}\hat{x}_4\hat{x}_6 - a_2\hat{x}_4\Omega + b_1U_2 + (z_{h11} - \hat{x}_1) + 25(z_{h12} - \hat{x}_2) \\ \dot{\hat{x}}_3 = \hat{x}_4 + 25(z_{h21} - \hat{x}_3) \\ \dot{\hat{x}}_4 = a_3\hat{x}_2\hat{x}_6 + a_4\hat{x}_2\Omega + b_2U_3 + b_2U_3 + (z_{h21} - \hat{x}_3) + 25(z_{h22} - \hat{x}_4) \\ \dot{\hat{x}}_5 = \hat{x}_6 + 25(z_{h31} - \hat{x}_5) \\ \dot{\hat{x}}_6 = a_5\hat{x}_2\hat{x}_4 + b_3U_4 + (z_{h31} - \hat{x}_5) + 25(z_{h32} - \hat{x}_6) \\ \dot{\hat{x}}_7 = \hat{x}_8 + 25(z_{h41} - \hat{x}_7) \\ \dot{\hat{x}}_8 = -g + \frac{U_1}{m}\cos\hat{x}_3\cos\hat{x}_1 + (z_{h41} - \hat{x}_7) + 25(z_{h42} - \hat{x}_8) \\ \dot{\hat{x}}_9 = \hat{x}_{10} + 25(z_{h51} - \hat{x}_9) \\ \dot{\hat{x}}_{10} = \frac{U_1}{m}(\cos\hat{x}_5\sin\hat{x}_3\cos\hat{x}_1 + \sin\hat{x}_1\sin\hat{x}_5) + (z_{h51} - \hat{x}_9) + 25(z_{h52} - \hat{x}_{10}) \\ \dot{\hat{x}}_{11} = \hat{x}_{12} + 25(z_{h61} - \hat{x}_{11}) \\ \dot{\hat{x}}_{12} = \frac{U_1}{m}(\sin\hat{x}_5\sin\hat{x}_3\cos\hat{x}_1 - \cos\hat{x}_5\sin\hat{x}_1) + (z_{h61} - \hat{x}_{11}) + 25(z_{h62} - \hat{x}_{12}) \end{cases} \quad (44)$$

The estimated parameter $\hat{\theta}$ is updated online following the adaptation law (40).

## 6.2 *Numerical Simulations for the Case of Quadrotor Systems Subject to Constant Unknown Parameters (Study Case 2)*

In order to illustrate the effectiveness of the proposed high gain observer based adaptive estimation approach, we consider again the nonlinear system consisting in a quadrotor system described by the state model (3).

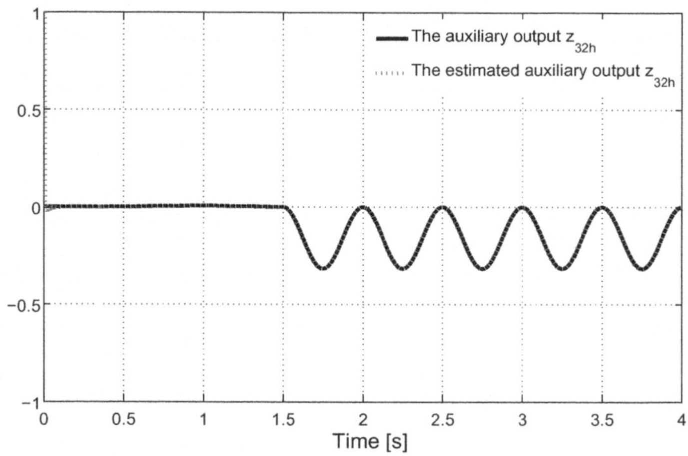


**Fig. 6** The auxiliary output $z_{32}$ and its estimate $z_{32h}$ by the high gain observer 20

The numerical values of the physical parameters, the control inputs as well as the initial conditions of the quadrator system are taken as in the numerical simulations section of the study case 1 (see Sect. 5.2).

As it was mentioned in Sect. 4.1, the observer matching assumption is not satisfied for the couple of matrices $(D, C)$ of the quadrotor system. We recall that the generated auxiliary output is given by:

$$z = Hx = \begin{bmatrix} z_{11} = C_1 x \\ z_{12} = C_1 A x \\ z_{12} = C_2 x \\ z_{12} = C_2 A x \end{bmatrix}. \tag{45}$$

where $H = I_{12}$. The observer matching condition is now clearly satisfied since $Rank(HB) = Rank(B)$.

We design a robust adaptive observer of the form (44) with the adaptation law (40) and the high gain observer (20). The initial states of both high-gain and adaptive observers are set to zero and the adaptive parameter is initialized as $\hat{\theta}(0) = 0$. The design parameters are set to $\epsilon = 0.006$, $\sigma = 0.001$, $\rho = 2.10^8$. We solve the system of linear matrix equations (41) and (42) by solving the same convex optimization problem considered in Sect. 5.2.

The numerical values of the design matrices $M$ and $L$ are selected as in Sect. 5.2.

Simulation tests are carried out without disturbances (i.e. with $d(t) = 0$) for the state model of the quadrotor system (3) and the adaptive observer (44) associated to the high gain observer (20) and the adaptation law (40). We get the following simulation results. In Fig. 6, we show the estimation of the auxiliary output $z_{32}$ by the high gain observer (20). Figure 7 depicts the convergence of the the estimation

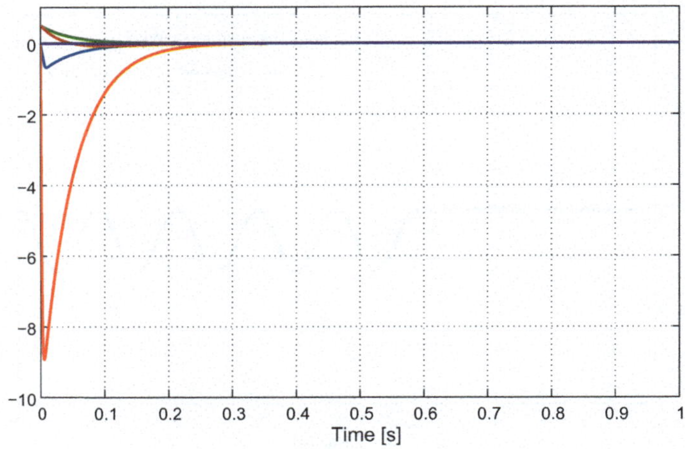


**Fig. 7** The estimation errors: $e_1 = x_1 - \hat{x}_1, \ldots, e_{12} = x_{12} - \hat{x}_{12}$ of the adaptive observer (44)

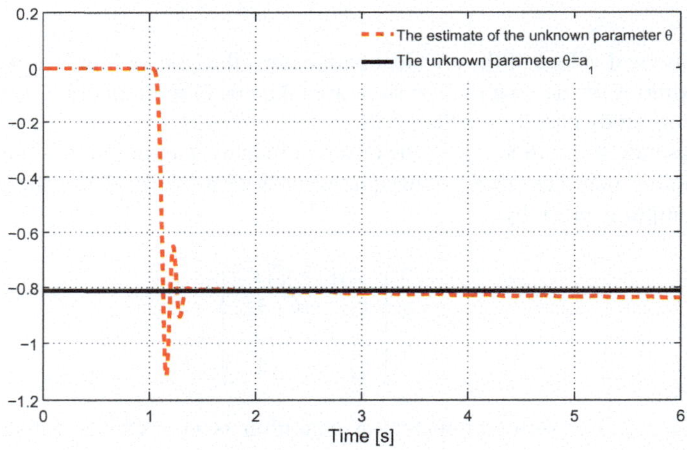


**Fig. 8** The unknown parameter $\theta = a_1 = -0.81$ and its estimate and its estimate by the adaptive observer (44) with adaptation law (40) and the associated high gain observer (20)

errors of the adaptive observer (39). Finally, in Fig. 8, we illustrate the parametric convergence as well as the estimation of the unknown parameter $\theta = a_1 = -0.81$.

## 7 Concluding Remarks

In this chapter, we have solved the problem of robust estimation as well as adaptive estimation for a quadrotor state model. Based on a high-gain approximate differentiator, we have shown that both the states and the unknown inputs of the quadrotor state model may be reconstructed by an appropriate sliding-mode observer using only positions and angular measurements despite that the observer matching condition is not satisfied. Similarly, we have considered the case of existence of constant unknown parameters in the quadrotor model and we have used a high gain approximate differentiator employed in cascade with an adaptive observer to solve the problem of adaptive state estimation despite that the relative degree of the outputs with respect to the unknown parameters vector is not equal to one. Numerical simulations were realized to validate the good performances of the proposed robust and adaptive estimation approaches.

## References


1. Benallegue, A., Mokhtari, A., Fridman, L.: High-order sliding-mode observer for a quadrotor uav. Int. J. Robust Nonlinear Control **18**, 427–440 (2008)
2. Besançon, G.: Remarks on nonlinear adaptive observer design. Syst. Control Lett. **41**, 271–280 (2000)
3. Chang, J., Cieslak, J., Davila, J., Zolghadri, A., Zhou, J.: Adaptive second-order sliding mode observer for quadrotor attitude estimation. In: American Control Conference (ACC), Boston, pp. 2246–2252 (2016)
4. Cho, Y.M., Rajamani, R.: Systematic approach to adaptive observer synthesis for nonlinear systems. IEEE Trans. Autom. Control **42**(4), 534–537 (1997)
5. Corless, M., Tu, J.: State and input estimation for a class of uncertain systems. Automatica **34**, 757–764 (1998)
6. Dimassi, H., Loria, A.: Adaptive unknown-input observers-based synchronization of chaotic systems for telecommunication. IEEE Trans. Circuits Syst. I **58**(4), 800–812 (2011)
7. Farza, M., Bouraoui, I., Ménard, T., Abdennour, R., M'Saad, M.: Adaptive observers for a class of uniformly observable systems with nonlinear parametrization and sampled outputs. Automatica **50**, 2951–2960 (2014)
8. Farza, M., M'saad, M., Maatoug, T., Kamoun, M.: Adaptive observers for nonlinearly parameterized class of nonlinear systems. Automatica **45**, 2292–2299 (2009)
9. Farza, M., M'Saad, M., Ménard, T., Ltaief, A., Maatoug, T.: Adaptive observer design for a class of nonlinear systems. Application to speed sensorless induction motor. Automatica **90**, 239–247 (2018)
10. Fethalla, N., Saady, M., Michalskaz, H., Ghommam, J.: Robust observer-based backstepping controller for a quadrotor UAV. In: IEEE 30th Canadian Conference on Electrical and Computer Engineering (CCECE), Canada (2011)
11. Floquet, T., Edwards, C., Spurgeon, S.K.: On sliding mode observers for systems with unknown inputs. Int. J. Adapt. Control Signal Process. **21**, 63–656 (2007)
12. Grip, H.F., Saberi, A., Johansen, T.A.: Estimation of states and parameters for linear systems with nonlinearly parameterized perturbations. Syst. Control Lett. **60**, 771–777 (2011)
13. Kalsi, K., Lian, J., Hui, S., Zak, S.H.: Sliding-mode observers for systems with unknown inputs: a high-gain approach. Automatica **46**, 347–353 (2010)
14. Liu, C., Chen, W.H., Andrews, J.: Tracking control of small-scale helicopters using explicit nonlinear mpc augmented with disturbance observers. Control Eng. Pract. **20**, 258–268 (2012)
15. Maatoug, T., Farza, M., M'Saad, M., Koubaa, Y., Kamouni, M.: Adaptive output feedback controller for a class of uncertain nonlinear systems. In: Proceedings of the 17th IFAC world congress, Seoul, pp. 13139–13144 (2008)

16. Morales, J.D.L., Huerta-Guevara, O., Besançon, G.: On adaptive observers for state affine systems. Int. J. Control **79**, 581–591 (2006)
17. Shim, H., Son, Y.I., Seo, J.H.: Semi-global observer for multi-output nonlinear systems. Syst. Control Lett. **42**, 233–244 (2001)
18. Tyukin, I.Y., Steur, E., Nijmeijer, H., Leeuwen, C.V.: Adaptive observers and parameter estimation for a class of systems nonlinear in the parameters. Automatica **49**, 2409–2423 (2013)
19. Walcott, B., Zak, S.H.: State observation of nonlinear uncertain dynamical systems. IEEE Trans. Autom. Control **32**, 166–170 (1987)
20. Zhang, Q.: Adaptive observer for multiple-input-multiple-output (mimo) linear time-varying systems. IEEE Trans. Autom. Control **47**(3), 525–529 (2002)

# Part IV

# Compound Fractional Integral Terminal Sliding Mode Control and Fractional PD Control of a MEMS Gyroscope

**Mehran Rahmani, Mohammad Habibur Rahman and Jawhar Ghommam**

**Abstract** This paper proposes a compound fractional order integral terminal sliding mode control (FOITSMC) and fractional order proportional-derivative control (FOPD-FOITSMC) for the control of a MEMS gyroscope. In order to improve the robustness of the conventional integral terminal sliding mode control (ITSMC), a fractional integral terminal sliding mode surface is applied. The chattering problem in FOITSMC, which is usually generated by the excitation of fast un-modelled dynamic is the main drawback. A fractional order proportional-derivative controller (FOPD) is employed in order to eliminate chattering phenomenon. The stability of the FOPD-FOITSMC is proved by Lyapunov theory. The performance of the proposed control method is compared with FOITSMC. Numerical simulations clearly verified the effectiveness of the proposed control approach.

**Keywords** MEMS gyroscope · Fractional integral terminal sliding mode control · Fractional PD controller

## 1 Introduction

An accelerometer is an electromechanical device that measures acceleration forces. These forces may be static, like the constant force Gestures provide a rich and intuitive form of interaction for controlling robots. Accelerometer Converts acceleration from motion (dynamic acceleration) or gravity (static acceleration) to either analog or digital electrical signals. Micro-electromechanical systems (MEMS) Gyro-

M. Rahmani (✉) · M. H. Rahman
Department of Mechanical Engineering, University of Wisconsin, Milwaukee, USA
e-mail: mrahmani@uwm.edu

M. H. Rahman
e-mail: rahmanmh@uwm.edu

J. Ghommam
Control and Management Lab, Sultan Qaboos University, Al Khoudh, Muscat, Oman
e-mail: jawhar.ghommam@gmail.com



J. Ghommam et al. (eds.), *New Trends in Robot Control*, Studies in Systems, Decision and Control 270, https://doi.org/10.1007/978-981-15-1819-5_18

scope accelerometers are one of the simplest but also most applicable micro-electromechanical systems.

The performance of the MEMS gyroscope usually encounters two problems with the results of external disturbances and time-varying parameters. The effects of these phenomena generate a frequency of oscillation that mismatch between the two vibrating axes [1]. In order to solve these problems, an advance control method is needed.

Fractional order sliding mode control (FOSMC) is a robust control method that is able to suppress external disturbances and has high tracking performance. Rabah and Ladaci [2] proposed an alternative fractional order adaptive sliding mode control (FOASMC) method for the synchronization of a class of nonlinear fractional order systems with chaotic behavior. The proposed method provided a set of fractional adaptive laws that guarantee asymptotic stability of fractional-order chaotic systems in the terms of the Lyapunov stability theorem. Tianyi et al. [3] FOSMC and design a controller which is used into the spacecraft attitude system so that it can be with more quickness and strong robustness. A novel sliding mode surface is proposed and it is proved that the system can be stable and can convergence in a finite time under this sliding mode surface. Tian et al. [4] proposed FOSMC for the active four-wheel steering vehicle. Optimal control is adopted to design the sliding mode in order to eliminate the state error caused by the different initial states. On this basis, FOSMC is designed to eliminate the state error caused by the different parameters. Kuang et al. [5] proposed a discrete-time fractional-order sliding mode (DFSM) control method based on the single-input second-order system. The general methods of designing a fractional-order sliding surface and the DFSM controller are introduced. The parametric uncertainties of the system are taken into consideration, which makes the conclusion suitable to the cases in practical industrial applications. Lin et al. [6] proposed synchronization for fractional master-slave Lorenz chaotic systems coupled with switched modified. Based on the Lyapunov stability theory and using the sliding mode control technique, a control scheme containing a new fraction-integer integral (FII) switching surface is developed to guarantee the synchronization of fractional Lorenz chaotic systems even when the switched modified is present.

Based on considered researches, FOSMC can be applied to different systems due to its robustness against external disturbances. The main drawbacks of FOSMC is creating chattering phenomenon, which is not appropriate for a control system. As a result of this problem, scholars have been proposed a compound control system that is able to eliminate chattering phenomenon [7–11]. Efe [12] proposed a new parameter adjustment method to enhance the robustness of fuzzy sliding-mode control obtained by the use of an adaptive neuro-fuzzy inference system (ANFIS) architecture. The proposed method uses fractional-order integration in the parameter tuning stage. The controller parameters are tuned such that the system under control is driven toward the sliding regime in the traditional sense. After a comparison with the classical integer-order counterpart, it is seen that the control system with the proposed adaptation scheme displays better tracking performance, chattering reduction, and a very high degree of robustness and insensitivity to disturbances are observed. Liu and Fei [13] proposed an adaptive FOSMC method based on dual radial basis function (RBF) neural networks (NNs) to improve the performance of a three-phase shunt active power filter (APF), where a conventional integer-order sliding surface is changed into a fractional-order one to speed up the system response and optimize the control performance. Furthermore, the control method adopts a class of dual RBF NNs, in which the network weights can be updated online to approximate the nonlinear system functions and the upper bound of estimated disturbances, respectively, improving the system stability and robustness, which reduce chattering phenomenon. Meanwhile, the adaptive control laws obtained by Lyapunov analysis can guarantee the system a stable operation. Rahmani and Rahman [14] proposed a new robust control method for the control of a 7-DOF exoskeleton robot. The external disturbances and unknown dynamics in the form of friction forces, different upper-limb's mass, backlash, and input saturation make robot unstable, which prevents the robot from correctly following the defined path. A new FOSMC is designed, which is robust against unknown dynamic and external disturbances. Fractional PID controller (FPID) has high trajectory tracking, but it is not robust against external disturbances. Therefore, by combining FOSMC and FPID controllers, a new compound fractional PID sliding mode controller (NCFPIDSMC) is proposed, which benefits high trajectory tracking of FPID and robustness of FOSMC. The stability of the proposed control method is verified by Lyapunov theory. A random noise is applied in order to confirm the robustness of the proposed control method. Rahmani et al. [15] proposed a novel adaptive fractional order PID sliding mode controller (AFOPIDSMC) using a Bat algorithm to control of a Caterpillar robot manipulator. A fractional order PID (FOPID) control is applied to improve both trajectory tracking and robustness. Sliding mode controller (SMC) is one of the control methods which provides high robustness and low tracking error. Using hybridization, a new combined control law is proposed for chattering reduction by means of FOPID controller and high trajectory tracking through using SMC. Then, an adaptive controller design motivated from the SMC is applied for updating FOPID parameters. A metaheuristic approach, the Bat search algorithm based on the echolocation behavior of bats is applied for optimal design of the Caterpillar robot in order to tune the parameter AFOPIDSMC controllers (BA-AFOPIDSMC). Numerical simulation results completely indicate the advantage of BA-AFOPIDSMC for trajectory tracking and chattering reduction. This chapter proposes a novel control approach to eliminate the chattering phenomenon. A sliding surface based on fractional order integral terminal sliding mode is applied in order to improve the robustness of the control system. The chattering phenomenon which is produced by the excitation of fast unmodelled dynamic is the major problem in FOITSMC method. It will be eliminated by FOPD controller. The rest of this chapter is organized as follows. In Sect. 2, the summary has described the dynamic equation of the MEMS gyroscope. In Sect. 3, the FOITSMC is included. In Sect. 4, FOPD-FOITSMC has been delineated. Section 5 presents simulation results and provide the conclusion and contributions of the work.

## 2 Dynamics of MEMS Gyroscope

A $z$-axis MEMS gyroscope is shown in Fig. 1. The conventional MEMS vibratory gyroscope comprises a proof mass suspended by springs, sensing mechanisms and an electronics actuation for forcing an oscillatory motion and velocity of the proof mass and sensing the position.

Oscillatory motion and velocity of the proof mass and sensing the position. The frame that the proof mass is mounted moves with a constant velocity and the gyroscope rotates at a slowly changing angular velocity $\Omega_z$. The centrifugal forces $m\Omega_z^2 x$ and $m\Omega_z^2 y$ are presumed to be negligible because of small displacements. The Coriolis force is generated in a direction perpendicular to the drive and rotational axes. The dynamics of gyroscope according to assumptions which presented above becomes as follows [1]:

$$m\ddot{x} + d_{xx}^{\star}\dot{x} + d_{xy}^{\star}\dot{y} + k_{xx}^{\star}x + k_{xx}^{\star}y = u_x^{\star} + 2m\Omega_z^{\star}\dot{y} \tag{1}$$

$$m\ddot{y} + d_{xy}^{\star}\dot{x} + d_{xy}^{\star}\dot{y} + k_{xy}^{\star}x + k_{yy}^{\star}y = u_y^{\star} - 2m\Omega_z^{\star}\dot{x} \tag{2}$$

The origin for $x$, $y$ coordinates is at the center of the proof mass without force being employed. Fabrication imperfections is supposed to be of help basically to the asymmetric spring and damping terms, $k_{xy}^{\star}$ and $d_{xy}^{\star}$. The $x$ and $y$ axes spring and damping terms $k_{xx}^{\star}$, $k_{yy}^{\star}$, $d_{xx}^{\star}$ and $d_{yy}^{\star}$ are often recognized, but may have small unknown variations from their nominal values [1]. The mass of the proof mass $m$ can be obtained accurately. The positive direction of the control force is same as the $(x, y)$ coordinate. $u_x^{\star}$ and $u_y^{\star}$ are the control forces in the $x$ and $y$ direction. Dividing

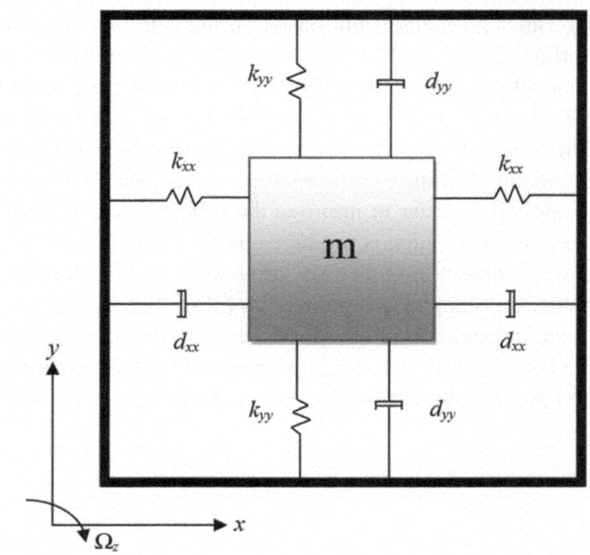


**Fig. 1** Structure of a MEMS gyroscope

gyroscope dynamics (1) and (2) by the reference mass results in the following vector forms as:

$$\ddot{q}^\star + \frac{D^\star}{m}\dot{q}^\star + \frac{K_a}{m}q^\star = \frac{1}{m}u^\star - 2\Omega^\star\dot{q}^\star \tag{3}$$

where:

$$q^\star = \begin{bmatrix} x^\star \\ y^\star \end{bmatrix},\ u^\star = \begin{bmatrix} u_x^\star \\ u_y^\star \end{bmatrix},\ \Omega^\star = \begin{bmatrix} 0 & -\Omega_z^\star \\ \Omega_z^\star & 0 \end{bmatrix}$$
$$D^\star = \begin{bmatrix} d_{xx}^\star & d_{xy}^\star \\ d_{xy}^\star & d_{yy}^\star \end{bmatrix},\ K_a = \begin{bmatrix} k_{xx}^\star & k_{xy}^\star \\ k_{xy}^\star & k_{yy}^\star \end{bmatrix}$$

The final form of the nondimensional equation of motion is as follows [1]:

$$\frac{1}{q_0}\ddot{q}^\star + \frac{D^\star}{m\omega_0 q_0}\dot{q}^\star + \frac{K_a}{m\omega_0^2 q_0}q^\star = \frac{1}{m\omega_0^2 q_0}u^\star - \frac{2}{\omega_0 q_0}\Omega^\star\dot{q}^\star \tag{4}$$

We determine a set of new parameters as follows:

$$q = \frac{q^\star}{q_0},\ d_{xy} = \frac{d_{xy}^\star}{m\omega_0},\ \Omega_z = \frac{\Omega_z^\star}{\omega_0} \tag{5}$$
$$u_x = \frac{u_x^\star}{m\omega_0^2 q_0},\ u_y = \frac{u_y^\star}{m\omega_0^2 q_0} \tag{6}$$
$$\omega_x = \sqrt{\frac{k_{xx}^\star}{m\omega_0^2}},\ \omega_{xy} = \sqrt{\frac{k_{xy}^\star}{m\omega_0^2}},\ \omega_y = \sqrt{\frac{k_{yy}^\star}{m\omega_0^2}} \tag{7}$$

Consequently, the nondimensional representation of (1) and (2) is written as follows:

$$\ddot{q} + D\dot{q} + K_b q = u - 2\Omega\dot{q} \tag{8}$$

where

$$q = \begin{bmatrix} x \\ y \end{bmatrix},\ u = \begin{bmatrix} u_x \\ u_y \end{bmatrix},\ \Omega = \begin{bmatrix} 0 & -\Omega_z \\ \Omega_z & 0 \end{bmatrix},\ D = \begin{bmatrix} d_{xx} & d_{xy} \\ d_{xy} & d_{yy} \end{bmatrix},\ K_b = \begin{bmatrix} \omega_x^2 & \omega_{xy}^2 \\ \omega_{xy}^2 & \omega_y^2 \end{bmatrix}$$

Evidently, Eq. (8) can be defined as:

$$\ddot{q} = -(D + 2\Omega)\dot{q} - K_b q + u + E \tag{9}$$

where $E$ is external disturbance. From Eq. (9), the dynamic equations for a MEMS gyroscope becomes as:

$$\ddot{q} = -Y\dot{q} - Pq + u + E \tag{10}$$

where $Y = D + 2\Omega$ and $P = K_b$. $\Delta Y$ and $\Delta P$ determine some uncertainties of parameter variations. Therefore, Eq. (10) can be denoted as:

$$\ddot{q} = -(Y + \Delta Y)\dot{q} - (P + \Delta P)q + u + E \tag{11}$$

Equation (11) can be denoted as follows:

$$\ddot{q} = -Y\dot{q} - Pq + u - d(t) \tag{12}$$

where $d(t) = \Delta Y\dot{q} + \Delta Pq - E$.

## 3 Fractional Integral Terminal Sliding Mode Control

The most important part of design of a FOITSMC is choosing suitable fractional integral terminal sliding mode surface. Therefore, the fractional integral terminal sliding mode surface can be defined as:

$$s(t) = \dot{e}(t) + \lambda D^{\mu-1}e(t) + \beta \int_0^t \left[\dot{e}^{\frac{m_1}{n_1}}(\tau) + e^{\frac{m_1}{n_1}}(\tau)\right] d\tau \tag{13}$$

where $\lambda$ and $\beta$ are positive constants, $\mu$ is fractional order operator, $m_1, n_1, m_2$ and $n_2$ are positive constants with $m_1 > n_1$ and $m_2 > n_2$. The tracking error can be defined as:

$$e(t) = q(t) - q_d(t) \tag{14}$$

Take derivative of Eq. (13) generates

$$\dot{s}(t) = \ddot{e}(t) + (\mu - 1)\lambda D^{\mu}e(t) + \beta \left[\dot{e}^{\frac{m_1}{n_1}}(t) + e^{\frac{m_1}{n_1}}(t)\right] \tag{15}$$

Equation (15) can be arranged as

$$\dot{s}(t) = \ddot{q}(t) - \ddot{q}_d(t) + (\mu - 1)\lambda D^{\mu}e(t) + \beta \left[\dot{e}^{\frac{m_1}{n_1}}(t) + e^{\frac{m_1}{n_1}}(t)\right] \tag{16}$$

Substituting (12) into (16) generates

$$\dot{s}(t) = -Y\dot{q} - Pq + u - d(t) - \ddot{q}_d(t) + (\mu - 1)\lambda D^{\mu}e(t) + \beta \left[\dot{e}^{\frac{m_1}{n_1}}(t) + e^{\frac{m_1}{n_1}}(t)\right] \tag{17}$$

Equivalent control can be obtained when $\dot{s}(t) = 0$ as

$$u_{eq}(t) = Y\dot{q} + Pq + \ddot{q}_d(t) - (\mu - 1)\lambda D^{\mu}e(t) - \beta \left[\dot{e}^{\frac{m_1}{n_1}}(t) + e^{\frac{m_1}{n_1}}(t)\right] \tag{18}$$

The equivalent control effort cannot guarantee the favorable control performance if unpredictable perturbations from the external disturbance or parameter variations occur. Therefore, in order to eliminate the effect of the unpredictable perturbations, auxiliary control effort should be designed. To do this, the Lyapunov function can be selected as [1]:

$$L(t) = \frac{1}{2}s^T(t)s(t) \tag{19}$$

The condition that control method is stable can be defined as, for $s(t) \neq 0$:

$$\dot{L}(t) = s^T(t)\dot{s}(t) < 0 \tag{20}$$

The control input can be defined as

$$u = u_{FOITSMC} = u_{eq} + u_s(t) \tag{21}$$

Substituting (17) into (20) generates

$$\dot{L}(t) = s^T(t)\Big[ - Y\dot{q} - Pq + u - \ddot{q}_d(t) + (\mu - 1)\lambda D^\mu e(t) + \beta\left(\dot{e}^{\frac{m_1}{n_1}}(t) + e^{\frac{m_1}{n_1}}(t)\right)\Big] \tag{22}$$

Substituting (21) into (22) produces

$$\dot{L}(t) = s^T(t)\Big[ - Y\dot{q} - Pq + u_{eq} + u_s - \ddot{q}_d(t) + (\mu - 1)\lambda D^\mu e(t) + \beta\left(\dot{e}^{\frac{m_1}{n_1}}(t) + e^{\frac{m_1}{n_1}}(t)\right)\Big] \tag{23}$$

Substituting (18) into (23) generates

$$\begin{aligned}\dot{L}(t) = s^T(t)\Big[ & - Y\dot{q} - Pq + Y\dot{q} + Pq + \ddot{q}_d(t) - (\mu - 1)\lambda D^\mu e(t) \\ & -\beta\left(\dot{e}^{\frac{m_1}{n_1}}(t) + e^{\frac{m_1}{n_1}}(t)\right) \\ & + u_s - \ddot{q}_d(t) + (\mu - 1)\lambda D^\mu e(t) + \beta\left(\dot{e}^{\frac{m_1}{n_1}}(t) + e^{\frac{m_1}{n_1}}(t)\right)\Big]\end{aligned} \tag{24}$$

which gives:

$$\dot{L}(t) = s^T(t)u_s \tag{25}$$

To ensure that $\dot{L}(t)$ is negative, the reaching control law should be chosen as:

$$u_s = -K_s s \tag{26}$$

where $K_s$ is a positive constant. By substituting (25) into (26), $\dot{L} < 0$ will be observed.

## 4 Compound Controller

Proportional derivative controller (PD) is a common control method which has been used in different structure. FOPD will improve the performance of the PD controller in terms of trajectory tracking and accuracy. The FOPD controller can be defined as:

$$u_{FOPD} = k_p e(t) + k_d D^{\mu} e(t) \tag{27}$$

where $k_p$ and $k_d$ are FOPD controller parameters.

The proposed FOPD-FOITSMC can be defined as:

$$u(t) = u_{FOITSMC}(t) - u_{FOPD}(t) \tag{28}$$

This gives:

$$\dot{L}(t) = s^T(t)\big[u_s - u_{FOPD}(t)\big] \tag{29}$$

When time goes to infinity ($t \longrightarrow +\infty$), the tracking error goes to zero ($e(t) \longrightarrow 0$). Equation (29) can be denoted as:

$$\dot{L}(t) = s^T(t)\big[u_s - u_{FOPD}(t)\big] \tag{30}$$

To guarantee that Eq. (30) is less than zero, the reaching control law should be chosen as:

$$u_s = -K_s s \tag{31}$$

where $K_s$ is a positive constant. By substituting Eq. (31) into Eq. (30), $\dot{L} < 0$ will be observed.

## 5 Simulation Results

The FOITSMC parameters are selected as $K_s = 10, \beta = 5, \lambda = 5, m_1 = 1.5, n_1 = 1$, $m_2 = 1.5, n_2 = 1$ and $\mu = 1.5$. The desired motion trajectory is determined by $q_{d1} = \sin 4.17t$, and $q_{d2} = 1.2 \sin 5.11t$. The FOPD controller parameters are as chosen as $k_{p1} = 20$, $k_{p2} = 30$, $k_{d1} = 10$, and $k_{d2} = 10$.

The initial values of the system are selected as: $q_1(0) = 0.4$, $q_2(0) = 0.6$, and $\dot{q}_1(0) = \dot{q}_2(0) = 0$.

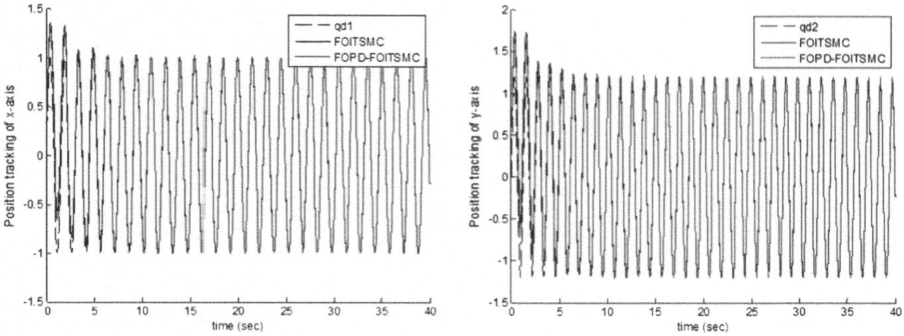


**Fig. 2** Position tracking of $x$-axis and $y$-axis under FOITSMC and FOPD-FOITSMC

The parameters of the MEMS gyroscope are selected as:

$$\begin{array}{lll} k_{xx} = 63.955 \text{ N/m} & k_{xy} = 12.779 \text{ N/m} & k_{yy} = 95.92 \text{ N/m} \\ d_{xx} = 18 \times 10^{-7} \text{ Ns/m} & d_{xy} = 3.6 \times 10^{-7} \text{ Ns/m} & d_{yy} = 18 \times 10^{-7} \text{ Ns/m} \\ m = 1.8 \times 10^{-7} \text{ kg} & & \end{array}$$

When the displacement range of the MEMS gyroscope in each axis is in sub-micrometer level, it is convenient to choose $1\mu$m as the reference length $q_0$ [1]. When the $\omega_0$ is selected as 1 kHZ, the common natural frequency of each axis of a MEMS gyroscope is in the kHZ range. The unknown angular velocity is assumed as $\Omega_z$ =100 rad/s. Therefore, the non-dimensional values of the MEMS gyroscope parameters are chosen as:

$$\omega_x^2 = 355.3\ ,\ \omega_y^2 = 532.9\ ,\ \omega_{xy} = 70.99$$
$$d_{xx} = d_{yy} = 0.01\ ,\ d_{xy} = 0.002\ ,\ \Omega_z = 0.1$$

Figure 2 illustrates the MEMS gyroscope along the x and y axes tracking trajectories using the FOITSMC, and FOPD-FOITSMC, respectively. It can be clearly observed that the actual motion trajectory of the MEMS gyroscope is consistent with the desired reference trajectory in a limited time, showing that the tracking performance of FOPD-FOITSMC is better in comparison with FOITSMC.

According to Fig. 3, a comparable investigation is accomplished between the two proposed novel controller FOITSMC, and FOPD-FOITSMC law applied to the MEMS gyroscope in order to show the advantages of the FOPD-FOITSMC approach, which improves the dynamic behavior of the MEMS gyroscope and verifies that the proposed control scheme can guarantee the asymptotical stability of the control system. In addition, the chattering problem exists in NFSMC according to Fig. 3. As Fig. 3 states, the chattering phenomenon is eliminated, maximum overshoot is reduced and settling time converged at zero in a limited time for $x$ and $y$-axes. As observed in Figs. 2 and 3, FOPD-FOITSMC can obtain a faster and more

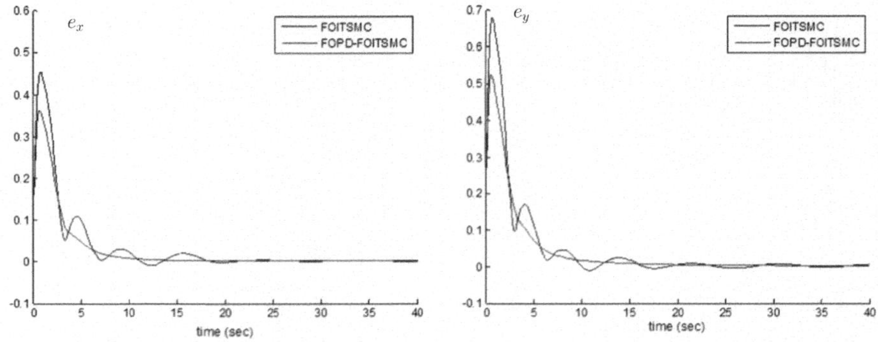


**Fig. 3** Position tracking error of $x$-axis and $y$-axis under FOITSMC and FOPD-FOITSMC

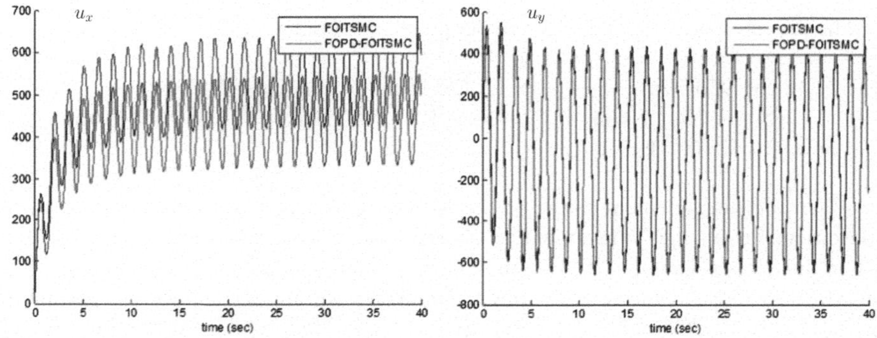


**Fig. 4** Control effort using FOITSMC and FOPD-FOITSMC

efficient performance of the reference signal than FOITSMC. Figure 4 illustrates the control efforts of FOITSMC and FOPD-FOITSMC. Clearly, the control input with the FOPD-FOITSMC is smoother than FOITSMC.

Therefore, the chattering phenomenon has been reduced in FOPD-FOITSMC. Figure 5 shows sliding surface under FOITSMC and FOPD-FOITSMC.

## 6 Conclusion

In this paper, a new compound FOPD and FOITSMC for the control of a MEMS gyroscope was proposed. The FOPD-FOITSMC approach is less sensitive to the parameter variations. It provides a model-free control when standard SMC required information of the dynamic system, which has to be controlled. Also, the proposed controller combines to eliminate their drawbacks, and has an excellent tracking trajectory performance following these conditions. Chattering phenomenon is the main

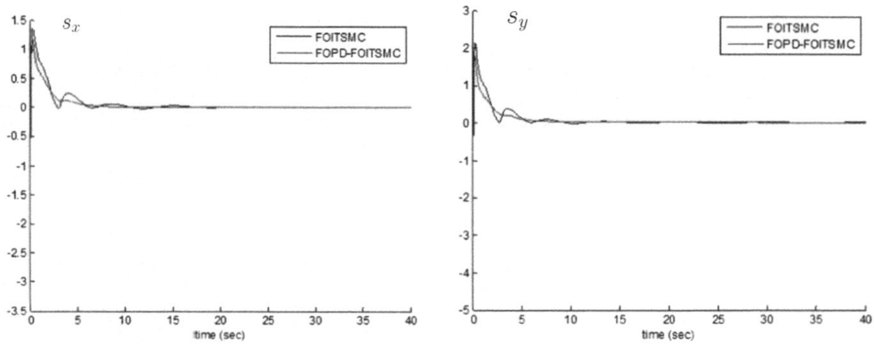


**Fig. 5** Sliding surface under FOITSMC and FOPD-FOITSMC

problem of FOITSMC. FOPD-FOITSMC removed chattering in MEMS gyroscope. It described an effective robustness when it kept low tracking errors. The numerical simulations confirm that we have a control system with excellent performances.

## References


1. Rahmani, M.: MEMS gyroscope control using a novel compound robust control. ISA Trans. **72**, 37–43 (2018)
2. Rabah, K., Ladaci, S.: Fractional adaptive sliding mode control laws for fractional order chaotic systems synchronization. In: 17th International Conference on Sciences and Techniques of Automatic Control and Computer Engineering (STA), pp. 293–302. Tunisia (2016)
3. Tianyi, Z., Xuemei, R., Yao, Z.: A fractional order sliding mode controller design for spacecraft attitude control system. In: 34th Chinese Control Conference (CCC), pp. 3379–3382 (2015)
4. Tian, J., Chen, N., Yang, J., Wang, L.: Fractional order sliding model control of active four-wheel steering vehicles. In: International Conference on Fractional Differentiation and Its Applications (ICFDA), pp. 1–5 (2014)
5. Kuang, Z., Shao, X., Li, X., Sun, G.: High-precision analysis of discrete-time fractional-order sliding mode control. In: Chinese Control and Decision Conference (CCDC), pp. 3083–3087 (2018)
6. Lin, J.S., Yan, J.J., Yang, Y.S., Liao, T.L.: Based on sliding mode control to synchronize of switched fractional Lorenz systems. International Symposium on Computer Communication Control and Automation (3CA), vol. 2, pp. 278–281 (2010)
7. Rahmani, M., Ghanbari, A., Ettefagh, M.M.: Hybrid neural network fraction integral terminal sliding mode control of an Inchworm robot manipulator. Mech. Syst. Signal Process. **80**, 117–136 (2016)
8. Rahmani, M., Ghanbari, A., Ettefagh, M.M.: A novel adaptive neural network integral sliding-mode control of a biped robot using bat algorithm. J. Vib. Control **24**(10), 2045–2060 (2018)
9. Wu, H., Wang, L., Niu, P., Wang, Y.: Global projective synchronization in finite time of nonidentical fractional-order neural networks based on sliding mode control strategy. Neurocomputing **235**, 264–273 (2017)
10. Hu, T., Zhang, X., Zhong, S.: Global asymptotic synchronization of nonidentical fractional-order neural networks. Neurocomputing (2018)
11. Ding, Z., Shen, Y.: Projective synchronization of nonidentical fractional-order neural networks based on sliding mode controller. Neural Netw. **76**, 97–105 (2016)

12. Efe, M.O.: Fractional fuzzy adaptive sliding-mode control of a 2-DOF direct-drive robot arm. IEEE Trans. Syst. Man Cybern. Part B **38**(6), 1561–1570 (2008)
13. Liu, N., Fei, J.: Adaptive fractional sliding mode control of active power filter based on dual RBF neural networks. IEEE Access **5**, 27590–27598 (2017)
14. Rahmani, M., Rahman, M.H.: Novel robust control of a 7-DOF exoskeleton robot. PloS one **13**(9), e0203440 (2018)
15. Rahmani, M., Ghanbari, A., Ettefagh, M.M.: Robust adaptive control of a bio-inspired robot manipulator using bat algorithm. Expert. Syst. Appl. **56**, 164–176 (2016)

# Highly Sensitive Polymer/Multiwalled Carbon Nanotubes Based Pressure and Strain Sensors for Robotic Applications

**Rajarajan Ramalingame, Ayda Bouhamed, Dhivakar Rajendran, Renato da Veiga Torres, Zheng Hu and Olfa Kanoun**

**Abstract** In the last decades, significant advances have been reached in the robotic field by the development and implementation of novel tactile sensors for robot hand and body attached sensors, in order to achieve high robot performance. Because of the unique and fascinating properties of polymer/carbon nanotubes (CNTs) nanocomposites, they were chosen for the design of highly sensitive and stable tactile sensors to detect pressure and strain. In this chapter we propose two sensor structures for robotic applications. Poly-Dimethylsiloxane/Multiwalled carbon nanotubes (MWCNTs) soft pressure sensors were prepared using solution processing method and molded in different shapes depending on the application requirements. The developed pressure sensors show promising piezocapacitive performance, very high sensitivity of 45%/N at low forces of 0–1 N and a wide pressure sensing a range of 0.5 Pa–570 kPa. These properties are essential for robotic applications like touch sensing, grasping and gait analysis. In the other part, conductive thermoplastic polyurethane (TPU)/MWCNTs filaments strain sensors were firstly made using an extrusion process. The sensors were subjected to strain loads. The results demonstrate excellent piezoresistive responses a gauge factor around 26 and stretchability more than 50% that was suitable for measuring finger bending.

**Keywords** Carbon nanotubes · Polymers · Nanocomposites · Pressure sensors · Strain sensors · Robotics · Grasping · Gait · Gesture recognition

## 1 Introduction

Nowadays, a lot of advancement is occurring in the field of robotics to meet the new requirements of the diverse and complex applications such as in industries, hospitals, homes, etc. Therefore, the use of tactile sensors in robotics is becoming a necessity to provide the robot to become cautious and capable to interact with environments

R. Ramalingame (✉) · A. Bouhamed · D. Rajendran · R. da Veiga Torres · Z. Hu · O. Kanoun
Faculty of Electrical Engineering and Information Technology, Technische Universität Chemnitz, 09126 Chemnitz, Germany
e-mail: rajarajan.ramalingame@etit.tu-chemnitz.de



J. Ghommam et al. (eds.), *New Trends in Robot Control*, Studies in Systems, Decision and Control 270, https://doi.org/10.1007/978-981-15-1819-5_19

and to operate safely in diverse locations. In fact, tactile sensors are used to provide quantitative pressure mapping due to the external contact, which can be helpful for robot grasping and handling different objects with different shapes, surface texture and material stiffness without slipping. Therefore, the key challenge of today is the development of artificial skin interfaces with fully distributed tactile sensors that should be wearable, with non-complex structure, superior sensing performance as well as to mimic the properties of human skin. Up to now, many attempts are exciting to develop flexible tactile sensors based on diverse transducing mechanisms such as capacitive [16], piezoresistive [5], piezoelectric [30], inductive [19] and optical principle [11]. Piezoresistive and capacitive tactile sensors are often preferred due to their simplicity and high performances as shown in Table 1.

Recently, conductive polymers are frequently used for tactile sensing in order to avoid the limitations of traditional sensors based on metal plates such as narrow sensing range ($\epsilon < 5\%$) and low sensitivity. Conductive polymer sensors consists usually of a binder polymer filled with conductive particles. Various conductive nanoparticles have been used as filler to form a piezoresistive material, e.g. carbon black [22], graphene [13], carbon nanotubes [18], and various metal powders [6, 15] as illustrated in Table 1. These sensing materials can change the resistivity under applied load and their sensitivity can be changed depending on the type of nanoparticle, the used polymer and the processing conditions. Comparing to the other type of sensors, these sensing materials have usually a wide range of sensitivity, good mechanical resistance, and low power consumption caused by the simple signal conditioning electronics. Based on this working principle, [31] developed a highly stretchable, sensitive and durable graphene-based strain sensor using a simple dry coating process. The developed sensor shows a stretchability up to 750%, high durability of 1500 stretching-releasing cycles and gauge factors ranging from 0.78 to 52.53 vary depending on the preparation conditions and strain state. The strain sensors can detect bending fingers and the pulse of the radial artery on the wrist. In another work, [32] fabricated a flexible and reproducible pressure sensor based on polyurethane (PU) foam coated with the multiwalled carbon nanotubes (MWCNTs) and reduced graphene oxide (rGO) ink (MWNT−rGO@PU foam). The MWNT−rGO@PU foam-based has the potential to detect both small-scale and large-scale movements over 5000 cycles.

In addition, capacitive sensors based on the use of conductive polymers are gaining attention due to their high sensitivity, good frequency response, less temperature effect and power consumption.

In this kind of sensors, the sensitivity is correlated to the compressibility of the dielectric material. Therefore, most of the works are introducing super-low modulus materials such as ecoflex silicone, poly-Dimethylsiloxane (PDMS) [20]. In addition, for maximum sensitivity, the sensor has to have an air gap between electrodes. Therefore, [15] developed a fabrication process to realize this air gap structure using a flexible polymer for a large-area capacitive tactile sensor array. The fabricated sensor array shows a sensitivity of 3%/mN within the full-scale range of 40 mN (250 kPa).

This work is dedicated to illustrate the recent progress and development in wearable tactile sensors based on polymer nanocomposites for robotic applications. In

**Table 1** Comparison of previous works on tactile pressure sensor. $P$-Pressure, $S$-Strain, $Ph.p$-Proximal phalange, $D.p$-Distal phalange

| Sensing principle | Qty | Material | Sensor characteristics | | Cells | Spatial resolution (mm) | Area of use | | | | Year/Author |
|---|---|---|---|---|---|---|---|---|---|---|---|
| | | | Sensitivity | Range | | | Ph.p | D.p | Palm | Arm | |
| Resistive | P | CNT/ZnO NPs | GF ≈ 306 | 1 Pa–25 kPa | – | – | | | | | [17] |
| Resistive | P | Nano-graphene | 4.3 $Pa^{-1}$ | 0.11–80 kPa | 10 | – | ✓ | ✓ | | | [24] |
| Pneumatic | P | Air bladder | – | 0–10 kPa | – | – | | ✓ | | | [9] |
| Resistive | S | Nanographene | GF:12.35–52.53 | 150–270% | 5 | – | ✓ | | | | [31] |
| Microfluidic | P | Galinstan | 0.0835 $kPa^{-1}$ | – | 1 | – | | | | | [7] |
| Capacitive | S | CNT/PDMS | – | – | – | – | | ✓ | | | [23] |
| Resistive | P | Flex | – | 1.41–214 kPa | 16 | 4 | ✓ | | | | [29] |
| Capacitive | P | Graphene | 0.002 $kPa^{-1}$ | 0.5–500 kPa | 6×6 | – | | ✓ | | | [12] |
| Resistive | P | Graphene | 0.96 $kPa^{-1}$ | 5–100 kPa | – | | ✓ | | | | [33] |
| Resistive | P | r-GO | 5.50 $kPa^{-1}$ | 0.0015–0.1 kPa | – | – | | | | | [36] |
| Capacitive | P | PDMS, Au, Cr | 0.08 pf/kPa | 250 kPa | 5 | – | ✓ | ✓ | | | [8] |
| Resistive | P | Rubber | – | 15 kPa | – | – | ✓ | ✓ | | | [2] |
| Capacitive | S | Silicon | 0.6 $\mu m^{-1}$ | 0.66% | 12 | – | | ✓ | | | [28] |
| OFET | P | FET | 0.8 kPa | 0–20 kPa | – | – | ✓ | | ✓ | | [21] |
| Resistive | P | Elastomer | – | 20–300 kPa | 32×32 | 5 | ✓ | | ✓ | | [34] |
| Optical | P | – | 0–05 N | 0.05–0.5 N | – | 5 | | ✓ | | | [4] |
| Resistive | P | Polymer | 0.1%/kPa | 0–30 kPa | 32×32 | 1.9 | ✓ | ✓ | | | [35] |
| Resistive | P | Polymer | 300 Ω/kPa | 20–300 kPa | 8×8 | 3 | ✓ | ✓ | | | [3] |
| Optical | P | Fiber | 0.023 V | 0.01–40 N | 4×4 | 2.6 | ✓ | | ✓ | | [10] |
| Capacitive | P | PDMS | 3%/250 kPa | 250 kPa | 16×16 | 1 | | | | ✓ | [14] |

this chapter, capacitive pressure sensor and high sensitive and flexible piezoresistive strain sensor were realized. A capacitive pressure sensor based on PDMS/MWCNTs were prepared in different shapes. Because of the high flexibility of these sensors, they were placed in different locations in the hand like fingers and palm for precision and power grasping in addition to gait analysis in a humanoid robot. Furthermore, a conductive thermoplastic polyurethane filled with MWCNTs was chosen to fabricate the filament via extrusion and then tested to sense finger bending.

## 2 Polymer/Carbon Nanotubes Tactile Sensors

Functional materials formed by the combination of polymer materials and conductive fillers are widely used for tactile sensing because of the tunability for different load ranges in addition to the possibility to adjust the properties depending on the application requirements. Among all polymer nanocomposites, polymer filled with carbon nanotubes is frequently used due to its robustness caused from the excellent mechanical properties of carbon nanotubes (CNTs) in addition to the excellent electrical properties at very low CNTs concentrations due to their high aspect ratio.

For tactile sensing, polymers with high stretchability and flexibility are usually used such as elastomers and thermoplastic polymers because they are ideal candidates for application requiring a high degree of bending and grasping. However, the main challenging aspect in the use of polymer/CNTs is to obtain a homogenous sensing material. In fact, CNTs tends to agglomerates caused by the Van der Walls attractive forces. For better CNTs distribution, many approaches have been adopted in literature such as solution mixing, in-situ polymerization, melt mixing via an extruder, etc. The selection of the approach depends usually on the used polymer. In this chapter, two different sensors were fabricated using two low cost fabrication techniques, which are solution mixing and melt mixing that it will be described in the Sect. 3.

## 3 Fabrication and Properties of Nanocomposite Sensors

### *3.1 Mold Cast, Soft Pressure Sensors*

The synthesis process of nanocomposite-based pressure sensor deals with the dispersion of MWCNTs in PDMS using an organic solvent such as Tetrahydrofuran (THF) as a pre-dispersion medium [25]. By an optimized sonication and mechanical stirring of MWCNTs in THF followed by in PDMS, a homogeneous dispersion of MWCNTs in PDMS is achieved [26]. This homogeneity is essential to realize a sensor with high sensitivity to pressure, at a low concentration of MWCNTs ($<1$ wt.%), which leads to a cost-effective sensor. The complete fabrication process is graphically represented in Fig. 1. To realize a sensor with predefined shapes and sizes, the

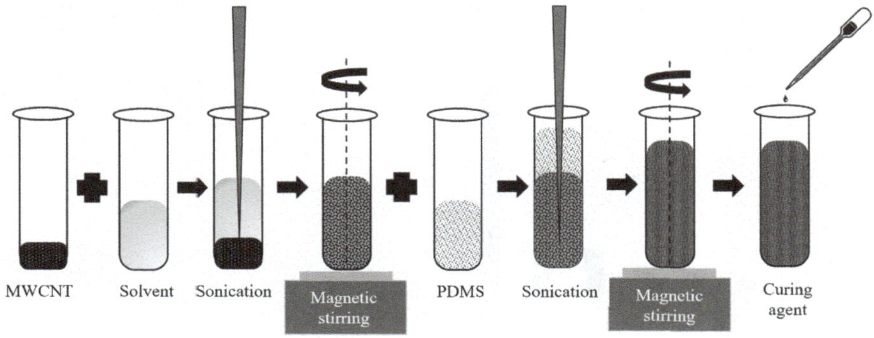


**Fig. 1** Nanocomposite synthesis process steps

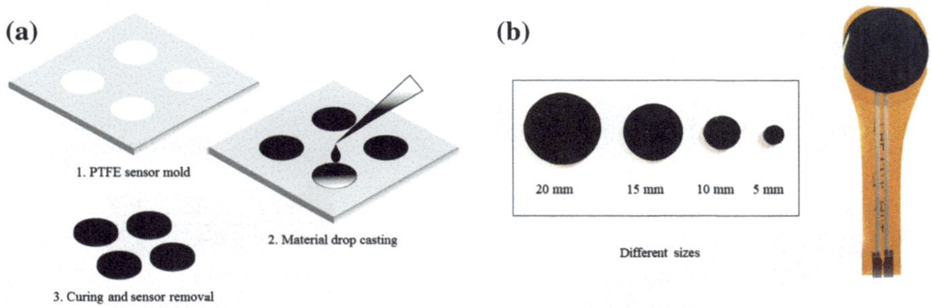


**Fig. 2** **a** Illustration of the sensor mold cast process steps, **b** different shapes of the diced sensor and **c** circular inter-digital electrodes, screen printed with silver ink for electrical contact to the sensor and sensor placed on the electrode structure

nanocomposite material can be drop casted in any mold structure and then heated in an oven for 2 h at 120 °C. As a best practice, before the curing process, the mold can be placed in a degassing chamber for 15 min to remove any trapped bubbles during the synthesis process. Following the curing process, the sensors can be lifted off the mold and placed over circular inter-digital electrode structures, screen printed using silver ink on a Kapton substrate. The complete fabrication process is graphically represented in Fig. 2. Figure 3a shows the ability of the sensor to sense low weights starting from 1 g and Fig. 3b shows the sensor response at higher forces up to 180 N.

## *3.2 Extruded Filament-Based Strain Sensors*

In order to prepare the filament sensor, conductive TPU masterbatch (10% of carbon in MWCNT form content) was purchased from Nanocyl (Belgium). CTPU was provided in form of pellets with diameter between 1 to 3 mm and length of 5 mm and mixed with SBS (styrene-butadiene-styrene rubber) in a glass flask. This mixture

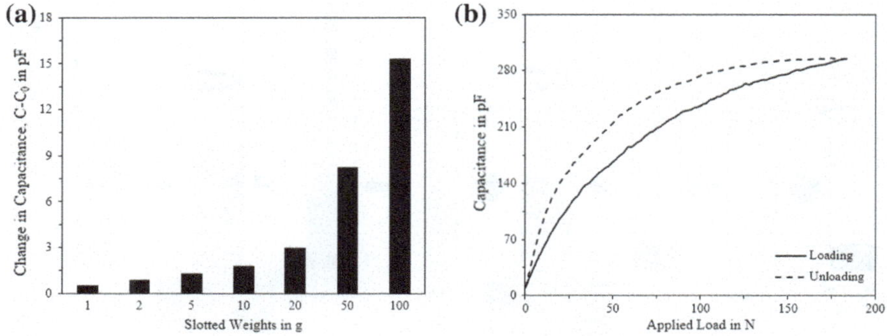


**Fig. 3** **a** Piezocapacitive response of the sensor at low forces and **b** Piezocapacitive response of the sensor at high forces

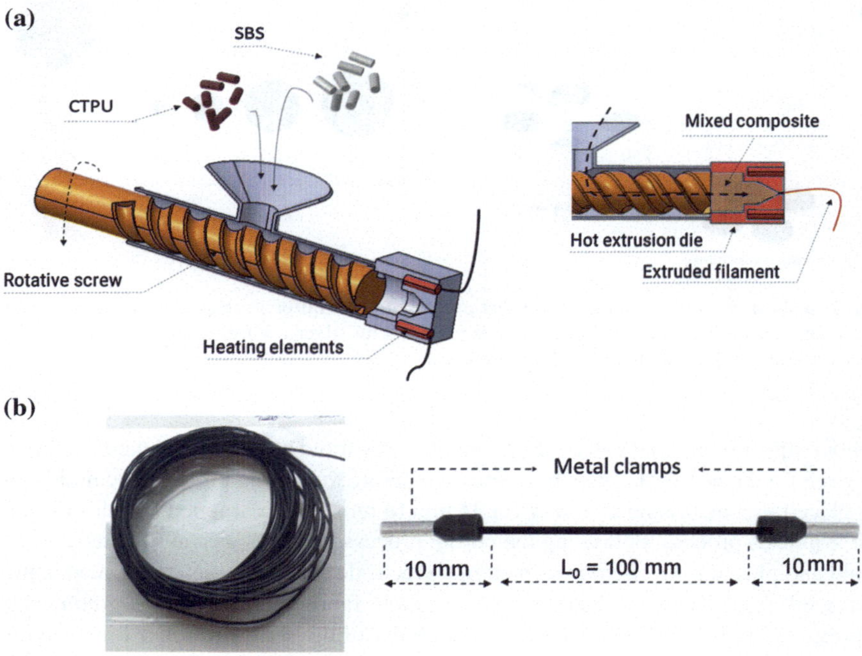


**Fig. 4** **a** Cross section of extruder device showing the material feeding and extruder parts and **b** obtained CTPU/SBS filaments with uniform diameter

is then added to an extruder device where the pellets are melted at a temperature up to 250 °C under a constant pressure exerted by the screw rotation as shown in Fig. 4a. The mixture is pushed towards a heated metal die where the filaments form by passing through the die hole of 1 mm. The resulting composite is a filament blend made of SBS and C-TPU. The filament sensor with 12 cm in length was attached

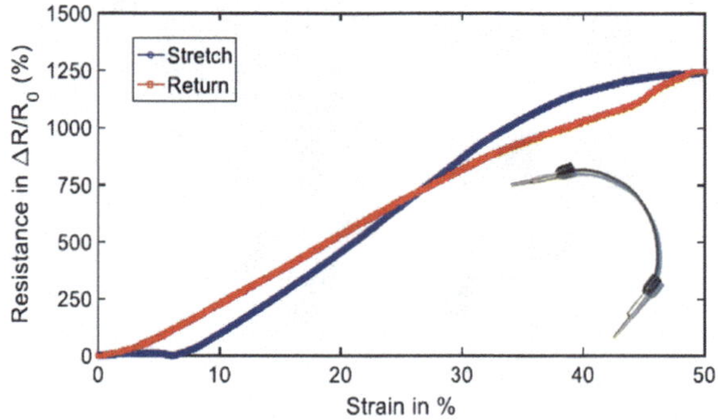


**Fig. 5** Relative resistance change of the filament sensor performance with the respect to strain

from both sides with metal clamps as illustrated in Fig. 4b. The initial resistance of obtained filaments was approximately 1 MΩ. According to the Fig. 5, a linear behavior is observed for the electrical resistance by increasing the strain till 50%. In addition, it can be seen that the sensor is highly sensitive to load change with a gauge factor (GF) around 26 demonstrating the high potential of this filament for detecting gesture change.

# 4 Case Study

## *4.1 Gait Analysis for Humanoid Robots*

Humanoid robots are designed to mimic human-like movements and their power-driven legs enable motion in addition to balancing the weight of the entire robot during motion. Thereby, such humanoid robot bears a gait that is much different from that of humans and hence it is essential to monitor and analyse it. The synthesized nanocomposite pressure sensor is utilized to analyze the gait of a humanoid robot. Custom-Build humanoid robot legs supported by a framework yet capable of moving from one place to other is the test subject [27]. The sensor is placed on the foot of the robot and the piezocapacitive response was recorded in real time using a Python controlled RCL meter Fluke PM6304. The robot was programmed to walk 1 m which corresponds to 3 steps in each leg and one such response is shown in Fig. 6a. The numbers along the curve in the Fig. 6b denote the following:

0—the right leg is above the ground and the robot is supported by the left leg
1—the right leg has established contact with the ground
2—both right and left leg is firmly placed in the ground
3—left leg starts to lift-up and the force exerted on the right leg increases
4—left leg approaches the ground and the robot starts to hang
0—right leg is above the ground until the walking cycle of the left leg is complete.

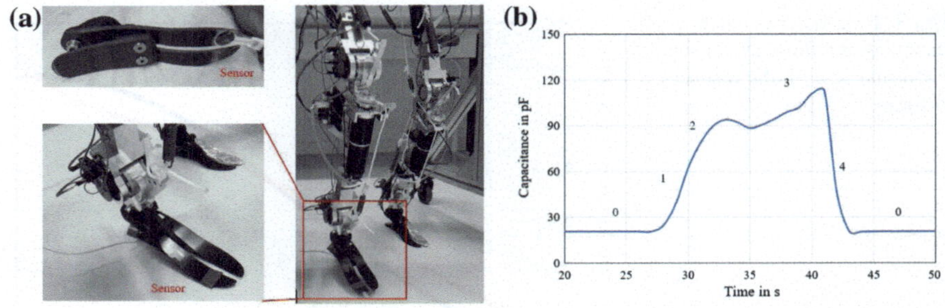


**Fig. 6** **a** Sensor placement in the foot of the humanoid robot for gait analysis and **b** Pressure response curve obtained by one walk cycle of the robot

The experiment shows that the sensor can track the physical movements of the robot and has the potential in analyzing the gait of the robot. Such an analysis can provide critical information to ensure optimal functioning of the robot in its desired application.

## 4.2 Tactile Pressure Sensors for Grasping

Robotic hands facilitate complex tasks that are challenging for human hands, either to perform and not easily accessible due to harsh environments. This robotic hands can perform both power and precision grasping. By combining and controlling both an accurate grasping of the desired object can be accomplished. Force feedback to the robot hand can enhance power and grasping accuracy. Hence, a sensor matrix with 12 sensor elements of 5 mm in diameter can be placed in each segment of the robotic finger as shown in Fig. 7a.

A slotted weight of 10 g and diameter of 8 mm was placed on the sensor matrix at different positions. The response of the sensors was acquired in real-time by means of a microcontroller-based data acquisition system equipped with a 16-channel multiplexer system. Figure 7b shows one such position where the 10 g weight was placed and the corresponding graphical representation of the visualized data. This sensor matrix could enable robots to perform grasping with a high level of precision.

## 4.3 Gesture Recognition

Gesture recongnition can be achieved by the electronic glove, which has nanocomposite filament strain sensor in each finger, and pressure sensors are on inside the hand. Sensor placement and a number of sensors depend on the finger movement and maximum pressure distribution formed during the object handling and this can

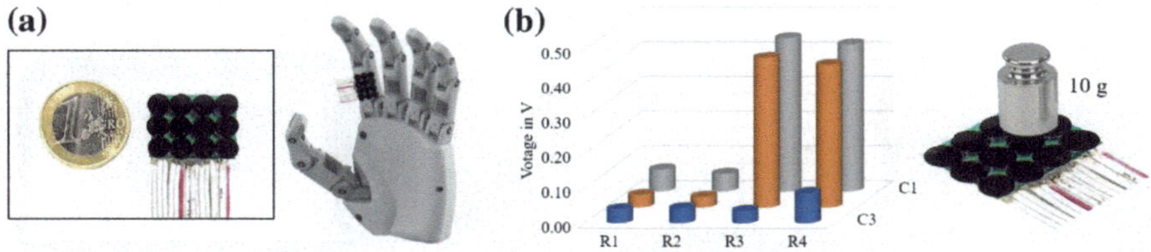


**Fig. 7** **a** Cross section of extruder device showing the material feeding and extruder parts and **b** obtained CTPU/SBS filaments with uniform diameter

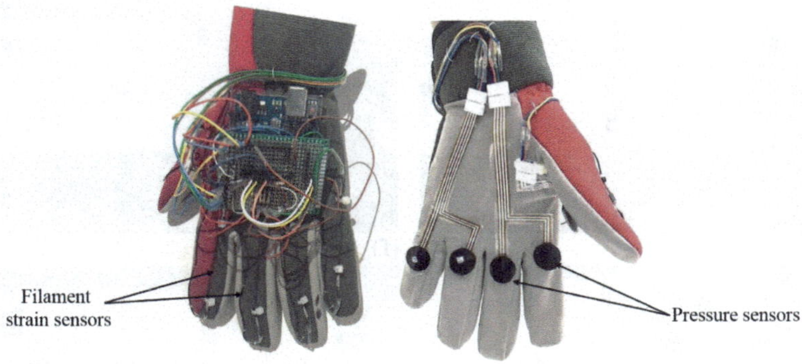


**Fig. 8** Electronic glove with sensors and acquisition system

be changed relying upon the patients need as well as doctors' advice. The hardware for this setup is placed over the glove as shown in Fig. 8.

In order to evaluate the working of the glove, we have recorded the movement of the hand while holding a glass bottle. From the Fig. 9b, it can be inferred that while holding a bottle, the pressure range is in the order: thumb > fore finger > middle finger > ring finger > baby finger which is similar to the measurements done in the previous works [1]. Figure 9b shows the pressure distribution while holding the object in fore and middle finger. Individual sensors were evaluated which can be seen in Fig. 9c. Depends on the measured resistance value of the strain sensors, the state of each finger can be classified into two classes: stretch or flexure, i.e., if the resistance increases with strain in the nanocomposites strain sensors is increased and so high resistance value means that the corresponding to the finger is in flexure position, while low resistance value means that the corresponding finger is in stretch position. Therefore, five fingers can have 32 different state combinations. Considering the fact that English has 26 letters, 26 finger combination states are selected out, and each finger combination state represents one English letter. The user can spell the English word letter by letter by controlling his/her fingers then. For the rest 5 fingers state combinations, they can be used to indicate different marks like period, question, exclamation, and so on. A predefined sign language library has been created for the alphabet "U" and words "can, control", which is shown in Fig. 9d, e, f.

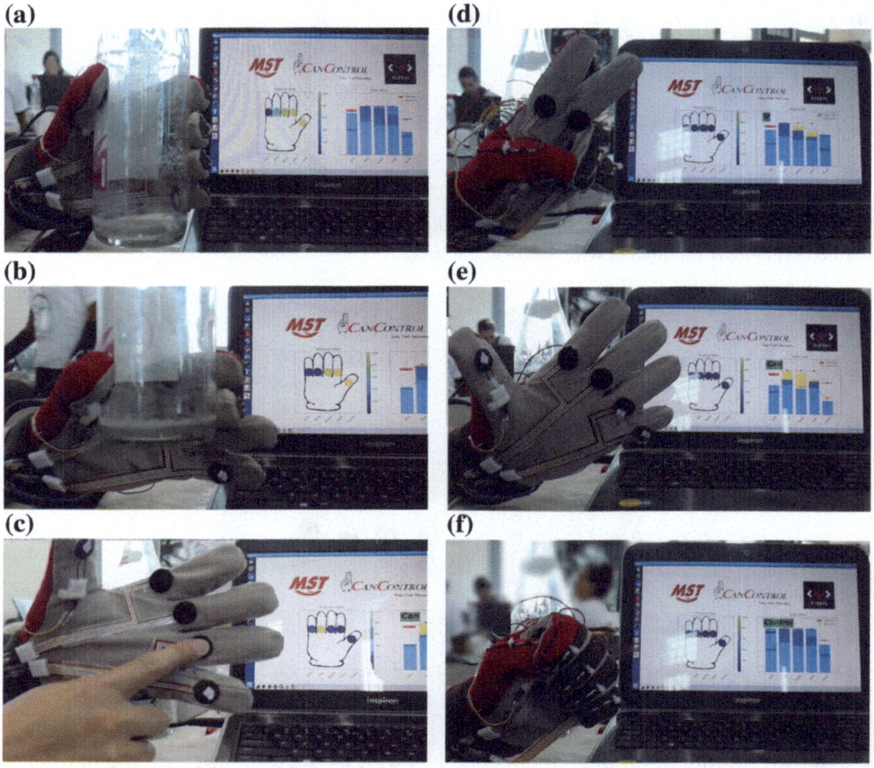

**Fig. 9** Left: **a** Holding the bottle with all fingers **b** Holding the bottle with thumb, fore and middle finger **c** Pressing the individual sensor in the ring finger. Converting sign language for the words: **d** U **e** Can **f** Control

## 5 Conclusions

In this chapter, we have firstly highlighted the recent progress in the field of tactile sensors for robotic applications. Then, two sensors based on the use of polymer/CNTs nanocomposite were proposed to realize highly flexible pressure and strain sensor. In fact, these two sensing materials are based on two different working principles; piezocapacitive and piezoresistive, respectively.

The developed sensors show excellent performance in terms of high flexibility, good adaptability to be in different size and to be implemented on different surfaces and good sensitivity to pressure and strain. In addition, the fabrication processes of these sensors were simple. Due to all these advantages, these sensors demonstrate high capability to detect different hand gesture and ability to be used for grasping and gait analysis for humanoid robots.

**Acknowledgements** The research work is carried out under the "Landesinnovationsstipendium (100284169)", funded by the Sachsische Aufbaubank (SAB) and the European Social Fund (ESF). In addition, a part of this work was funded by the Brazilian CSF (Ciencia sem Fronteiras) program from CAPES/CNPq under the contract number 207319/2014-6. We thank Prof. Dr.-Ing. Ulrike Thomas, Chair of Robotics and Human Machine Interaction for providing facilities to analysis the gait in humanoid robots.

## References

1. Ahmed, S.F., Ali, S.M.B., Qureshi, S.S.M.: Electronic speaking glove for speechless patients, a tongue to a dumb. In: 2010 IEEE Conference on Sustainable Utilization and Development in Engineering and Technology, pp. 56–60. IEEE (2010)
2. Aqilah, A., Jaffar, A., Bahari, S., Low, C.Y., Koch, T.: Resistivity characteristics of single miniature tactile sensing element based on pressure sensitive conductive rubber sheet. In: 2012 IEEE 8th International Colloquium on Signal Processing and Its Applications, pp. 223–227. IEEE (2012)
3. Cheng, M.-Y., Tsao, C.-M., Lai, Y.-T., Yang, Y.-J.: A novel highly-twistable tactile sensing array using extendable spiral electrodes. In: 2009 IEEE 22nd International Conference on Micro Electro Mechanical Systems, pp. 92–95. IEEE (2009)
4. Chorley, C., Melhuish, C., Pipe, T., Rossiter, J.: Development of a tactile sensor based on biologically inspired edge encoding. In: 2009 International Conference on Advanced Robotics, pp. 1–6. IEEE (2009)
5. Dao, D.V., Sugiyama, S., Hirai, S., et al.: Analysis of sliding of a soft fingertip embedded with a novel micro force/moment sensor: simulation, experiment, and application. In: 2009 IEEE International Conference on Robotics and Automation, pp. 889–894. IEEE (2009)
6. Fiorillo, A.S.: A piezoresistive tactile sensor. IEEE Trans. Instrum. Meas. **46**(1), 15–17 (1997)
7. Gao, Y., Ota, H., Schaler, E.W., Chen, K., Zhao, A., Gao, W., Fahad, H.M., Leng, Y., Zheng, A., Xiong, F., et al.: Wearable microfluidic diaphragm pressure sensor for health and tactile touch monitoring. Adv. Mater. **29**(39), 1701985 (2017)
8. Gerratt, A.P., Sommer, N., Lacour, S.P., Billard, A.: Stretchable capacitive tactile skin on humanoid robot fingers? first experiments and results. In: 2014 IEEE-RAS International Conference on Humanoid Robots, pp. 238–245. IEEE (2014)
9. Gong, D., He, R., Yu, J., Zuo, G.: A pneumatic tactile sensor for co-operative robots. Sensors **17**(11), 2592 (2017)
10. Hasegawa, Y., Shikida, M., Ogura, D., Suzuki, Y., Sato, K.: Fabrication of a wearable fabric tactile sensor produced by artificial hollow fiber. J. Micromechanics Microengineering **18**(8), 085014 (2008)
11. Heo, J.-S., Kim, J.-Y., Lee, J.-J.: Tactile sensors using the distributed optical fiber sensors. In: 2008 3rd International Conference on Sensing Technology, pp. 486–490. IEEE (2008)
12. Ho, D.H., Sun, Q., Kim, S.Y., Han, J.T., Kim, D.H., Cho, J.H.: Stretchable and multimodal all graphene electronic skin. Adv. Mater. **28**(13), 2601–2608 (2016)
13. Jeong, Y.R., Park, H., Jin, S.W., Hong, S.Y., Lee, S.-S., Ha, J.S.: Highly stretchable and sensitive strain sensors using fragmentized graphene foam. Adv. Funct. Mater. **25**(27), 4228–4236 (2015)
14. Lee, H.-K., Chang, S.-I., Yoon, E.: A flexible polymer tactile sensor: fabrication and modular expandability for large area deployment. J. Microelectromechanical Syst. **15**(6), 1681–1686 (2006)
15. Lee, J., Kim, S., Lee, J., Yang, D., Park, B.C., Ryu, S., Park, I.: A stretchable strain sensor based on a metal nanoparticle thin film for human motion detection. Nanoscale **6**(20), 11932–11939 (2014)
16. Li, T., Luo, H., Qin, L., Wang, X., Xiong, Z., Ding, H., Gu, Y., Liu, Z., Zhang, T.: Flexible capacitive tactile sensor based on micropatterned dielectric layer. Small **12**(36), 5042–5048 (2016)

17. Li, J., Orrego, S., Pan, J., He, P., Kang, S.H.: Ultrasensitive, flexible, and low-cost nanoporous piezoresistive composites for tactile pressure sensing. Nanoscale **11**(6), 2779–2786 (2019)
18. Liu, C.-X., Choi, J.-W.: Patterning conductive pdms nanocomposite in an elastomer using microcontact printing. J. Micromechanics Microengineering **19**(8), 085019 (2009)
19. Liu, Y., Han, H., Liu, T., Yi, J., Li, Q., Inoue, Y.: A novel tactile sensor with electromagnetic induction and its application on stick-slip interaction detection. Sensors **16**(4), 430 (2016)
20. Ma, L., Shuai, X., Hu, Y., Liang, X., Zhu, P., Sun, R., Wong, C.-P.: A highly sensitive and flexible capacitive pressure sensor based on a micro-arrayed polydimethylsiloxane dielectric layer. J. Mater. Chem. C **6**(48), 13232–13240 (2018)
21. Manunza, I., Sulis, A., Bonfiglio, A.: Pressure sensing by flexible, organic, field effect transistors. Appl. Phys. Lett. **89**(14), 143502 (2006)
22. Mattmann, C., Clemens, F., Tröster, G.: Sensor for measuring strain in textile. Sensors **8**(6), 3719–3732 (2008)
23. Nag, A., Mukhopadhyay, S., Kosel, J.: Transparent biocompatible sensor patches for touch sensitive prosthetic limbs. In: 2016 10th International Conference on Sensing Technology (ICST), pp. 1–6. IEEE (2016)
24. Núñez, C.G., Navaraj, W.T., Polat, E.O., Dahiya, R.: Energy-autonomous, flexible, and transparent tactile skin. Adv. Funct. Mater. **27**(18), 1606287 (2017)
25. Ramalingame, R., Chandraker, P., Kanoun, O.: Investigation on the influence of solvents on MWCNT-PDMS nanocomposite pressure sensitive films. In: Multidisciplinary Digital Publishing Institute Proceedings, vol. 1, p. 384 (2017)
26. Ramalingame, R., Hu, Z., Gerlach, C., Rajendran, D., Zubkova, T., Baumann, R., Kanoun, O.: Flexible piezoresistive sensor matrix based on a carbon nanotube PDMS composite for dynamic pressure distribution measurement. J. Sens. Sens. Syst. **8**(1), 1–7 (2019)
27. Ramalingame, R., Lakshmanan, A., Müller, F., Thomas, U., Kanoun, O.: Highly sensitive capacitive pressure sensors for robotic applications based on carbon nanotubes and pdms polymer nanocomposite. J. Sens. Sens. Syst. **8**(1), 87–94 (2019)
28. Schmitz, A., Maggiali, M., Natale, L., Bonino, B., Metta, G.: A tactile sensor for the fingertips of the humanoid robot iCub. In: 2010 IEEE/RSJ International Conference on Intelligent Robots and Systems, pp. 2212–2217. IEEE (2010)
29. Seo, S., Kim, S., Jung, J., Ma, R., Baik, S., Moon, H.: Flexible touch sensors made of two layers of printed conductive flexible adhesives. Sensors **16**(9), 1515 (2016)
30. Spanu, A., Pinna, L., Viola, F., Seminara, L., Valle, M., Bonfiglio, A., Cosseddu, P.: A high-sensitivity tactile sensor based on piezoelectric polymer PVDF coupled to an ultra-low voltage organic transistor. Org. Electron. **36**, 57–60 (2016)
31. Tadakaluru, S., Kumpika, T., Kantarak, E., Sroila, W., Panthawan, A., Sanmuangmoon, P., Thongsuwan, W., Singjai, P.: Highly stretchable and sensitive strain sensors using nano-graphene coated natural rubber. Plast. Rubber Compos. **46**(7), 301–305 (2017)
32. Tewari, A., Gandla, S., Bohm, S., McNeill, C.R., Gupta, D.: Highly exfoliated MWNT-rGO ink-wrapped polyurethane foam for piezoresistive pressure sensor applications. ACS Appl. Mater. Interfaces **10**(6), 5185–5195 (2018)
33. Tian, H., Shu, Y., Wang, X.-F., Mohammad, M.A., Bie, Z., Xie, Q.-Y., Li, C., Mi, W.-T., Yang, Y., Ren, T.-L.: A graphene-based resistive pressure sensor with record-high sensitivity in a wide pressure range. Sci. Rep. **5**, 8603 (2015)
34. Yang, Y.-J., Cheng, M.-Y., Shih, S.-C., Huang, X.-H., Tsao, C.-M., Chang, F.-Y., Fan, K.-C.: A 32 × 32 temperature and tactile sensing array using pi-copper films. Int. J. Adv. Manuf. Technol. **46**(9–12), 945–956 (2010)
35. Yu, S.-L., Chang, D.-R., Tsao, L.-C., Shih, W.-P., Chang, P.-Z.: Porous nylon with electro-active dopants as flexible sensors and actuators. In: 2008 IEEE 21st International Conference on Micro Electro Mechanical Systems, pp. 908–911. IEEE (2008)
36. Zhu, B., Niu, Z., Wang, H., Leow, W.R., Wang, H., Li, Y., Zheng, L., Wei, J., Huo, F., Chen, X.: Microstructured graphene arrays for highly sensitive flexible tactile sensors. Small **10**(18), 3625–3631 (2014)

The manufacturer's authorised representative in the EU is Springer Nature Customer Service Centre GmbH, Europaplatz 3, 69115 Heidelberg, Germany. If you have any concerns regarding our products, please contact ProductSafety@springernature.com

Printed and bound by CPI Group (UK) Ltd, Croydon, CR0 4YY
25/09/2026
02222886-0004